A Historical Anthology from **Technology & Culture**

4—

Technology

A Historical Anthology from **Technology & Culture**

& the West

Edited by Terry S. Reynolds and Stephen H. Cutcliffe

The University of Chicago Press

CHICAGO AND LONDON

The essays in this volume originally appeared in various issues of *Technology and Culture*.

The University of Chicago Press, Chicago 60637
The University of Chicago Press, Ltd., London

Printed in the United States of America
01 00 99 98 97 5 4 3 2 1

Library of Congress Cataloging-in-Publication Data

Technology and the west : a historical anthology from technology and culture / edited by Terry S. Reynolds and Stephen H. Cutcliffe.
p. cm.
Includes bibliographical references and index.
ISBN 0-226-71033-5 (cloth). — ISBN 0-226-71034-3 (pbk.)
1. Technology—History. I. Reynolds, Terry S. II. Cutcliffe, Stephen H. III. Technology and culture.
T15.T42 1997
609—dc21 97-5926
CIP

The paper used in this publication meets the minimum requirements of American National Standard for Information Science—Permanence of Paper for Printed Library Materials, ANSI Z39.48-1984.

Technology and the West: A Historical Anthology from Technology and Culture

For Linda and Katie

Introduction

Technology has always played an important role in human affairs. Even the most primitive human societies had to have at least a modicum of technical activity to achieve the essentials of food, clothing, and shelter. The civilizations that emerged in Mesopotamia and Egypt after 5000 B.C. built on that modicum to establish irrigation networks, cities, and massive monuments and, in turn, passed their technological knowledge down to the immediate ancestors of Western civilization, the classical Greeks and Romans. Drawing not only from classical Greek and Roman technology but also from contemporary Islamic, Indian, and Chinese technologies, medieval Europe laid the foundations for contemporary Western civilization's heavy dependence on technology and for its preference for seeking technological solutions to its social problems.

Technology, however, is simply one element of culture. It influences and, in turn, is influenced by the other components of the cultures in which it is enmeshed. Technological developments present societies with new options, but it does not compel them to take these options. Which of those options are taken, or whether any are taken at all, is determined by cultural context. In other words, society shapes technology as often and as radically as technology shapes society. Thus, even though the West embraced technology and technological solutions to a greater extent than other civilizations, the history of technology in the West cannot be viewed in isolation. Western cultural beliefs, Western geography, and other elements of the Western social context influenced the way technology evolved in the West.

This volume, *Technology and the West,* contains eighteen high-quality essays. These essays place the history of Western technology in its cultural context and are suitable for both the general reader and the

TERRY S. REYNOLDS is professor of history and chair of the Department of Social Sciences at Michigan Technological University. He has written several books on aspects of the history of water power and edited *The Engineer in America* (Chicago, 1991). STEPHEN H. CUTCLIFFE is a member of the Department of History and the director of Lehigh University's Science, Technology and Society Program, where he edits their *Science, Technology and Society Curriculum Development Newsletter.* He is also the co-editor with Robert C. Post of *In Context: History and the History of Technology—Essays in Honor of Melvin Kranzberg* (Lehigh University Press, 1989).

college student.[1] It is one-half of a two-volume set of anthologies consisting of articles selected from *Technology and Culture*, the official journal of the Society for the History of Technology (SHOT).[2] Its companion volume (*Technology and American History*) looks at technology in the context of American history. The eighteen essays selected for *Technology and the West* cover the scope of the history of technology in Western civilization from the early irrigation civilizations, through the medieval period and early modern Europe, to late-20th-century America and Europe. We have organized the essays in roughly chronological order, making selections representative of key time periods, key themes, and key technologies without trying or claiming to be all-inclusive.

We found the task of selecting eighteen articles to cover the entire scope of the history of technology from the early irrigation-based civilizations to the present extraordinarily difficult. Often the best article from the viewpoint of academic scholarship was not the best article from the viewpoint of the general reader or college student of the history of technology. Often we reluctantly eliminated very good pieces simply because we could afford only one article in a particular time frame, if we were to cover the entire history of Western technology in a single volume.

Our opening selection in this volume provides a broad overview of some of the major themes and issues in the history of technology. For this purpose, we selected Mel Kranzberg's 1985 presidential address: "Kranzberg's Laws." Although the "laws" may cause professional scholars to squirm, they provide informative, entertaining, and sometimes thought provoking ideas to readers who have never been ex-

[1] Faculty teaching undergraduate survey courses in the history of Western technology have long been plagued by the limited number of published classroom materials available. This problem was noted by Svante Lindqvist, *The Teaching of History of Technology in USA—a Critical Survey in 1978,* Stockholm Papers in History and Philosophy of Technology, TRITA-HOT-5003 (Stockholm: Royal Institute of Technology Library, 1981), p. 51; Darwin Stapleton with Liz Paley, "Its Own Reward: Three Decades of Teaching and Scholarship in the History of Technology," in *In Context: History and the History of Technology,* ed. Stephen H. Cutcliffe and Robert C. Post (Bethlehem, Pa.: Lehigh University Press, 1989), pp. 206, 207; and Lars O. Olsson, *Undergraduate Teaching of History of Technology: A Survey of the Teaching at Some Universities in the USA in 1993* (Gothenburg: Chalmers University of Technology, Department of History of Technology and Industry, 1995), p. 37.

[2] Stapleton and Paley, "Its Own Reward," p. 208, noted in 1988 that journal articles and chapters excised from books were the staple of most history of technology courses in America and that *Technology and Culture* was the most-used scholarly journal for undergraduate course readings.

posed to the history of technology. Because "Kranzberg's Laws" was originally a presidential address and contains sections suitable for that occasion but of little interest otherwise, we have slightly shortened the original article. In all other cases the essays are complete as published.

The bulk of this volume is divided into two large sections. The first section covers technology through the traditional "Industrial Revolution." The second part deals with technology in the industrial era (ca. 1850 to present). Internally, the two broad chronological divisions of this anthology have parallel organizations. An editors' overview introduces each section. These overviews set the *Technology and Culture* selections that follow in broad historical context, identify common theses and issues, and provide a brief synopses of each article. They are meant to be read as essays in their own right. The selected essays within each division then follow in roughly the chronological order of their subject matter.

Although the primary criterion used in selecting essays was appropriateness for the general reader and college student and "fit" with the focus of this anthology, a significant number of the sections have been singled out for recognition for the quality of their scholarship by the Society for the History of Technology. The Society awards its Usher Prize annually for the best essay appearing in *Technology and Culture,* selected from the past three years' issues of the journal. Among the Usher Prize–winning pieces found in this anthology are Thomas Esper's "The Replacement of the Longbow by Firearms in the English Army" (1966); Lynwood Bryant's "The Development of the Diesel Engine" (1970); and William H. Te Brake's "Air Pollution and Fuel Crises in Preindustrial London, 1250–1650" (1977). In addition, we have used Ruth Cowan's "Industrial Revolution in the Home," which was to grow into her 1984 Dexter Prize–winning book *More Work for Mother: The Ironies of Household Technology from the Open Hearth to the Microwave* (New York: Basic, 1983) and Walter A. McDougall's "Space-Age Europe: Gaullism, Euro-Gaullism, and the American Dilemma," an article stemming from research conducted on his 1986 Dexter Prize–winning book . . . *the Heavens and the Earth: A Political History of the Space Age* (New York: Basic, 1985).[3] Finally, we have included Brett Steele's essay on the ballistics revolution of the 18th century, based on his Joan Cahalin Robinson Prize paper

[3] The Dexter Prize is awarded annually by the Society for the History of Technology to the author of an outstanding book in the history of technology published during any of the three years preceding the award.

of 1991, and J. Samuel Walker's essay on nuclear power and the environment, winner of the 1990 IEEE Life Members' Prize.[4]

In compiling this anthology, we have accumulated a number of debts. Our original stimulus came from the Executive Council of the Society for the History of Technology, which suggested this project in 1993. We owe particular thanks to Bob Post, former editor of *Technology and Culture,* for his encouragement. Without it, we might very well have given up on this project. We also thank Charles Hyde of Wayne State University, Richard Hirsh of Virginia Tech, Bruce Seely of Michigan Tech, and Roger Simon and John Smith of Lehigh University for reviewing and commenting on our proposal for this reader and our article selections. Finally, we are indebted to the fine staff at the University of Chicago Press Journals Division for help in bringing our early vision to final reality.

TERRY S. REYNOLDS AND STEPHEN H. CUTCLIFFE

[4]The Joan Cahalin Robinson Prize is awarded annually to the best presented paper at a meeting of the Society for the History of Technology by an individual giving a first paper. The IEEE Life Members' Prize is awarded by the Society for the best paper in electrical history published in the previous year.

Overview

TECHNOLOGY AND HISTORY: "KRANZBERG'S LAWS"

MELVIN KRANZBERG

A few months ago I received a note from a longtime collaborator in building the Society for the History of Technology, Eugene S. Ferguson, in which he wrote, "Each of us has only one message to convey." Ferguson was being typically modest in referring to an article of his in a French journal[1] emphasizing the hands-on, design component of technical development, and he claimed that he had been making exactly the same point in his many other writings. True, but he has also given us many other messages over the years.

However, Ferguson's statement of "only one message" might indeed be true in my case. For I have been conveying basically the same message for over thirty years, namely, the significance in human affairs of the history of technology and the value of the contextual approach in understanding technical developments.

Because I have repeated that same message so often, utilizing various examples or stressing certain elements to accord with the interests of the different audiences I was attempting to reach, my thoughts have jelled into what have been called "Kranzberg's Laws." These are not laws in the sense of commandments but rather a series of truisms deriving from a longtime immersion in the study of the development of technology and its interactions with sociocultural change.

* * *

DR. KRANZBERG, now deceased, was one of the pioneers of the history of technology in the United States. He was the founding editor of *Technology & Culture,* recipient of the Society for the History of Technology's Leonardo da Vinci Medal (1967), and the Society's president (1983–84). He was Callaway Professor of the History of Technology at the Georgia Institute of Technology at the time this paper was presented as his delayed presidential address on October 19, 1985, at the Henry Ford Museum in Dearborn, Michigan.

[1]Eugene S. Ferguson, "La Fondation des machines modernes: des dessins," *Culture technique* 14 (June 1985): 182–207. *Culture technique* is the publication of the Centre de Recherche sur la Culture Technique, located in Paris under the direction of Jocelyn de Noblet. The June 1983 edition of *Culture technique,* dedicated to *Technology and Culture,* contained French translations of a number of articles from the SHOT journal.

We historians tend to think of historical change in terms of cause and effect and of means and ends. Although it is not always easy to find causative elements and to distinguish ends from means in the interactions between technology and society, that has not kept scholars from trying to do so.

Indeed one of the intellectual clichés of our time, whose scholarly statement is embodied in the writings of Jacques Ellul and Langdon Winner, is that technology is pursued for its own sake and without regard to human need.[2] Technology, it is said, has become autonomous and has outrun human control; in a startling reversal, the machines have become the masters of man. Such arguments frequently result in the philosophical doctrine of technological determinism, namely, that technology is the prime factor in shaping our life-styles, values, institutions, and other elements of our society.

Not all scholars accept this version of technological omnipotence. Lynn White, jr., has said that a technical device "merely opens a door, it does not compel one to enter."[3] In this view, technology might be regarded as simply a means that humans are free to employ or not, as they see fit—and White recognizes that many nontechnical factors might affect that decision. Nevertheless, several questions do arise. True, one is not compelled to enter White's open door, but an open door is an invitation. Besides, who decides which doors to open—and, once one has entered the door, are not one's future directions guided by the contours of the corridor or chamber into which one has stepped? Equally important, once one has crossed the threshold, can one turn back?

Frankly, we historians do not know the answer to this question of technological determinism. Ours is a new discipline; we are still working on the problem, and we might never reach agreement on an answer—which means that it will provide employment for historians of technology for decades to come. Yet there are several things that we do know, and that I summarize under the label of Kranzberg's First Law.

Kranzberg's First Law reads as follows: Technology is neither good nor bad; nor is it neutral.

By that I mean that technology's interaction with the social ecology is such that technical developments frequently have environmental, social, and human consequences that go far beyond the immediate purposes of the technical devices and practices themselves, and the same

[2]Jacques Ellul, *The Technological Society* (New York, 1964), and Langdon Winner, *Autonomous Technology: Technics Out-of-Control as a Theme in Political History* (Cambridge, Mass., 1977).

[3]Lynn White, jr., *Medieval Technology and Social Change* (Oxford, 1962), p. 28.

technology can have quite different results when introduced into different contexts or under different circumstances.

Many of our technology-related problems arise because of the unforeseen consequences when apparently benign technologies are employed on a massive scale. Hence many technical applications that seemed a boon to mankind when first introduced became threats when their use became widespread. For example, DDT was employed to raise agricultural productivity and to eliminate disease-carrying pests. Then we discovered that DDT not only did that but also threatened ecological systems, including the food chain of birds, fishes, and eventually man. So the Western industrialized nations banned DDT. They could afford to do so, because their high technological level enabled them to use alternative means of pest control to achieve the same results at a slightly higher cost.

But India continued to employ DDT, despite the possibility of environmental damage, because it was not economically feasible to change to less persistent insecticides—and because, to India, the use of DDT in agriculture was secondary to its role in disease prevention. According to the World Health Organization, the use of DDT in the 1950s and 1960s in India cut the incidence of malaria in that country from 100 million cases a year to only 15,000, and the death toll from 750,000 to 1,500 a year. Is it surprising that the Indians viewed DDT differently from us, welcoming it rather than banning it? The point is that the same technology can answer questions differently, depending on the context into which it is introduced and the problem it is designed to solve.

Thus while some American scholars point to the dehumanizing character of work in a modern factory,[4] D. S. Naipaul, the great Indian author, assesses it differently from the standpoint of his culture, saying, "Indian poverty is more dehumanizing than any machine."[5] Hence in judging the efficacy of technological development, we historians must take cognizance of varying social contexts.

It is also imperative that we compare short-range and long-range impacts. In the 19th century, Romantic writers and social critics condemned industrial technology for the harsh conditions under which the mill workers and coal miners labored. Yet, according to Fernand Braudel, conditions on the medieval manor were even worse.[6] Certain

[4]E.g., Christopher Lasch, *The Minimal Self: Psychic Survival in Troubled Times* (New York, 1984).

[5]Quoted in Dennis H. Wrong, "The Case against Modernity," *New York Times Book Review*, October 28, 1984, p. 7.

[6]Fernand Braudel, *The Structures of Everyday Life*, vol. 1 of *Civilization and Capitalism, 15th–18th Century* (New York, 1981).

economic historians have pointed out that, although the conditions of the early factory workers left much to be desired, in the long run the worker's living standards improved as industrialization brought forth a torrent of goods that were made available to an ever-wider public.[7] Of course, those long-run benefits were small comfort to those who suffered in the short run; yet it is the duty of the historian to show the differences between the immediate and long-range implications of technological developments.

Although our technological advances have yielded manifold benefits in increasing food supply, in providing a deluge of material goods, and in prolonging human life, people do not always appreciate technology's contributions to their lives and comfort. Nicholas Rescher, citing statistical data on the way people perceive their conditions, explains their dissatisfaction on the paradoxical ground that technical progress inflates their expectations faster than it can actually meet them.[8]

Of course, the public's perception of technological advantages can change over time. A century ago, smoke from industrial smokestacks was regarded as a sign of a region's prosperity; only later was it recognized that the smoke was despoiling the environment. There were "technological fixes," of course. Thus, one of the aims of the Clean Air Act of 1972 was to prevent the harmful particulates emitted by smokestacks from falling on nearby communities. One way to do away with this problem was to build the smokestacks hundreds of feet high; then a few years later we discovered that the sulfur dioxide and other oxides, when sent high into the air, combined with water vapor to shower the earth with acid rain that has polluted lakes and caused forests to die hundreds of miles away.

Unforeseen "dis-benefits" can thus arise from presumably beneficent technologies. For example, although advances in medical technology and water and sewage treatment have freed millions of people from disease and plague and have lowered infant mortality, these have also brought the possibility of overcrowding the earth and producing, from other causes, human suffering on a vast scale. Similarly, nuclear technology offers the prospect of unlimited energy resources, but it has also brought the possibility of worldwide destruction.

That is why I think that my first law—Technology is neither good nor bad; nor is it neutral—should constantly remind us that it is the historian's duty to compare short-term versus long-term results, the

[7]E.g., T. S. Ashton, *The Industrial Revolution, 1760–1830* (Oxford, 1948), and David S. Landes, *The Unbound Prometheus: Technological Change and Industrial Development in Western Europe from 1750 to the Present* (Cambridge, 1969).

[8]Nicholas Rescher, *Unpopular Essays on Technological Progress* (Pittsburgh, 1980).

utopian hopes versus the spotted actuality, the what-might-have-been against what actually happened, and the trade-offs among various "goods" and possible "bads." All of this can be done only by seeing how technology interacts in different ways with different values and institutions, indeed, with the entire sociocultural milieu.[9]

* * *

Whereas my first law stresses the interactions between technology and society, my second law starts with internalist elements in technology and then stretches to include many nontechnical factors. Kranzberg's Second Law can be simply stated: Invention is the mother of necessity.

Every technical innovation seems to require additional technical advances in order to make it fully effective. If one invents a lathe that can cut metal faster than existing machines, this necessitates improvements in the lubricating system to keep the mechanism running efficiently, improved grinding materials to stand up under the enhanced speed, and new means of taking away quickly the waste material from the item being turned.

Many major innovations have required further inventions to make them completely effective. Thus, Alexander Graham Bell's telephone spawned a variety of technical improvements, ranging from Edison's carbon-granule microphone to central-switching mechanisms. A variation on this same theme is described in Hugh Aitken's book on the origins of radio, in which he indicates the various innovative steps whereby the spark technology that produced radio waves was tuned into harmony (syntonized) with the receiver.[10] In more recent times, the design of a more powerful rocket, giving greater thrust, necessitates innovation in chemical engineering to produce the thrust, in materials to withstand the blast, in electronic control mechanisms, and the like.

A good case of invention mothering necessity can be seen in the landmark textile inventions of the 18th century. Kay's "flying shuttle" wove so quickly that it upset the usual ratio of four spinners to one weaver; either there had to be many more spinners or else spinning had to be similarly quickened by application of machinery. Thereupon Hargreaves, Cartwright, and Crompton improved the spinning pro-

[9]The "New Directions" program session at the 1985 SHOT annual meeting indicated that historians of technology are continuing to broaden their concerns and are indeed investigating new areas of the sociocultural context in relation to technological developments.

[10]Hugh G. J. Aitken, *Syntony and Spark: The Origins of Radio* (New York, 1976).

cess; then Cartwright set about further mechanizing the weaving operation in order to take full advantage of the now-abundant yarn produced by the new spinning machines.

Thomas P. Hughes would refer to the phenomenon that I have just described as a "reverse salient";[11] but I prefer to call it a "technological imbalance," a situation in which an improvement in one machine upsets the previous balance and necessitates an effort to right the balance by means of a new innovation. No matter what one calls it, Hughes and I are talking about the same thing. Indeed, Hughes has gone further in discussing technological systems, for he shows how, as a system grows, it generates new properties and new problems, which in turn necessitate further changes.

The automobile is a prime example of how a successful technology requires auxiliary technologies to make it fully effective, for it brought whole new industries into being and turned existing industries in new directions by its need for rubber tires, petroleum products, and new tools and materials. Furthermore, large-scale use of the auto demanded a host of auxiliary technological activities—roads and highways, garages and parking lots, traffic signals, and parking meters.

While it might be said that each of these other developments occurred in response to a specific need, I claim that it was the original invention that mothered that necessity. If we look into the internal history of any mechanical device, we find that the basic invention required other innovative changes to make it fully effective and that the completed mechanism in turn necessitated changes in auxiliary and supporting technological systems, which, taken all together, brought many changes in economic and sociocultural patterns.

* * *

What I have just said is virtually a statement of my Third Law: Technology comes in packages, big and small.

The fact is that today's complex mechanisms usually involve several processes and components. Radar, for example, is a very complicated system, requiring specialized materials, power sources, and intricate devices to send out waves of the proper frequency, detect them when they bounce off an object, and then interpret them and place the results on a screen.

That might explain why so many different people have laid claim to inventing radar. Each is perfectly right in pointing out that he pro-

[11]Thomas P. Hughes, "Inventors: The Problems They Choose, the Ideas They Have, and the Inventions They Make," in *Technological Innovation: A Critical Review of Current Knowledge*, ed. Patrick Kelly and Melvin Kranzberg (San Francisco, 1978), pp. 166–82.

vided an element essential to the final product, but that final product is composed of many separate elements brought together in a system that could not function without every single one of the components. Thus radar is the product of a packaging process, bringing together elements of different technologies into a single device.

In his fascinating account of the development of mass production, David A. Hounshell tells how many different experiments and techniques were employed in bringing Ford's assembly line into being.[12] Although many of the component elements were already in existence, Ford put these together into a comprehensive system—but not without having to develop additional technical capabilities, such as conveyor lines, to make the assembly process more effective.

My third law has been extended even further by Thomas P. Hughes's 1985 Dexter Prize–winning book *Networks of Power*. What I call "packages" Hughes more precisely and accurately calls "systems," which he defines as coherent structures composed of interacting, interconnected components.[13] When one component changes, other parts of the system must undergo transformations so that the system might continue to function. Hence the parts of a system cannot be viewed in isolation but must be studied in terms of their interrelations with the other parts.

Although Hughes concentrates on electric power systems, what he provides is a paradigm that is applicable to other systems—transportation, water supply, communications, and the like. And because entire systems interact with other systems, a system cannot be studied in isolation any more than can its component parts; hence one must also look at the interaction of these systems with the entire social, political, economic, and cultural environment. Hughes's book thus provides excellent case studies proving the validity of the first three of Kranzberg's Laws, and also of my fourth dictum.

* * *

Unfortunately, Kranzberg's Fourth Law cannot be stated so pithily as the first three. It reads as follows: Although technology might be a prime element in many public issues, nontechnical factors take precedence in technology-policy decisions.

Engineers claim that their solutions to technical problems are not

[12]David A. Hounshell, *From the American System to Mass Production 1800–1932: The Development of Manufacturing Technology in the United States* (Baltimore, 1984), chap. 6.

[13]Thomas P. Hughes, *Networks of Power: Electrification in Western Society, 1880–1930* (Baltimore, 1983), p. ix.

based on mushy social considerations; instead, they boast that their decisions depend on the hard and measurable facts of technical efficiency, which they define in terms of input-output factors such as cost of resources, power, and labor. However, as Edward Constant has shown in studying the Kuhnian paradigm's applicability to technological developments, many complicated sociocultural factors, especially human elements, are involved, even in what might seem to be "purely technical" decisions.[14]

Besides, engineers do not always agree with one another; different fields of engineering might have different solutions to the same problem, and even within the same field they might disagree on what weight to assign to different trade-off factors. Indeed, as Stuart W. Leslie demonstrated in his Usher Prize article on "Charles F. Kettering and the Copper-cooled Engine,"[15] the most efficient device does not always win out even in what we might regard as a narrowly technical decision within a single industrial corporation. Although Kettering regarded his copper-cooled engine as a technical success, it never went into production. Why not? True, it had some technical "bugs," but these could not be successfully ironed out because of divisions between the research engineers and the production people—and because of the overall decision that the copper-cooled engine could not meet the corporate demand for immediate profit. So technical worth, or at least potential technical capability and efficiency, was not the decisive element in halting the copper-cooled engine.

In *Networks of Power* Hughes likewise demonstrates how nontechnical factors affected the efficient growth of electrical networks by comparing developments in Chicago, Berlin, and London. Private enterprise in Chicago, in the person of Samuel Insull, followed the path of the most efficient technology in seeking economies of scale. In Berlin and London, however, municipal governments were more concerned about their own authority than about technical efficiency, and political infighting meant that they lagged behind in developing the most economical power networks.

Technologically "sweet" solutions do not always triumph over political and social forces.[16] The debate a dozen years ago over the super-

[14]Edward W. Constant, *The Origins of the Turbojet Revolution* (Baltimore, 1980). This book was awarded the Dexter Prize by SHOT in 1982.

[15]Stuart W. Leslie, "Charles F. Kettering and the Copper-cooled Engine," *Technology and Culture* 20 (October 1979): 752–76.

[16]Eugene B. Skolnikoff states, "Technology alters the physical reality, but is not the key determinant of the political changes that ensue," in *The International Imperatives of Technology: Technological Development and the International Political System* (Berkeley, Calif.: University of California Institute of International Studies, n.d.), p. 2.

sonic transport (SST) provides an example. Although the SST offered potential advantages, its development to the point where its feasibility and desirability could be properly determined was never allowed to take place. Economic factors might have underlain the decision to cut R&D funds for the SST, but the public decision seems also to have been based on a fear of the environmental hazards posed by the supersonic aircraft in commercial aviation.

Environmental concerns have indeed assumed a major place in public decisions regarding technical initiatives. These concerns are not groundless, for we have seen how certain technologies, employed without awareness of potential environmental effects, have boomeranged to present hazardous problems, despite their early beneficial effects. Many engineers believe that hysterical fear about technological development has so gripped our nation that people overlook the benefits provided by technology and concentrate on the dangers presented either by ill-conceived technological applications or by human error or oversight in technical operations. But who can blame the public, with Love Canal and Bhopal crowding the headlines?[17]

American politics has now become the battleground of special-interest groups, and few of these groups are willing to make the trade-offs required in many engineering decisions. In the case of potential environmental hazards, Daniel A. Koshland has stated that we can satisfy one or the other of the different groups, but only at a cost of something undesirable to the others.[18]

Especially politicized has been the question of nuclear power. The nuclear industry itself has been partly to blame for technological deficiencies, but the presumption of risk by the public, especially following the Three Mile Island and Chernobyl accidents, has affected the future of what was once regarded as a safe and inexhaustible source of power. The public fears possible catastrophic consequences from nuclear generators.

Yet the historical fact is that no one has been killed by commercial nuclear power accidents in this country. Contrast this with the 50,000 Americans killed each year by automobiles. But although antinuclear protestors picket nuclear power plants under construction, we never see any demonstrators bearing signs saying "Ban the Buick"!

[17]Speaking of the Bhopal tragedy, President John S. Morris of Union College has said: "Methyl isocyanate makes it possible to grow good crops and feed millions of people, but it also involves risks. And analyzing risks is not a simple matter" (*New York Times*, April 14, 1985).

[18]Daniel A. Koshland, "The Undesirability Principle," *Science* 229 (July 5, 1985): 9.

Partly this is due to the public's perception of risk, rather than to the actual risks themselves.[19] People seek a zero-risk society. But as Aaron Wildavsky has so aptly put it, "No risk is the highest risk of all."[20] For it would not only petrify our technology but also stultify developmental growth in society along any lines.

Nevertheless, the fact that political considerations take precedence over purely technical considerations should not alarm us. In a democracy, that is as it should be. To deal with questions involving the interactions between technology and the ecology, both natural and social, we have devised new social instruments, such as "technology assessment," to evaluate the possible consequences of the applications of technologies before they are applied.

Of course, political considerations often continue to take precedence over the commonsensible results of comprehensive and impartial technological assessments. But at least there is the recognition that technological developments frequently have social, human, and environmental implications that go far beyond the intention of the original technology itself.

* * *

The fact that historians of technology must be aware of outside forces and factors affecting technology—from the human personality of the inventor to the larger social, economic, political, and cultural milieu—has led me to Kranzberg's Fifth Law: All history is relevant, but the history of technology is the most relevant.

In her presidential address to the Organization of American Historians several years ago, Gerda Lerner pointed out how history satisfies a variety of human needs, serving as a cultural tradition that gives us personal identity in the continuum of the past and future of the human enterprise.[21] Other apologists for the profession point out that history is one of the fundamental liberal arts and is essential as a key to an understanding of the future.

No one would quarrel with such worthy sentiments, but, to repeat questions raised by Eugene D. Genovese, "If so, how can we explain the

[19]See Dorothy Nelkin, ed., *Controversy: The Politics of Ethical Decisions* (Santa Monica, Calif., 1984).

[20]Aaron Wildavsky, "No Risk Is the Highest Risk of All," *American Scientist* 67 (1979): 32–37.

[21]Gerda Lerner, "The Necessity of History and the Professional Historian," *Journal of American History* 69 (June 1982): 7–20.

dangerous decline in the teaching of history in our schools; the cynical taunt, 'What is history good for anyway?' "[22] Although historians might write loftily of the importance of historical understanding by civilized people and citizens, many of today's students simply do not see the relevance of history to the present or to their future. I suggest that this is because most history, as it is currently taught, ignores the technological element.

Two centuries ago the great German philosopher Immanuel Kant stated that the two great questions in life are (1) What can I know? and (2) What ought I do?

To answer Kant's first question, we can learn the history of the past. I look on history as a series of questions that we ask of the past in order to find out how our present world came into being. We call ours a "technological age." How did it get to be that way? That indeed is the major question that the history of technology attempts to answer. Our students know that they live in a technological age, but any history that ignores the technological factor in societal development does little to enable them to comprehend how their world came into being.

True, economic and business historians have perforce taken cognizance of those technological elements that had a mighty effect on their subject matter. Similarly, social historians of the *Annales* school have stressed how technology set the patterns of daily life for the vast majority of people throughout history, and Brooke Hindle, in a fine historiographical article, has indicated how some of our fellow historians have begun to see how technology impinges on their special fields of study.[23] But for the most part, social, political, and intellectual historians have been oblivious to the technological parameters of their own subjects.

Perhaps most guilty of neglecting technology are those concerned with the history of the arts and with the entire panoply of humanistic concerns. Indeed, in many cases they are disdainful of technology, regarding it as somehow opposed to the humanities. This might be because they regard technology solely in terms of mechanical devices and do not even begin to comprehend the complex nature of techno-

[22]Eugene D. Genovese, "To Celebrate a Life—Biography as History," *Humanities* 1 (January–February 1980): 6. An analysis of today's low state of the history profession is to be found in Richard O. Curry and Lawrence D. Goodheart, "Encounters with Clio: The Evolution of Modern American Historical Writing," *OAH Newsletter* 12 (May 1984): 28–32.

[23]Brooke Hindle, "'The Exhilaration of Early American Technology': A New Look," in *The History of American Technology: Exhilaration or Discontent?* ed. David A. Hounshell (Wilmington, Del., 1984).

logical developments and their direct influences on the arts, to say nothing of their indirect influence on mankind's humanistic endeavors.

Yet anyone familiar with Cyril Stanley Smith's writings would be aware of the importance of the aesthetic impulse in technical accomplishments and of how these in turn amplified the materials and techniques available for artistic expression.[24] And any historian of art or of the Renaissance should perceive that such artistic masters as Leonardo and Michelangelo were also great engineers. That relationship continues today, as David Billington has shown in stressing the relationship of structural design and art.[25]

Today's technological age provides new technical capabilities to enlarge the horizons and means of expression for artists in every field. Advances in musical instruments have given larger scope to the imagination of composers and to musical interpretation by performers. The advent of photography, the phonograph, radio, movies, and television have not only given artists, composers, and dramatists new tools with which to exercise their vision and talents but have also enlarged the audience for music, drama, and the whole panoply of the arts. They also extend our audio and visual memory, enabling us to see, hear, and preserve the great works of the past and present.

In the field of learning and education, there is little point in belaboring the impact of writing tools, paper, the printing press, and, nowadays, radio and TV. But there is also an indirect influence of technology on education, one that makes it more possible than ever before in human history for larger numbers of people in the industrialized nations to take advantage of formal schooling.

Let me give a brief example drawn from American history. Thomas Jefferson was very proud of the educational system that he devised for the state of Virginia. But in his educational scheme, only a very small percentage could ever hope to ascend to the heights of a university education.

This is not because Jefferson was an elitist. Far from it! But the fact is that the agrarian technology of his time was not productive enough to allow large numbers of youth to participate in the educational process. From a very early age, children worked in the fields alongside

[24]See especially Cyril Stanley Smith's Usher Prize article, "Art, Technology, and Science: Notes on Their Historical Interaction," *Technology and Culture* 11 (October 1970): 493–549.

[25]See David Billington's Dexter Prize–winning book, *Robert Maillart's Bridges: The Art of Engineering* (Princeton, N.J., 1979), and "Bridges and the New Art of Structural Engineering," *American Scientist* 72 (January–February 1984): 22–31.

their parents or, if they were town dwellers, were apprenticed to craftsmen. Only when great increases in agricultural and industrial productivity were made possible by revolutionary developments in technology did society acquire sufficient wealth to keep children out of the work force and enable them to attend school. As the 19th century progressed, first elementary education was made compulsory, then secondary education, and by the mid-20th century, America had grown so wealthy that it could afford a college education for all its citizens. True, some students drop out of high school before completing it, and not everyone going to college takes full advantage of the educational opportunities. But the fact is that the majority of Americans today have the equivalent education of the small segment of the upper-class elite in preindustrial society. In brief, technology has been a significant factor, not only in the pattern of our daily lives and in our workaday world, but also in democratizing education and the intellectual realm of the arts and humanities.

However, such vast generalizations might do little to convince the public of the wisdom of Stanley N. Katz's vision of scholars participating "in public discourse in order to recover the traditional role of the humanist as a public figure."[26] But the relevance of the history of technology to today's world can be spelled out in very specific terms. For example, because we live in a "global village," made so by technological developments, we are conscious of the need to transfer technological expertise to our less fortunate brethren in the less developed nations. And the history of technology has a great deal to say about the conditions, complexities, and problems of technology transfer.

Likewise, we are faced with public decisions regarding global strategy, environmental concerns, educational directions, and the ratio of resources to the world's burgeoning population. Technological history can cast light on many parameters of these very specific problems confronting us now and in the future—and that is why I say that the history of technology is more relevant than other histories.

One proof of this is that the outside world, especially the political community, is becoming increasingly cognizant of the contributions that historians of technology can make to public concerns. Whereas several decades ago historians were rarely called on to provide information to Congress on matters other than historical archives, memorials, and national celebrations, nowadays it is almost commonplace for historians of technology to testify before congressional committees dealing with scientific and technological expenditures, aerospace developments, transportation, water supplies, and other

[26]Stanley N. Katz, "The Scholar and the Public," *Humanities* 6 (June 1985): 14–15.

problems having a technological component. Congressmen obviously think that the information provided by historians of technology is relevant to coping with the problems of today and tomorrow.

Leaders in all fields are increasingly turning to historians of technology for expertise regarding the nature of the sociotechnical problems facing them. Let me give a few more specific examples. SHOT is an affiliate of the American Association for the Advancement of Science (AAAS), and there was a time when historians of technology appeared only on the program sessions of Section L of the AAAS, the History and Philosophy of Science. But historians of technology also have important things to say to a public larger than that composed of their historical colleagues. Hence it was a source of great personal pride to me—almost paternal pride—when, at the 1985 AAAS meeting, Carroll Pursell appeared on a program session with a congressman and a former assistant secretary of commerce; the program dealt with certain social and economic problems affecting the United States today, and Pursell's historical account of the technological parameters was truly germane to the thrust of the discussion. Similarly, at a recent conference, at my own Georgia Tech, on the problems expected to affect the workplace in the future, David Hounshell provided a meaningful technological historical context for a discussion that involved top labor leaders, political figures, and corporate executives. (I took family pride in that too!)

I regard this entrance of historians of technology into the public arena as empirical evidence of the true relevance of the history of technology to the worlds of today and tomorrow. To reiterate, all history is relevant, but the history of technology is most relevant. The rest of the world realizes that, and SHOT is working to make our historical colleagues from other fields recognize it too.

* * *

This brings me to my final law, Kranzberg's Sixth Law: Technology is a very human activity—and so is the history of technology.

Anthropologists and archaeologists studying primate evolution tell us of the importance of purposive toolmaking in the formation of *Homo sapiens.* The physical development of our species is apparently inextricably bound up with cultural developments, so that technology is classed as one of the earliest and most basic of human cultural characteristics, one helping to develop language and abstract thinking. Or, to put it another way, man could not have become *Homo sapiens*, "man the thinker," had he not at the same time been *Homo faber*, "man the maker."

Man is a constituent element of the technical process. Machines are made and used by human beings. Behind every machine, I see a face—indeed, many faces: the engineer, the worker, the businessman or businesswoman, and, sometimes, the general and admiral. Furthermore, the function of the technology is its use by human beings—and sometimes, alas, its abuse and misuse.

To those who identify technology simply with the machines themselves, I use the computer as a metaphor to show the importance of the interaction of human and social factors with the technical elements—for computers require both the mechanical element, the "hardware," and the human element, the "software"; without the software, the machine is simply an inert device, but without the hardware, the software is meaningless. We need both, the human and the purely technical components, in order to make the computer a usable and useful piece of technology.

Those of you who were at our Silver Anniversary meeting in 1983 will recall that I told an anecdote, which I sometimes use to quiet my most voluble antitechnological humanistic colleagues. A lady came up to the great violinist Fritz Kreisler after a concert and gushed, "Maestro, your violin makes such beautiful music." Kreisler held his violin up to his ear and said, "I don't hear any music coming out of it."

You see, the instrument, the hardware, the violin itself, was of no use without the human element. But then again, without the instrument, Kreisler would not have been able to make music. The history of technology is the story of man and tool—hand and mind—working together. If the hardware is faulty or if the software is deficient, the sounds that emerge will be discordant; but when man and machine work together, they can make some beautiful music.

People sometimes speak of the "technological imperative," meaning that technology rules our lives. Indeed, they can point to many technical elements, such as the clock, that determine the character and pace of our daily existence. Likewise, the automobile determines where and how we Americans live, work, think, play, and pray.

But this does not necessarily mean that the "technological imperative," usually based on efficiency or economy, necessarily directs all our thoughts and actions. We can point to many technical devices that would make life simpler or easier for us but which our social values and human sensibilities simply reject. Thus, for example, Ruth Schwartz Cowan has shown in her Dexter Prize–winning book, *More Work for Mother*, how communal kitchens would be feasible and save the mother from much drudgery of food preparation. But our adherence to the concept of the home has made that technical solution unworkable; instead, we have turned to other technologies to ease the housework and cooking chores, albeit requiring more time and attention from mother.[27]

In other words, technological capabilities do not necessarily determine our actions. Indeed, how else can we explain why we have spent billions of dollars on nuclear power plants that we have had to abandon before they were completed? Obviously, other human factors proved more powerful than the combined technical and economic pressures.

[27]Ruth S. Cowan, *More Work for Mother: The Ironies of Household Technology from the Open Hearth to the Microwave* (New York, 1983), chap. 5.

Technology and the West through Britain's Industrial Revolution (to ca. 1850)

Technology in the Preindustrial West

TERRY S. REYNOLDS AND
STEPHEN H. CUTCLIFFE

From at least the Renaissance, Europeans and their cultural offspring in the Americas and elsewhere have regarded change, particularly technological change, as normal and desirable. For millennia this was not the case. The value systems of early cultures were radically different. They often had no interest in change, either social or technological, and did not regard it as especially desirable. This was particularly the case among prehistoric peoples, where traditional ways of doing things were very important. In this context, technology (the ability to manipulate or control nature) evolved slowly and sporadically.

This is not to say that technology did not have an important role in prehistoric cultures. It did. In the selection that opens this volume, "Technology and History: 'Kranzberg's Laws'," Melvin Kranzberg expresses certain truisms about the history of technology in the form of "laws." For example, Kranzberg's fifth law states that "All history is relevant, but the history of technology is the most relevant," and his sixth law asserts that "Technology is a very human activity." These two "laws" can be applied to humanity's prehistory, for, as Kranzberg notes, anthropologists and archaeologists looking at the prehistoric period have observed that technology, in the form of tool making, was one of the earliest and most basic of human cultural characteristics. In fact, one of the most widely used organizational tools for prehistory—the chronological division into Bone Age, Stone Age, Bronze Age, Iron Age—reflects the importance of technology to early peoples.[1]

Even though technology evolved very slowly among prehistoric peoples, by the dawning of the historical era (around 4000–3000 B.C.) humans had accumulated a considerable arsenal of technical instruments—plows, axes, sickles, saws, needles, looms, spindles, and spears, for example.[2] The next few thousand years would see the

[1] Leslie White's *The Evolution of Culture* (New York: McGraw-Hill, 1959) is a classic exposition of the central role of technology in cultural development. White develops an abstract theory of cultural evolution that makes technology the determinant of a society's social organization, which, in turn, determines that society's ideologies.

[2] Robert F. G. Spier's *From the Hand of Man: Primitive and Preindustrial Technologies* (Boston: Houghton Mifflin, 1970) provides a good overview of prehistoric technologies.

pace of change, both social and technological, increase significantly. The trigger was a closely related set of technological innovations—the development of settled, permanent agriculture relying on artificial irrigation in the river valleys of, first, Mesopotamia and, shortly after, Egypt. The changes that these innovations stimulated can be seen as a reflection of Kranzberg's second "law" ("Invention is the mother of necessity"), which suggests that every technological innovation requires additional innovations—both technical and social—to make it work effectively. Change begets change.

The notion that change begets change and the notion that close ties exist between technical and social change are probably nowhere better demonstrated than in the transformations wrought by the adoption of irrigated agriculture. Peter Drucker in his essay on the "First Technological Revolution and Its Lessons" argues that many of humanity's most fundamental social and political institutions emerged directly out of the establishment of the irrigation city, including law, standing armies, government as a distinct and permanent institution, social classes, schools, and teachers. Drucker, in fact, argues that major technological changes *require* subsequent social and political changes. Nonetheless, he points out, societies still have considerable control over the specifics of how they adjust to massive technological change. For instance, the coming of settled, permanent agriculture required the establishment of a system of organized defense (a standing military) to protect the food surpluses being produced by increasingly pacific farmers from more warlike hunters and nomads. Reacting to different geographical situations and cultural value systems, however, different civilizations adopted radically different means of meeting this need. The Egyptians used mercenaries, the Mesopotamians developed a warrior class, the ancient Greeks relied on the citizen soldier.

The irrigation civilizations that first developed many of the institutions still common to all civilized societies dominated the ancient Near East for several millennia. In the second and first millennia B.C., however, newer and younger civilizations emerged further to the west (hence the term "Western" civilization) and struck out in new directions.

Many of the values of the modern West emerged from the civilization established around the Aegean Sea in the area of modern Greece and the western seaboard of Turkey (Asia Minor) in the first millennium B.C. Greek tribes first migrated into the modern Greek peninsula around 2000 B.C., establishing a loose civilization closely related to earlier Minoan civilization, centered on the island of Crete. Internecine warfare and a new wave of invasions by other Greek-speaking tribes around 1200–1100 B.C. destroyed Mycenean civilization, initiating a "dark age" that lasted several centuries. By the 8th century B.C., however, the Greeks had begun to develop a new civilization based on the *polis* (city-state), maritime commerce, a system of writing derived from the Phoenician alphabet, and a unique set of beliefs and values. Many Western concepts like democracy, the importance of the individual, and the value of competition, logic, and reason have their origins in Hellenic Greece (ca. 750–336 B.C.).

Hellenic civilization borrowed heavily from the Near Eastern civilizations in the area of technology, including the wheel, smelting, building forms, sailing ships, brewing, glass making, and the potter's wheel and kiln, to name just a few. While innovative in the arts, in literature, in philosophy, in mathematics, and in what would later come to be called science, classical Greece added little to the arsenal of technology, especially mechanical technology. Classical Greece valued ideas rather more than applications. Technology was not regarded as a pursuit of importance, and the intellectual and political elites tended to hold manual and technical work in relatively low esteem.[3] Artisans and architects, nonetheless, played an essential role in Greek civic and commercial life, and Hellenic Greece made some significant contributions to technology in the way its craftsmen and craftswomen unified aesthetics with practicality in the crafts, a reflection of the values of Hellenic Greek society.[4]

The context of the Greek world changed dramatically in the 4th century B.C. after Philip II of Macedonia and his son Alexander, later to be called "the Great," invaded and conquered much of Greece. After Philip's assassination, Alexander led Greek armies against the Persian Empire. In eleven years he was able to conquer an empire stretching from Greece to India. One of the implements that enabled him to conquer such a vast empire in such a short time was the torsion catapult.

[3]For good overviews of the link between social conditions and the slow pace of technological progress in the ancient world, see M. I. Finley, "Technical Innovation and Economic Progress in the Ancient World," *Economic History Review*, 2d. ser., 18 (1965): 29–45; H. W. Pleket, "Technology and Society in the Graeco-Roman World," *Acta Historiae Neerlandica* 2 (1967): 1–25; and David W. Reece, "The Technological Weakness of the Ancient World," *Greece and Rome*, 2d. ser., 16 (1969): 32–47.

[4]An excellent example of a major Greek development in technical aesthetics is the exterior decoration of pottery. For details see, esp., Joseph-Veach Noble, "Pottery Manufacture," in *The Muses at Work: Arts, Crafts, and Professions in Ancient Greece and Rome*, ed. Carl Roebuck (Cambridge, Mass.: MIT Press, 1969), pp. 118–46. The important historic links between art and technology are reviewed by Cyril Stanley Smith, "Art, Technology and Science: Notes on Their Historical Interaction," *Technology and Culture* 11 (1970): 493–549.

Barton Hacker's essay "Greek Catapults and Catapult Technology: Science, Technology, and Work in the Ancient World" outlines the varieties of catapults used in antiquity, their operation, and their origins around 398 B.C. as a state-patronized technology in the Greek city-state of Syracuse in Sicily. There Dionysios, the ruler of Syracuse, gathered artisans from all over the Greek world to develop and fabricate siege engines to use against his Carthagenian enemies. Government support of missile research thus has very ancient roots.

By late in the 4th century B.C., catapults had spread to mainland Greece and, in improved form, were an important component of Alexander the Great's army. Before effective and accurate catapults, all the advantages lay with the defenders of city walls; few could hold back many. With the new technology, Alexander took walled towns with "a regularity and rapidity unknown in the . . . world before his time."

Even though his empire died with him, Alexander's conquests spread Greek rule over much of the Near East. The kingdoms established by Alexander's generals after his death maintained, with slowly diminishing effectiveness, this rule and shifted the center of Greek culture away from mainland Greece, displaced the *polis* (or city-state) in political importance, and, generally, altered the context of Greek civilization, ushering in what has been called the Hellenistic period (336–31 B.C.). The larger political units of Hellenistic Greek civilization, the expanded nature of political and military conflict between these units, and the concentration of more political power in the state proved to be more fertile ground for technological development than Hellenic Greek civilization, at least in the sphere of mechanical technology.

The perfection of the torsion catapults described in Hacker's essay provides one example of accelerated Hellenistic technological development. Hellenistic improvements to the torsion catapult, like the implement's beginnings in Syracuse, evolved from state support. In the Hellenistic world, as Hacker notes, military-related industries were "lavishly supported." Rulers "competed keenly for the services of the most able engineers" to improve their siege artillery and other military implements. In fact, the standardization of the torsion catapult in the Hellenistic era provides the only known example of the use of systematic experimentation and geometrical concepts in mechanical design in the ancient world.

To some extent, states had always supported technology. The early irrigation civilizations were dependent on state supervision of irrigation canal construction and operation, and the earliest states almost inevitably used their organizational powers to build large monumental or religious structures, such as the pyramids of Egypt or temples and civic buildings of Athens. State patronage, however, had seldom gone beyond structures to other areas of technology, particularly mechanical technology.

The mechanicians of the Hellenistic period did more than develop improved military machinery. They also introduced several of the fundamental elements of mechanical technology, including the pulley and complex geared mechanisms.[5] Lynn White mentions another of their contributions in passing in his essay "The Act of Invention: Causes, Contexts, Continuities and Consequences": the development of the helix—which evolved into that ubiquitous device, the screw. And, lest we forget that technology involves knowledge as well as artifacts, White also points out that the first evidence of alphabetization, a key method for organizing knowledge, emerged in the same era. Finally, the first evidence of the use of water power appears in the Hellenistic world.

In the 1st century B.C., weakened by internecine warfare and struggles with their non-Greek subjects, the Hellenistic world fell under the control of a people even further to the west: the Romans. The Romans showed more interest than the Greeks in technical achievement. Focusing more on "practical" structures than "useless" monuments, the Romans built more than 186,000 miles of roads, numerous bridges and aqueducts, dams, arenas, and impressive public buildings (the dome of the Pantheon, built in Rome in the second century, had a span 142 feet 6 inches, not bettered anywhere until the mid-19th century). They were the first people to make large-scale use of the semicircular arch and concrete in construction. In Rome, as in earlier civilizations, large-scale engineering was principally state-sponsored engineering, accomplished with traditional technical means and excellent organization. In the field of structures, the results of Roman state engineering were very impressive.

Such was not the case in other areas. In the private sector (areas such as metallurgy, textiles, pottery, and power), the Roman Empire widely diffused existing techniques, but provided few new technological developments and even these seem to have only slowly disseminated.[6] Rome's limited contributions to mechanical technology were due to the empire's social and economic structure. Like most ancient

[5]The most readable account of the contributions of Hellenistic engineers is L. Sprague de Camp, *The Ancient Engineers* (New York: Dorset, 1990); 114–63. The book was originally published in 1960. For additional information on mechanical developments in the Hellenistic period, see A. G. Drachmann, *The Mechanical Technology of Greek and Roman Antiquity* (Copenhagen: Munksgaard, 1963).

[6]For additional details on Roman technology, see J. G. Landels, *Engineering in the Ancient World* (Berkeley: University of California Press, 1978); or K. D. White, *Greek and Roman Technology* (Ithaca, N.Y.: Cornell University Press, 1984).

civilizations, Rome was controlled by a small, propertied elite with values that belittled those involved in trade and manufacturing. Painstakingly produced handicraft products were more important to this elite than standardized, industrial products. Moreover, Roman civilization frequently enjoyed labor surpluses, whether in the form of unemployed or underemployed citizens or masses of slaves. Hence there was little incentive to develop laborsaving machinery or reward for doing so. Even Rome's massive public works relied on the labor of armies of slaves or soldiers little aided by machinery. Mechanical ingenuity went not into productive machinery but into providing *automata* (ingenious devices) for the entertainment of the upper classes.[7]

The western portion of the Roman Empire collapsed in the 5th century A.D. from a combination of inner weaknesses and pressure from the Germanic peoples outside. Western civilization emerged from the ashes of Rome in the West, fertilized by the Germanic people who succeeded them, and moderated by the one universal institution—the Roman Catholic Church—that survived the collapse of the Empire.

Europe's medieval period (ca. 500–1500) has often been pictured as a "dark age." Culturally it was. Early Western civilization was overwhelmingly rural, poor, and politically fragmented; literacy was rare and often restricted to the church and its agents. As a result, we know few details of the technological history of the Middle Ages. Indeed, as White points out in his essay on "The Act of Invention," we know little about how even such widely used things as coins, alphabetization, and buttons emerged.

Nonetheless, White's essay suggests that the first millennia of "Western" civilization was substantially more technologically innovative than commonly believed. He cites medieval Western development of the crankshaft, the weight-driven trebuchet, cannon, mechanical clocks and watches, and the button as evidence. Contributing to the medieval West's technological growth was its willingness to borrow from other civilizations. White's account indicates that the West took the stirrup from nomadic tribes to the east and used it to revolutionize its methods of waging war. It took manpowered Chinese projectile-throwing engines and improved them to operate using falling weights. The West probably learned of the rotary blower, gunpowder, compass, and paper from China and then significantly improved on all of them. While the medieval West may have lagged far

[7]See the works by Finley, Reece, and Pleket cited in n. 3 above for discussions of Rome's limited accomplishments in mechanical technology.

behind other civilizations in other elements of culture, White believes that it was a technologically precocious civilization.[8]

White's essay provides additional evidence of the importance of context to technological history. During the early medieval period the most popular weapon was the long, laminated slashing sword, a weapon developed in the Roman era but ignored by the Empire because its adoption would have required abandonment of a long-standing military system based on close formations, drill, and discipline. The fall of the Empire changed the social context, enabling the new technology to emerge and flourish, until, in the 10th century, further contextual changes caused it, too, to disappear. Similarly, numerous societies had encountered the stirrup before it reached western Europe in the 8th century. Only in Europe did social, political, and military conditions prompt the development of a new method of fighting based on the stirrup. In other words, a technological idea that fails to develop in one social context may, when transferred to another, suddenly expand and flourish.

One of the critical elements of medieval European technology was the watermill. It provides another good example of how a technology, when transported to a different social context, may suddenly flourish. The first documentary evidence of the use of waterpower comes in the Hellenistic period—around the 1st century B.C., but before the 3rd or 4th century A.D., the mill found limited application. The labor surpluses of the early Roman Empire probably account for this, for it is only in the labor-short late Roman Empire that archaeological evidence indicates growing use of waterpower.

The collapse of organized government in the West and several centuries of political, economic, and social chaos no doubt set back growth, but the watermill became one of the foundations of medieval European society, flourishing in labor-short medieval Europe as it never had during the Pax Romana. By 1086 England alone had over 5,000 water-driven mills, and by 1400 Europeans were applying water

[8]In other essays, White has argued for the medieval West's technological prowess, including *Medieval Technology and Social Change* (New York: Oxford University Press, 1962); "What Accelerated Technological Progress in the Western Middle Ages?" in Scientific Change, ed. A. C. Crombie (London: Heinemann, 1963), pp. 272–91; "Cultural Climates and Technological Advance in the Middle Ages," *Viator* 2 (1971): 171–201, and "Technology Assessment from the Stance of a Medieval Historian," *American Historical Review*, 79 (1974): 1–13. Other good reviews of technology in the medieval West include Jean Gimpel, *The Medieval Machine: The Industrial Revolution of the Middle Ages* (New York: Holt, Rinehart & Winston, 1976); and Frances and Joseph Gies, *Forge, Cathedral, and Waterwheel: Technology and Invention in the Middle Ages* (New York: HarperCollins, 1994).

to a host of tasks never imagined by either earlier or even contemporary civilizations.[9] "The chief glory of the later Middle Ages," White wrote elsewhere, "was not its cathedrals or its epics or its scholasticism: it was the building for the first time in history of a complex civilization which rested not on the backs of sweating slaves or coolies but primarily on non-human power."[10]

While water formed one foundation of medieval European technology, wood formed the other.[11] In medieval Europe wood was the construction material of choice and the predominant source of thermal energy. As William Te Brake notes in his essay "Air Pollution and Fuel Crises in Preindustrial London, 1250–1650," wood was used to construct most dwellings and all ships, and charcoal (charred wood) was essential for smelting, glassmaking, soapmaking, brewing, making plaster, and a host of other industries. "In almost every way imaginable," Te Brake concludes, "woodland was essential to preindustrial European society."

If medieval technology was a wood-and-water complex, wood was the first element of that complex to show signs of weakness. The height of the European medieval period was the period between 1000 and 1300. By the mid-13th century, industrial expansion (based on the use of wood as a fuel) and the expansion of cultivated lands had created, as Te Brake shows, significant wood shortages in parts of England. This pushed both industrial and domestic users of wood to a less desirable and much more noxious alternative fuel, sea coal, and led to an early instance of severe atmospheric pollution, often considered largely a 20th-century phenomenon.

The transition from wood to coal as a source of thermal energy—one of the technical characteristics of Europe's industrial revolution in the late 18th and early 19th centuries—might have occurred earlier had not medieval Europe's golden age come to an end in the 14th century. Overpopulation and an era of colder and wetter weather led to famine and malnutrition early in the century. When the Black Death reached Europe around 1350, it struck an already weakened population with disastrous effects. England's population, Te Brake notes, dropped from around 3.7 million in the 1340s to a low of around 2.1 million in the early 1400s and only very slowly rebounded.

[9] For medieval development of water power, see Terry S. Reynolds, *Stronger than a Hundred Men: A History of the Vertical Water Wheel* (Baltimore, Md.: Johns Hopkins University Press, 1983), pp. 47–121.

[10] Lynn White, jr., "Technology and Invention in the Middle Ages," *Speculum* 15 (1940): 155–56.

[11] Lewis Mumford described medieval European technology (which he referred to as "eotechnic technology") as a "wood-and-water complex" in his classic *Technics and Civilization* (New York: Harcourt, Brace & World, 1934), p. 110.

The population decline decreased the pressure on wood resources; only in the second half of the 16th century, with a revived population, did industrial and domestic demands for wood once again drive the English to the use of coal with its resultant problems.

The pressures population and industrial growth put on wood supplies in 13th-century Europe and the aggressive expansion of water power throughout the medieval period heralded the emergence of a technologically aggressive civilization. White and others have maintained that it was in the medieval West that a civilization embraced for the first time a set of values that emphasized man's domination over nature and had few reservations about environmental manipulation. Only in the Latin West were activities that emphasized domination over nature, such as manual labor, land reclamation, the damming of streams, and labor-saving invention, considered not only good, but even as acts of religious piety.[12]

Te Brake's study of English wood supply demonstrates the importance of context in technological choices relating to fuel. Thomas Esper's essay, "The Replacement of the Longbow by Firearms in the English Army," demonstrates the same point with military armaments. From at least the 13th century onward, English arms enjoyed considerable success in medieval Europe in large part because of the technological superiority of the English longbow over rival arms, such as the crossbow and the armored knight. As late as the 18th century, Esper notes, it was still possible for intelligent observers to argue that the longbow was technically superior to handheld firearms. Nonetheless, in the late 16th century, the English gave up the accurate, rapid-firing longbow in favor of relatively primitive, inaccurate, slow-firing, handheld firearms. In explaining why and how this transition occurred, Esper makes several key points. First, he demonstrates that the "superiority" of a technology is not simply a function of the artifact; it is also heavily dependent on the skills and training of the user. Without appropriately trained human operators, "superior" technologies can be displaced by "inferior" technologies which require less training. Second, Esper illustrates the importance of social context in technological history by pointing out how changing recreation patterns played a key role in the demise of the skills needed to keep the longbow a potent weapon.[13]

In terms of its relationship to the rest of the world, the period from

[12] See, e.g., Lynn White, jr., "The Historical Roots of Our Ecological Crisis," in White, *Machina ex deo: Essays in the Dynamics of Western Culture* (Cambridge, Mass., and London: MIT Press, 1969), pp. 75–94.

[13] For another good case study where social and geographical conditions enabled a seemingly "superior" technology to be replaced by an "inferior" one, see Richard W. Bulliet, *The Camel and the Wheel* (Cambridge, Mass.: Harvard University Press, 1975).

1450 to 1550 was decisive to Western civilization. By 1450 the West had recovered economically and socially from the devastations of the 14th century: climactic change, famine, malnutrition, plague, and war. In the next century, influenced by its continued political fragmentation and the rise of cities dominated by a dynamic merchant class, the West steadily improved a variety of technologies—especially in the transportation and military areas—that were to give it an advantage over other civilizations and to permit it to expand its influence far beyond its original European homeland.

One of the key Western innovations was the cannon-armed, oceangoing sailing vessel and the associated knowledge of how to navigate out of sight of land. Development of oceangoing navigation and large cannon-armed sailing vessels enabled Western nations in the late 15th century to develop trading routes around Africa to the Indies that bypassed the Muslims. It also enabled the West to discover two previously unknown (to Europeans, Africans, and Asians) continents: North America and South America. The inhabitants of these continents, long isolated from the bulk of humanity, proved extremely vulnerable to European diseases (e.g., smallpox) and weapons, and thus to conquest, plunder, and absorption into the orbit of Western civilization. The voyages of de Gama to India and Columbus to America in the 1490s, as John Law notes in his essay "On the Social Explanation of Technical Change: The Case of the Portuguese Maritime Expansion," signaled a change in the balance of power between Europe and the rest of the world.

Law is a sociologist, and his essay, written from the perspective of a sociologist rather than a historian, uses some sociological jargon. Nonetheless, the central portion of his piece is useful for its introduction of the concept of systems building, or heterogeneous engineering, another way of describing Kranzberg's Third Law: "Technology comes in packages, big and small."[14] Law argues that the maritime developments that enabled the Portuguese eventually to round Africa and dominate the Indian Ocean were examples of heterogeneous engineering. In other words, the Portuguese brought together both technological and social elements to create a maritime system capable of overcoming both natural and human opposition (e.g., the power of ocean storms and the resistance of Muslim traders). These elements included square rigging, reduced use of man power, greater ship storage space, lighter and cheaper cannons, navi-

[14] For a good example of the importance of looking at technology in terms of systems, see Thomas P. Hughes, *Networks of Power: Electrification in Western Society, 1880–1930* (Baltimore, Md.: Johns Hopkins University Press, 1983).

gation instruments, and knowledge of astronomy, currents, and prevailing winds. Portuguese systems building produced a sailing vessel that "was essentially a compact device that allowed a relatively small crew to master unparalleled masses of inanimate energy for movement and destruction."

Law notes that Portugual's maritime systems also relied on elements of "social engineering": the development of knowledge and the means of training people to use it. Portuguese officials convened a "scientific commission" in the 1480s to find improved navigation methods, that is, to transform esoteric scientific knowledge into applicable practices. They then reduced the commission's findings to a set of instructions that converted vessels and their instruments into astronomical observatories, took steps to secure new data on the location of coastal features, and developed institutions to train new navigators. Portuguese "heterogeneous" and "social" engineering worked. Within fifteen years of de Gama's first visit in 1498, Portugal had completely destroyed the naval and commercial power of the Arabs in Indian waters and had taken control of oceangoing trade in the region.[15]

Simultaneously with European maritime expansion, European natural philosophers revolutionized the nature of scientific inquiry, led by the work of Copernicus, Kepler, Galileo, and Newton. By the 16th century, European science had begun to focus on detailed narrow problems where measurements could be made and assumptions tested, instead of on the broad, universal questions of ancient and medieval philosophers where neither solutions nor proofs were possible. European scientists had begun to seek quantitative descriptions of how things happened rather than qualitative explanations of why. These changes in approach established the foundations of modern science. However, the simplifications early Western scientists made in order to establish basic principles (e.g., ignoring friction or air resistance), meant that the principles they established initially had little impact on technological practice. The problems encountered by practicing technicians and engineers in the real world were simply too complex for the scientific laws being derived. The use of mathematics and systematic measurement and the concentration on narrowly conceived problems, however, laid the base for future interactions between the new science and practical technology.

Near the beginning of the 18th century European engineers began

[15] For an extended discussion of Western maritime technological developments in this era, see Carlo M. Cipolla, *Guns, Sails, and Empires: Technological Innovation and the Early Phases of European Expansion, 1400–1700* (New York: Pantheon, 1965).

to attempt to apply theoretical mechanics and the newly discovered calculus to practical problems. For example, between 1700 and 1740 French engineers used mechanics and calculus to analyze the operation of the waterwheel, and by 1750 both French and English engineers were using systematic, quantitative experimentation to understand better and to improve the operation of this basic prime mover of industry. Not surprisingly, military needs also prompted early attempts to apply science to practical ends, as Brett Steele's "Muskets and Pendulums: Benjamin Robins, Leonhard Euler, and the Ballistic Revolution" demonstrates. War often dominated early modern European political life, and the state, the chief agent of civilized warfare, had supported the application of mathematical and theoretical knowledge to weapon design and operation as early as the Hellenistic catapult designers, as we have seen.

From Galileo's work in the 1630s onward, those familiar with theoretical mechanics recognized its potential application to artillery fire. Galileo even attempted to apply his new mechanics to that use. By the mid-18th century the potential of applying quantitative theory and systematic quantitative experimentation to practical problems was more widely recognized. Steele's study of the "ballistic revolution" of the 18th century contradicts the once prevalent idea that practical technology and the mathematical sciences remained essentially separate until the 19th century. He describes how Benjamin Robins, Leonhard Euler, and a host of other engineers and mathematicians in France, Prussia, Austria, Britain, and Sardinia applied mechanics and systematic, quantitative experimentation to revolutionize ballistics as a science and create knowledge that made the use of artillery a far more precise thing. In the process, they made calculus and mechanics an important element in the training of artillery and engineering officers all over Europe and influenced the future course of formal civilian engineering education, which often evolved from military precedents.[16]

The importance of government as patron of technology is clear from a number of the essays in this volume. State patronage played a role in the development of monumental architecture in the ancient civilizations, torsion artillery in the Hellenistic period, transcontinental oceanic navigation in the 15th and 16th centuries, and the ballistics

[16] For introductory information on the influence of military engineering education on the later evolution of civilian engineering education, see Frederick B. Artz, *The Development of Technical Education in France, 1500–1850* (Cambridge, Mass.: Society for the History of Technology and MIT Press, 1966); and Peter M. Molloy, "Technical Education and the Young Republic: West Point as America's École Polytechnique, 1802–1833" (Ph.D. dissertation, Brown University, 1975).

revolution of the 18th century. However, technological innovations also emerged from anonymous craftsmen and craftswomen practicing their trades. Certainly such widely used items as the button, the mechanical clock, improved plows, the crank, the rotary blower, and the screw, all mentioned in White's essay, and the fireplace and flue, mentioned in Brake's essay, emerged from inventors and craftsmen working without state patronage. The same can be said of the numerous improvements made in the watermill and its applications, the windmill and its applications, the blast furnace, and a host of other technologies.

This was certainly the case with the mechanization of textile production in late-18th-century England, a task carried out by craftsmen and entrepreneurs without government sponsorship and, in some sense, working against established laws and tradition. Before 1770 textile production was largely carried out inside homes by families working with hand-operated carding combs, spinning wheels, and looms. Between 1770 and 1800 a series of textile inventions revolutionized production by mechanizing one after another of the numerous tasks involved in producing cloth. The more complex, powerful, and expensive machinery emerging from these inventions required nonhuman power sources and capital investments far beyond those involved in handicraft production. They permitted the centralization of production in the factory and the imposition of much higher levels of control over the workforce—now stripped of its control over the means of production, the work site, and the pace of work.

Kranzberg's "First Law" asserts: "Technology is neither good, nor bad, nor is it neutral." Adrian J. Randall's essay, "The Philosophy of Luddism: The Case of the West of England Woolen Workers, ca. 1790–1809," readily demonstrates this with respect to the introduction of mechanized textile machinery, often considered the beginning of the Industrial Revolution and modern machine-oriented civilization.

Aggressive wool entrepreneurs viewed the new machine technology as good. They touted the "benefits" of centralized production, pointing to the larger volumes of standardized goods at lower prices produced by the new textile machinery. Some of their rhetoric, as elucidated by Randall, may sound familiar today. For example, they argued that any region failing to adopt machinery would "inevitably lose its trade to rivals in other regions and overseas, thereby ruining not only the capitalists but also the workers." Moreover, they warned local officials that interference with their operations would force them to pack up their operations and move. Such positions are paralleled today when manufacturers argue that, if regions do not provide them with tax breaks or labor unions do not grant wage and benefit concessions, they will have to move elsewhere to remain competitive.

While entrepreneurs viewed the adoption of a machine economy as both good and necessary, woolen workers, as Randall notes, viewed matters otherwise. They felt that the new machinery brought not salvation but ruin. Representing a traditional culture that placed a premium on stability, regulation, and custom, they quite rightly recognized that the coming machine economy represented an alien value system. Clearly what was "good" for one element of industrializing society was "bad" for another, and the new textile technologies were certainly not neutral.

The mechanization of the textile industry in late-18th-century Britain and the parallel development of the steam engine and coal-produced iron set off what is often referred to as Britain's "Industrial Revolution." This "revolution" consisted of a host of social, economic, and technological changes. Technologically, it saw the transition from wood to coal as the primary source of thermal energy, the centralization of production in mechanized factories, the replacement of water power by steam as a source of mechanical power, and the emergence of the steam-powered railroad as the primary means of transportation.[17] These technological changes were interconnected. In other words, just as Portugal's development of oceanic navigation can be considered as a system, so can Britain's Industrial Revolution. The expansion and centralization of production through the use of machines and the transition from wood to coal as a fuel source required the development of an adequate transportation system to bring raw materials to factories and to export the vastly increased volume of finished products. In a sense, this development illustrates Kranzberg's "Second Law": "Invention is the mother of necessity." Technical innovations require additional technical innovations to make them fully effective.

Francis Evans's "Roads, Railways, and Canals: Technical Choices in 19th-Century Britain" illustrates the struggles involved in developing an appropriate transportation system for Britain's expanding industrial economy. As Evans points out, in the mid-18th century the

[17] There is an extensive literature on the Industrial Revolution, including some questioning of the utility of the concept. The classic study of Britain's industrial revolution is T. S. Ashton, *The Industrial Revolution, 1760–1830* (London: Oxford University Press, 1948). Other good overviews include David S. Landes, *The Unbound Prometheus: Technological Change and Industrial Development in Western Europe from 1750 to the Present* (Cambridge: Cambridge University Press, 1969); Christine MacLeod, *Inventing the Industrial Revolution* (Cambridge: Cambridge University Press, 1988); and Peter N. Stearns, *The Industrial Revolution in World History* (Boulder, Colo.: Westview, 1993), pp. 17–40.

very idea that roads were intended to accommodate wheeled vehicles was in question. Wheeled vehicles were regarded as what destroyed roads, as what roads needed to be protected against.

Not surprisingly, Britain's fast-growing economy quickly outstripped the capabilities of existing transportation technologies, leading to improved roads and the construction of extensive systems of canals. Canals, however, required locks or a suitable substitute to overcome gradients, and the high water requirements of locks, Evans points out, put a practical limit on the volume of commerce canals could handle and, in turn, created problems for other users of water, notably, industry, which was still heavily dependent on water power. This permitted railroads, at first rightly regarded as merely supplements to the canal system, to emerge as the predominant form of transportation.

Evans's review of the competition between transportation technologies during Britain's industrial expansion also provides some insights into technical change. Often technical change is viewed simplistically. A superior technology—say, the steam engine—emerges, and the inferior technology against which it competes—the traditional wooden waterwheel—soon disappears. Evans's account makes it clear that this is often not what happens. Older technologies do not disappear. They steadily improve, often borrowing ideas from the newer technologies to remain competitive and, in improved form, remain in operation as alternatives or supplements to the newer technologies. Finally, the success or failure of technologies, as we have seen, is often dependent on social context. But social context cannot explain everything. Although social and political factors have often been blamed for the failure of the road steamer in the early 19th century, Evans argues convincingly that the key factors behind its failure were technological rather than social.

* * *

The West in A.D. 1000 had been among the smallest, poorest, and weakest of the world's civilizations. By the mid-19th century—building on social values that promoted the domination of nature, on the stimulus of labor shortages, and on a decentralized political system that permitted the rise of a strong middle class and encouraged military development—the West had developed an industrial and technological complex significantly more powerful than that of contemporary civilizations. Its military and nautical technologies had given it control over the world's oceans and had permitted it to expand its domination over two new continents. The stone-and-man complex of the classical civilizations and the wood-and-water complex of medieval European civilization had been displaced, or were in the process of being displaced, by the coal-and-iron complex that had emerged in Britain in the late 18th century. The decentralized production of goods by craftsmen—the dominant form of production for millennia—had been successfully challenged in the textile industry and would soon be challenged in other areas as well. The ramifications of the new technologies and the economic and value systems that accompanied them would shortly be felt throughout Western society and, indeed, throughout the world.

The First Technological Revolution and Its Lessons

PETER F. DRUCKER

Aware that we are living in the midst of a technological revolution, we are becoming increasingly concerned with its meaning for the individual and its impact on freedom, on society, and on our political institutions. Side by side with messianic promises of utopia to be ushered in by technology, there are the most dire warnings of man's enslavement by technology, his alienation from himself and from society, and the destruction of all human and political values.

Tremendous though today's technological explosion is, it is hardly greater than the first great revolution technology wrought in human life seven thousand years ago when the first great civilization of man, the irrigation civilization, established itself. First in Mesopotamia, and then in Egypt and in the Indus Valley, and finally in China there appeared a new society and a new polity: the irrigation city, which then rapidly became the irrigation empire. No other change in man's way of life and in his making a living, not even the changes under way today, so completely revolutionized human society and community. In fact, the irrigation civilizations were the beginning of history, if only because they brought writing.

The age of the irrigation civilization was pre-eminently an age of technological innovation. Not until a historical yesterday, the eighteenth century, did technological innovations emerge which were comparable in their scope and impact to those early changes in technology, tools, and processes. Indeed, the technology of man remained essentially unchanged until the eighteenth century insofar as its impact on human life and human society is concerned.

Peter Drucker has since 1971 been Clarke Professor of Social Science and Management at the Claremont Graduate School of the Claremont Colleges in Claremont, California. From 1950 until 1971 he was professor of management at the Graduate Business School of New York University. Among his more recent books are *The New Realities* (1989), *The Post Capitalist Society* (1993), and *Managing in a Time of Great Changes* (1995). This article is Dr. Drucker's presidential address to the Society for the History of Technology, presented on December 29, 1965, in San Francisco.

But the irrigation civilizations were not only one of the great ages of technology. They represent also mankind's greatest and most productive age of social and political innovation. The historian of ideas is prone to go back to ancient Greece, to the Old Testament prophets, or to the China of the early dynasties for the sources of the beliefs that still move men to action. But our fundamental social and political institutions antedate political philosophy by several thousand years. They all were conceived and established in the early dawn of the irrigation civilizations. Any one interested in social and governmental institutions and in social and political processes will increasingly have to go back to those early irrigation cities. And, thanks to the work of archeologists and linguists during the last fifty years, we increasingly have the information, we increasingly know what the irrigation civilizations looked like, we increasingly can go back to them for our understanding both of antiquity and of modern society. For essentially our present-day social and political institutions, practically without exception, were then created and established. Here are a few examples.

1. The irrigation city first established government as a distinct and permanent institution. It established an impersonal government with a clear hierarchical structure in which very soon there arose a genuine bureaucracy—which is of course what enabled the irrigation cities to become irrigation empires.

Even more basic: the irrigation city first conceived of man as a citizen. It had to go beyond the narrow bounds of tribe and clan and had to weld people of very different origins and blood into one community. This required the first super-tribal deity, the god of the city. It also required the first clear distinction between custom and law and the development of an impersonal, abstract, codified legal system. Indeed, practically all legal concepts, whether of criminal or of civil law, go back to the irrigation city. The first great code of law, that of Hammurabi, almost four thousand years ago, would still be applicable to a good deal of legal business in today's highly developed, industrial society.

The irrigation city also first developed a standing army—it had to. For the farmer was defenseless and vulnerable and, above all, immobile. The irrigation city which, thanks to its technology, produced a surplus, for the first time in human affairs, was a most attractive target for the barbarian outside the gates, the tribal nomads of steppe and desert. And with the army came specific fighting technology and fighting equipment: the war horse and the chariot, the lance and the shield, armor and the catapult.

2. It was in the irrigation city that social classes first developed. It needed people permanently engaged in producing the farm products on

which all the city lived; it needed farmers. It needed soldiers to defend them. And it needed a governing class with knowledge, that is, originally a priestly class. Down to the end of the nineteenth century these three "estates" were still considered basic in society.[1]

But at the same time the irrigation city went in for specialization of labor resulting in the emergence of artisans and craftsmen: potters, weavers, metal workers, and so on; and of professional people: scribes, lawyers, judges, physicians.

And because it produced a surplus it first engaged in organized trade which brought with it not only the merchant but money, credit, and a law that extended beyond the city to give protection, predictability, and justice to the stranger, the trader from far away. This, by the way, also made necessary international relations and international law. In fact, there is not very much difference between a nineteenth-century trade treaty and the trade treaties of the irrigation empires of antiquity.

3. The irrigation city first had knowledge, organized it, and institutionalized it. Both because it required considerable knowledge to construct and maintain the complex engineering works that regulated the vital water supply and because it had to manage complex economic transactions stretching over many years and over hundreds of miles, the irrigation city needed records, and this of course meant writing. It needed astronomical data, as it depended on a calendar. It needed means of navigating across sea or desert. It therefore had to organize both the supply of the needed information and its processing into learnable and teachable knowledge. As a result, the irrigation city developed the first schools and the first teachers. It developed the first systematic observation of natural phenomena, indeed, the first approach to nature as something outside of and different from man and governed by its own rational and independent laws.

4. Finally, the irrigation city created the individual. Outside the city, as we can still see from those tribal communities that have survived to our days, only the tribe had existence. The individual as such was neither seen nor paid attention to. In the irrigation city of antiquity, however, the individual became, of necessity, the focal point. And with this emerged not only compassion and the concept of justice; with it emerged the arts as we know them, the poets, and eventually the world religions and the philosophers.

This is, of course, not even the barest sketch. All I wanted to suggest is the scope and magnitude of social and political innovation that underlay the rise of the irrigation civilizations. All I wanted to stress is

[1] See the brilliant though one-sided book by Karl A. Wittvogel, *Oriental Despotism: A Comparative Study of Total Power* (New Haven, Conn., 1957).

that the irrigation city was essentially "modern," as we have understood the term, and that, until today, history largely consisted in building on the foundations laid five thousand or more years ago. In fact, one can argue that human history, in the last five thousand years, has largely been an extension of the social and political institutions of the irrigation city to larger and larger areas, that is, to all areas on the globe where water supply is adequate for the systematic tilling of the soil. In its beginnings, the irrigation city was the oasis in a tribal, nomadic world. By 1900 it was the tribal, nomadic world that had become the exception.

The irrigation civilization was based squarely upon a technological revolution. It can with justice be called a "technological polity." All its institutions were responses to opportunities and challenges that new technology offered. All its institutions were essentially aimed at making the new technology most productive.

* * *

I hope you will allow me one diversion.

The history of the irrigation civilizations has yet to be written. There is a tremendous amount of material available now, where fifty years ago we had, at best, fragments. There are splendid discussions available of this or that irrigation civilization, for instance of Sumer. But the very big job of recreating this great achievement of man and of telling the story of his first great civilization is yet ahead of us.

This should be pre-eminently a job for historians of technology such as we profess to be. At the very least the job calls for a historian with high interest in, and genuine understanding of, technology. The essential theme around which this history will have to be written must be the impacts and capacities of the new technology and the opportunities and challenges which this, the first great technological revolution, presented. The social, political, cultural institutions, familiar though they are to us today—for they are in large measure the institutions we have been living with for five thousand years—were all brand-new then, and were all the outgrowth of new technology and of attempts to solve the problems the new technology posed.

It is our contention in the Society for the History of Technology that the history of technology is a major, distinct strand in the web of human history. We believe that the history of mankind cannot be properly understood without relating to it the history of man's work and man's tools, that is, the history of technology. Some of our colleagues and friends—let me mention only such familiar names as Lewis Mumford, Fairfield Osborne, Joseph Needham, R. J. Forbes, Cyril Stanley Smith, and Lynn White—have in their own works brilliantly demonstrated the profound impact of technology on political, social, economic, and cultural history. But while technological change has always had impact on the way men live and work, surely at no other time has technology so literally influenced civilization and culture as during the first technological revolution, that is, during the rise of the irrigation civilizations of antiquity.

Only now, however, is it possible to tell the story. No longer can its neglect be justified. For the facts are available, as I stated before. And we now, because we live in a technological revolution ourselves, are capable of understanding what happened then—at the very dawn of history. There is a big job to be done: to show that the traditional approach to our history—the approach taught in our schools—in which "relevant" history really begins with the Greeks (or with the Chinese dynasties), is short-sighted and distorts the real "ancient civilization."

* * *

I have, however, strayed off my topic: the question I posed at the beginning, what we can learn from the first technological revolution regarding the impacts likely to result on man, his society, and his government from the new industrial revolution, the one we are living in. Does the story of the irrigation civilization show man to be determined by his technical achievements, in thrall to them, coerced by them? Or does it show him capable of using technology to his own, to human ends, and of being the master of the tools of his own devising?

The answer which the irrigation civilizations give us to this question is threefold.

1. Without a shadow of doubt, major technological change creates the need for social and political innovation. It does make obsolete existing institutional arrangements. It does require new and very different institutions of community, society, and government. To this extent there can be no doubt: technological change of a revolutionary character coerces; it *demands innovation.*

2. The second answer also implies a strong necessity. There is little doubt, one would conclude from looking at the irrigation civilizations, that specific technological changes demand equally specific social and political innovations. That the basic institutions of the irrigation cities of the Old World, despite great cultural difference, all exhibited striking similarity may not prove much. After all, there probably was a great deal of cultural diffusion (though I refuse to get into the quicksand of debating whether Mesopotamia or China was the original innovator). But the fact that the irrigation civilizations of the New World around the Lake of Mexico and in Maya Yucatán, though culturally

completely independent, millennia later evolved institutions which, in fundamentals, closely resemble those of the Old World (e.g., an organized government with social classes and a permanent military, and writing) would argue strongly that the solutions to specific conditions created by new technology have to be specific and are therefore limited in number and scope.

In other words, one lesson to be learned from the first technological revolution is that new technology creates what a philosopher of history might call "objective reality." And objective reality has to be dealt with on *its* terms. Such a reality would, for instance, be the conversion, in the course of the first technological revolution, of human space from "habitat" into "settlement," that is, into a permanent territorial unit always to be found in the same place—unlike the migrating herds of pastoral people or the hunting grounds of primitive tribes. This alone makes obsolete the tribe and demands a permanent, impersonal, and rather powerful government.

3. But the irrigation civilizations can teach us also that the new objective reality determines only the gross parameters of the solutions. It determines where, and in respect to what, new institutions are needed. It does not make anything "inevitable." It leaves wide open *how* the new problems are to be tackled, what the purposes and values of the new institutions are to be.

In the irrigation civilizations of the New World the individual, for instance, failed to make his appearance. Never as far as we know, did these civilizations get around to separating law from custom nor, despite a highly developed trade, did they invent money, and so on.

Even within the Old World, where one irrigation civilization could learn from the others, there were very great differences. They were far from homogeneous even though all had similar tasks to accomplish and developed similar institutions for these tasks. The different specific answers expressed above all different views regarding man, his position in the universe, and his society—different purposes and greatly differing values.

Impersonal bureaucratic government had to arise in all these civilizations; without it they could not have functioned. But in the Near East it was seen at a very early stage that such a government could serve equally to exploit and hold down the common man and to establish justice for all and protection for the weak. From the beginning the Near East saw an ethical decision as crucial to government. In Egypt, however, this decision was never seen. The question of the purpose of government was never asked. And the central quest of government in China was not justice but harmony.

It was in Egypt that the individual first emerged, as witness the many statues, portraits, and writings of professional men, such as scribes and administrators, that have come down to us—most of them superbly aware of the uniqueness of the individual and clearly asserting his primacy. It is early Egypt, for instance, which records the names of architects who built the great pyramids. We have no names for the equally great architects of the castles and palaces of Assur or Babylon, let alone for the early architects of China. But Egypt suppressed the individual after a fairly short period during which he flowered (perhaps as part of the reaction against the dangerous heresies of Ikhnaton). There is no individual left in the records of the Middle and New Kingdoms, which perhaps explains their relative sterility.

In the other areas two entirely different basic approaches emerged. One, that of Mesopotamia and of the Taoists, we might call "personalism," the approach that found its greatest expression later in the Hebrew prophets and in the Greek dramatists. Here the stress is on developing to the fullest the capacities of the person. In the other approach—we might call it "rationalism," taught and exemplified above all by Confucius—the aim is the molding and shaping of the individual according to pre-established ideals of rightness and perfection. I need not tell you that both these approaches still permeate our thinking about education.

Or take the military. Organized defense was a necessity for the irrigation civilization. But three different approaches emerged: a separate military class supported through tribute by the producing class, the farmers; the citizen-army drafted from the peasantry itself; and mercenaries. There is very little doubt that from the beginning it was clearly understood that each of these three approaches had very real political consequences. It is hardly coincidence, I believe, that Egypt, originally unified by overthrowing local, petty chieftains, never developed afterwards a professional, permanent military class.

Even the class structure, though it characterizes all irrigation civilizations, showed great differences from culture to culture and within the same culture at different times. It was being used to create permanent castes and complete social immobility, but it was also used with great skill to create a very high degree of social mobility and a substantial measure of opportunities for the gifted and ambitious.

Or take science. We now know that no early civilization excelled over China in the quality and quantity of scientific observations. And yet we also know that early Chinese culture did not point toward anything we would call "science." Perhaps because of their rationalism the Chinese refrained from generalization. And though fanciful and speculative, it is the generalizations of the Near East or the mathematics of Egypt which point the way toward systematic science. The Chinese, with their superb gift for accurate observation, could obtain an enormous amount of information about nature. But their view of the universe remained totally unaffected thereby—in sharp contrast to what we know about the Middle Eastern developments out of which Europe arose.

In brief, the history of man's first technological revolution indicates the following:

1. Technological revolutions create an objective need for social and political innovations. They create a need also for identifying the areas in which new institutions are needed and old ones are becoming obsolete.

2. The new institutions have to be appropriate to specific new needs. There are right social and political responses to technology and wrong social and political responses. To the extent that only a right institutional response will do, society and government are largely circumscribed by new technology.

3. But the values these institutions attempt to realize, the human and social purposes to which they are applied, and, perhaps most important, the emphasis and stress laid on one purpose as against another, are largely within human control. The bony structure, the hard stuff of a society, is prescribed by the tasks it has to accomplish. But the ethos of the society is in man's hands and is largely a matter of the "how" rather than of the "what."

* * *

For the first time in thousands of years, we face again a situation that can be compared with what our remote ancestors faced at the time of the irrigation civilization. It is not only the speed of technological change that creates a "revolution," it is its scope as well. Above all, today, as seven thousand years ago, technological developments from a great many areas are growing together to create a new human environment. This has not been true of any period between the first technological revolution and the technological revolution that got under way two hundred years ago and has still clearly not run its course.

We therefore face a big task of identifying the areas in which social and political innovations are needed. We face a big task in developing the institutions for the new tasks, institutions adequate to the new needs and to the new capacities which technological change is casting up. And, finally, we face the biggest task of them all, the task of insuring that the new institutions embody the values we believe in, aspire to the purposes we consider right, and serve human freedom, human dignity, and human ends.

If an educated man of those days of the first technological revolution —an educated Sumerian perhaps or an educated ancient Chinese— looked at us today, he would certainly be totally stumped by our technology. But he would, I am sure, find our existing social and political institutions reasonably familiar—they are after all, by and large, not fundamentally different from the institutions he and his contemporaries first fashioned. And, I am quite certain, he would have nothing but a wry smile for both those among us who predict a technological heaven and those who predict a technological hell of "alienation," of "technological unemployment," and so on. He might well mutter to himself, "This is where I came in." But to us he might well say, "A time such as was mine and such as is yours, a time of true technological revolution, is not a time for exultation. It is not a time for despair either. It is a time for work and for responsibility."

Greek Catapults and Catapult Technology

SCIENCE, TECHNOLOGY, AND WAR IN THE ANCIENT WORLD

BARTON C. HACKER

Scholars and laymen alike have long displayed a lively interest in ancient catapults.[1] Research has focused on the most obvious problem: How did catapults work? Beyond its inherent interest, an answer to this question could perhaps contribute to a fuller understanding of the mechanical technology of antiquity in general. The question itself has largely been answered. Reliable information on the mechanical characteristics of catapults comes primarily from three sources: the *Belopoiika* of Philon of Byzantium,[2] the tenth book of Vitruvius' *De architectura*,[3] and the *Belopoiika* of Heron of Alexandria.[4] Philon and Heron belonged to the school of mechanicians centered at Alexandria in Hellenistic times; the two treatises spring from a single tradition, though Philon flourished about 250 B.C., Heron A.D. 62.[5] Vitruvius, a Roman engineer, composed his treatise toward the end of the first cen-

DR. HACKER assumed the post of historian at the University of California Lawrence Livermore National Laboratory in 1992, continuing the pattern of research positions alternating with academic teaching that has marked his career throughout. His books and articles have won several awards, including SHOT's 1993 Usher Prize for *An Annotated Index to . . . "Technology and Culture" 1959–1984* (Chicago, 1991). At present, he is working not only on nuclear weapons history but also on *Ordering Society: A World History of Military Institutions* (Westview Press).

[1] I use "catapult" as a generic term for missile-throwing engines of war. Problems of nomenclature, endemic to any discussion of these machines, I shall consider in greater detail further on.

[2] *Philons Belopoiika (viertes Buch der Mechanik)*, trans. H. Diels and E. Schramm (*Abhandlungen der preussischen Akademie der Wissenschaften*, Philosophisch-historische Klasse, No. 16 [1918] [Berlin, 1919]).

[3] *Vitruvius De architectura libri decem*, ed. F. Krohn (Leipzig, 1912).

[4] *Herons Belopoiika (Schrift vom Geschützbau)*, trans. H. Diels and E. Schramm (*Abhandlungen der königlichen preussischen Akademie der Wissenschaften*, Philosophisch-historische Klasse, No. 2 [1918] [Berlin, 1918]).

[5] When Heron lived was long a subject of controversy. Heron presents data on an eclipse of the moon in one of his works, however, that will only fit an eclipse that took place in A.D. 62. See A. G. Drachmann, *Ktesibios, Philon and Heron: A Study in Ancient Pneumatics* (*Acta historica scientiarum naturalium et medicinalium*, Vol. IV [Copenhagen, 1948]), pp. 74–77.

tury B.C. Two other extant treatises on catapult construction, by Biton[6] and Athenaios,[7] are too obscure to be technically useful; their dates are as obscure as their data.[8]

Scholarly efforts to re-establish the principles of ancient catapult construction date from the mid-nineteenth century and culminated in the definitive investigation of Colonel Erwin Schramm in Germany, 1903–4.[9] Quite apart from such scholarly investigations, the literature on catapults has been swelled by the contributions of enthusiasts dating back at least to the eighteenth century.[10] Such minor obscurities as persist reflect shortcomings in the surviving texts rather than failures of scholarship. We may safely assume that no further research will

[6] *Bitons Bau von Belagerungsmaschinen und Geschützen*, trans. A. Rehm and E. Schramm (*Abhandlungen der bayerischen Akademie der Wissenschaften*, Philosophisch-historische Abteilung, N.S., No. 2 [Munich, 1929]).

[7] *Athenaios über Maschinen. Griechische Poliorketiker, mit den handschriftlichen Bildern*, III, trans. Rudolf Schneider (*Abhandlungen der königlichen Gesellschaft der Wissenschaften zu Göttingen*, Philologisch-historische Klasse, N.S., Vol. XII, No. 5 [Berlin, 1912]).

[8] Athenaios addressed his treatise to one Marcellus, who has usually been identified with the Roman conqueror of Syracuse (212 B.C.). But in *The Mechanical Technology of Greek and Roman Antiquity: A Study of the Literary Sources* (Copenhagen, 1963), p. 11, A. G. Drachmann says that the proper Marcellus was the son-in-law of Augustus, which means that Athenaios flourished at the beginning of the first century A.D. Dating Biton involves a similar problem. He addressed his treatise to King Attalos of Pergamum, who may have been any of the three bearing that name; accordingly, Biton flourished sometime between 241 and 133 B.C., the inclusive dates of their reigns.

[9] The best short account, which includes a complete bibliography, is E. Schramm, "Poliorketik," in Johannes Kromayer and Georg Veith (eds.), *Heerwesen und Kriegführung der Griechen und Römer* (*Handbuch der Altertumswissenschaft*, Sec. 4, Part III, Vol. II [Munich, 1928]), pp. 209–45. Schramm, *Die antiken Geschütze der Saalburg* (Berlin, 1919), is definitive. For a recent brief and authoritative survey, see Drachmann, *Mechanical Technology of Greek and Roman Antiquity* (n. 8 above), pp. 186–90. Other useful accounts include: Hermann Diels, *Antike Technik: Sieben Vorträge* (2d ed.; Leipzig and Berlin, 1920), pp. 91–120; W. W. Tarn, *Hellenistic Military and Naval Developments* (Cambridge, 1930), pp. 101–18; and A. R. Hall, "Military Technology," in Charles Singer, E. J. Holmyard, A. R. Hall, and Trevor I. Williams (eds.), *A History of Technology*, II (Oxford, 1956), 695–730.

[10] See, e.g., Francis Grose, *Military Antiquities respecting a History of the English Army from the Conquest to the Present Time* (London, 1786), II, 286–302; Ralph Payne-Gallwey, *The Crossbow, . . . with a Treatise on the Ballista and Catapult of the Ancients* (London, 1903); and Payne-Gallwey's *A Summary of the History, Construction, and Effects in Warfare of the Projectile-throwing Engines of the Ancients* (London, 1907).

substantially alter our present knowledge of the mechanical characteristics of catapults. In his recent study of the literary sources for the mechanical technology of ancient Greece and Rome, Aage G. Drachmann justifiably confined himself to a short survey of engines of war, noting, "Of all the technical writings of Antiquity those on engines of war have received by far the best treatment. The texts have been edited and translated; the figures have been reproduced and interpreted; the catapults have been reconstructed and tried."[11]

I shall briefly review the current state of knowledge concerning the mechanical characteristics of catapults. Were this my sole purpose, however, I could scarcely excuse adding to an already voluminous literature. But need the study of so successful an innovation as the catapult rest complete when its mechanical characteristics have been established? I think not. By looking at the catapult as a technological innovation, we are led to some thought-provoking questions that have scarcely been raised in the existing literature. Why, for example, were catapults so successful, so widely adopted? What role, if any, did Greek science play in the development of catapults? What were the consequences of a highly developed catapult technology?

Questions like these may be asked about any technological innovation. Because of the catapult's military significance, however, asking such questions about catapults may be more fruitful than asking them about other ancient inventions. We are not restricted to information drawn from surviving technical treatises and archeological findings;[12] we have, in addition, evidence on the development, spread, and use of catapults from military and political history and from military treatises. In this, as in all fields of ancient history, the historian must rely on scattered, fragmentary, and heterogeneous sources. Nevertheless, we may follow the introduction and diffusion of new weapons far more readily than of any other kind of technical advance. My main purpose, then, is to suggest some of these larger implications of the advanced catapult technology of antiquity.

* * *

First, however, I want to describe briefly the mechanical characteristics of the machines I shall be talking about and to indicate some

[11] Drachmann, *Mechanical Technology of Greek and Roman Antiquity* (n. 8 above), p. 186.

[12] The most important archeological contribution to our knowledge of catapults was the discovery, in 1912 at Ampurias in Spain, of a catapult dating from 150 B.C. See *ibid.*, p. 189.

of the major problems of nomenclature.[13] The earliest catapults were the product of relatively straightforward attempts to increase the range and penetrating power of missiles by strengthening the bow which propelled them. Since such a reinforced bow could not be spanned in the ordinary way, mechanical ingenuity supplemented human muscle. In the *gastraphetes* (Fig. 1), a hand weapon Heron attributed to Zopyros of Tarentum (fl. *ca.* 350 B.C.), the bow was mounted on a stock grooved on its upper surface. A slider, with a

[13] The basis for these remarks are the works by Schramm, Diels, Tarn, and Hall cited in n. 9 above and Drachmann's *Mechanical Technology of Greek and Roman Antiquity* (n. 8 above).

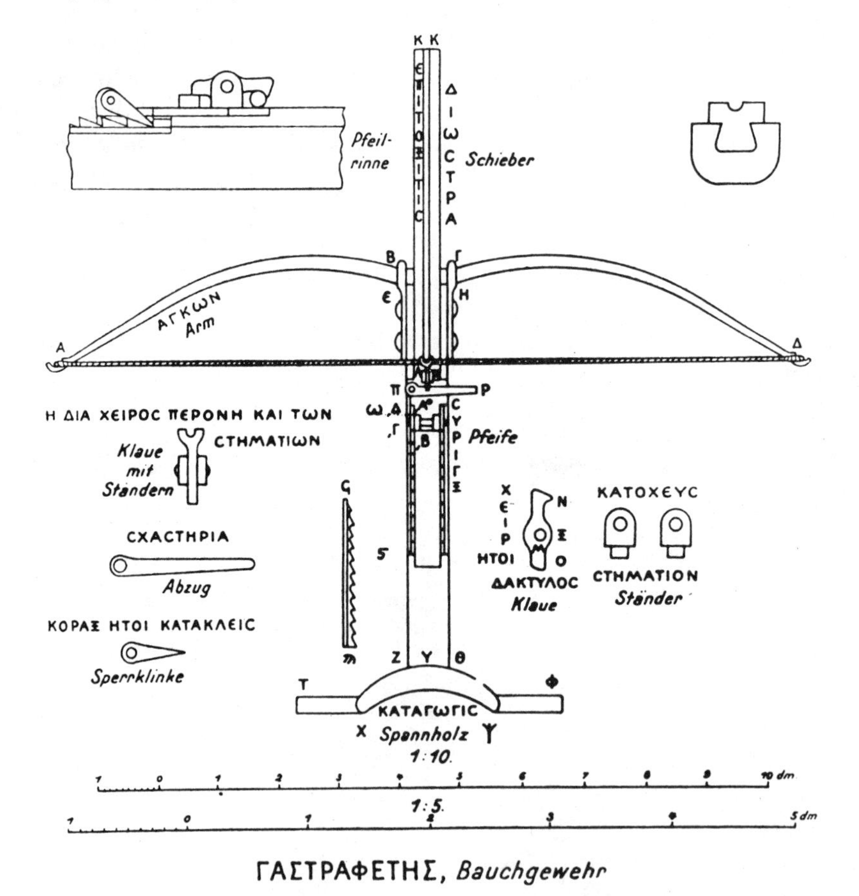

FIG. 1.–*Gastraphetes*. Reproduced from *Herons Belopoiika* (cited in n. 4), Fig. 4, p. 10.

hook at its back end to catch the bowstring, rested in the groove, its front projecting beyond the front of the stock. To span the weapon, the archer braced the curved back of the stock against his belly (hence the name) and forced the slider toward him by pressing it against the ground. Thus the archer could use not only his arm muscles but his more powerful back muscles as well. Once spanned, a rack-and-pawl device held the slider in position while the archer inserted an arrow before the bowstring, aimed, and fired. He could then move the slider forward and repeat the process. Larger and more powerful bow-catapults were standmounted and spanned by winch but otherwise were similar in design to the *gastraphetes.*

Dependence on the elasticity of the bow was a basic limitation of bow-catapults; the elastic strength of materials available to the Greeks was simply not great enough to permit much elaboration along these lines. Bow-catapults were supplanted during the course of the fourth century by torsion catapults. This new kind of catapult used the torsional force of thick, tightly twisted skeins of hair or sinew, a force considerably greater than the elastic force of strained horn or wood. Two forms of torsion catapult evolved; the *euthytonon* (Fig. 2), an arrow-shooting machine, and the *palintonon* (Fig. 3), a stone-thrower. Both types were standmounted and spanned by winch, supplemented in larger machines by pulleys; like earlier machines, they used a slider in a grooved stock. A frame at the front end, however, replaced the bow of earlier machines. It held two skeins, one on either side of the stock, into each of which a wooden arm was inserted. The string was attached to these arms. In the *euthytonon,* a single frame housed both skeins. But the *palintonon,* constructed for heavier missiles, had two separate cable frames.

The arms of both *euthytonon* and *palintonon* operated horizontally. Another machine, the *monankon,* used the torsion principle somewhat differently. It had but a single, massive arm, tipped by a sling, which moved in a vertical plane. The arm was winched down and a missile placed in the sling; when released, it flung a fairly heavy missile in a high trajectory. The *monankon* is described in none of the treatises from which we learn of other catapults. In particular, the failure of Vitruvius, who had been an official in charge of construction and repair of engines of war for his government,[14] to mention this machine strongly suggests that it was a later invention. Reliable evidence for its existence is no earlier than the late fourth century A.D.[15]

[14] Vitruvius *De architectura* i. 1. 2.

[15] Ammianus Marcellinus xxiii. 4. 4–7.

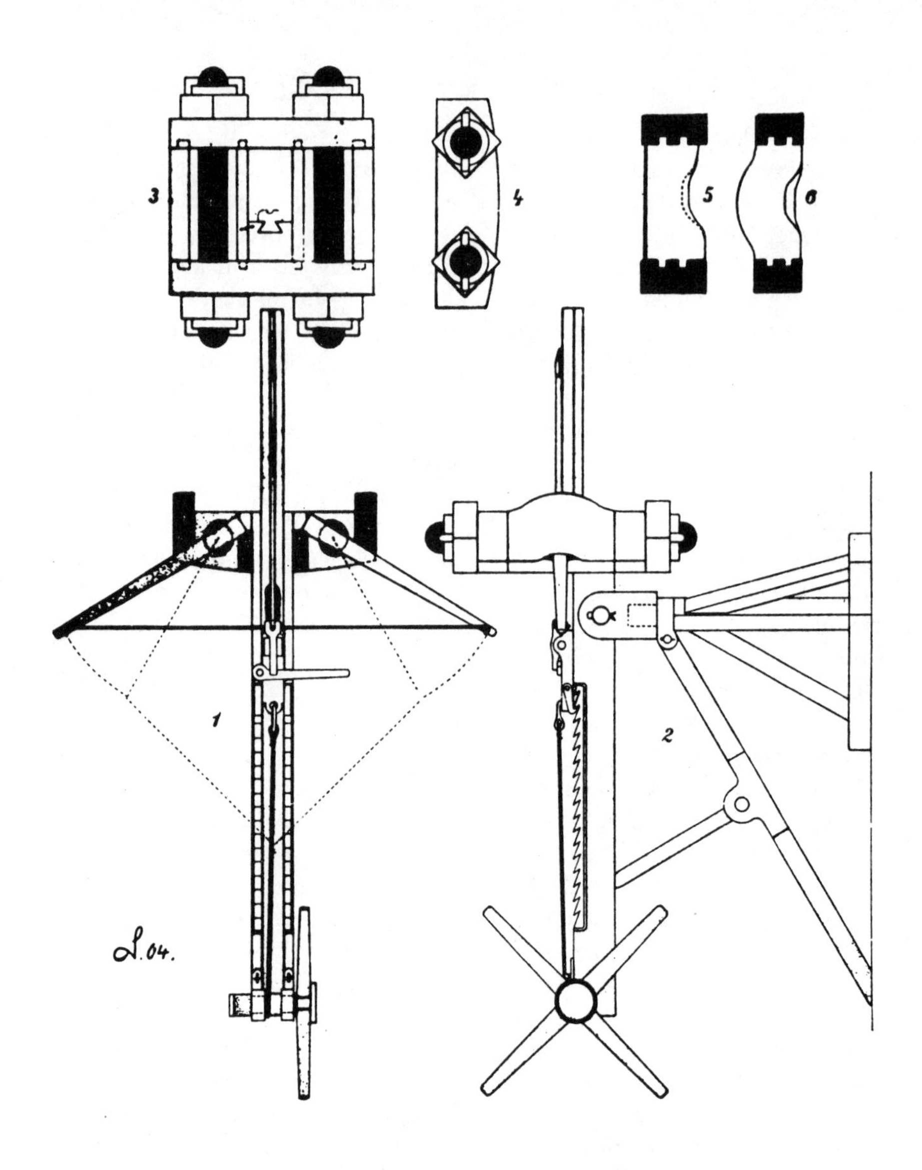

Euthytonon nach Philon

1 von oben, 2 von der Seite. 3 Spannrahmen von vorn. 4 von oben, 5 u. 6 Schnitte

Maßstab 1: 20

FIG. 2.—*Euthytonon*. Reproduced from *Philons Belopoiika* (cited in n. 2), Pl. 3

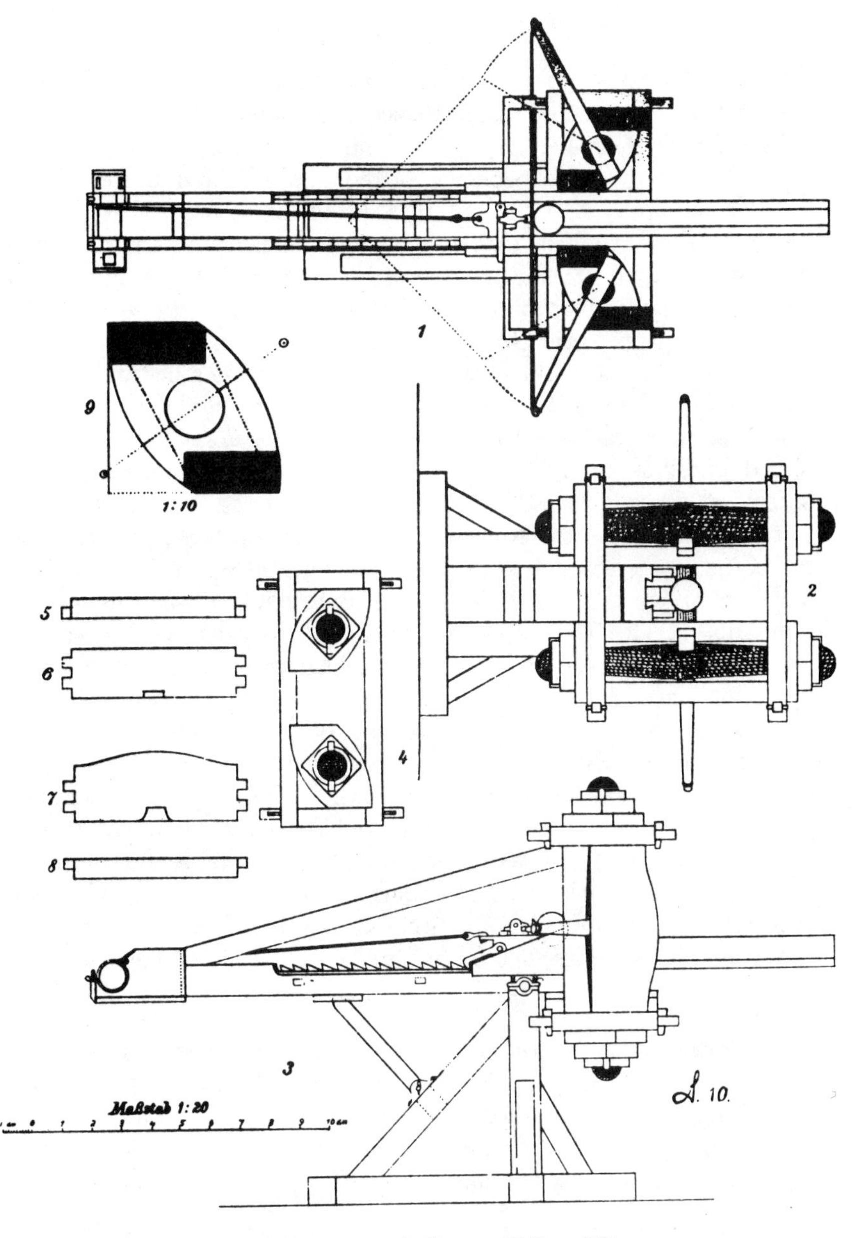

Palintonon nach Heron, Philon, Vitruv

1 von oben, 2 von vorn, 3 von der Seite, 4 Spannrahmen von oben.
5, 6, 7, 8 Mittel- und Seitenständer, 9 Construction der Peritreten.

FIG. 3.—*Palintonon.* Reproduced from *Philons Belopoiika* (cited in n. 2), Pl. 4

Euthytonon was the Greek name for the ordinary single-curved bow, *palintonon* for the double-curved (compound) bow. Why catapults received these names has been much debated. The most likely explanation is that the machines looked like the kind of bow they were named after. As I noted earlier, problems of nomenclature pervade any study of missile-throwing engines. Ancient writers sometimes seem to have used the same name for different machines, sometimes to have called similar machines by different names. The *euthytonon* was the catapult proper—the Greek *katapeltes* meant something that would pierce a shield, *pelte*—but it was also called *oxybolos* because it shot arrows. The *palintonon* was termed *ballista* (or *lithobolos* or *petrobolos*) because it threw stones. *Catapulta* and *ballista* eventually replaced *euthytonon* and *palintonon* as the standard names. This is Vitruvius' usage. Because they used twisted skeins, torsion machines were referred to by the Romans as *tormenta.* Still later, *ballista* came to be used generically for all torsion engines except the *monankon.* Vegetius (fl. *ca.* A.D. 384–95), in his *Epitoma rei militaris,* used *catapulta* not at all. The Romans termed the *monankon* "onager" or "scorpion" on the basis of fancied resemblances to those animals. But "scorpion" was also applied to what was apparently a kind of hand weapon, as, for example, those Archimedes was alleged to have built for the defense of Syracuse (212 B.C.).[16] A distinction based on whether the machine threw darts or stones has little meaning; with suitable, and relatively slight, modifications, a machine could do either. All this is from men who knew what they were writing about, for the most part engineers or soldiers. Later, during the Middle Ages, when we must rely on the chronicles of churchmen not particularly familiar with the subject, obscurities of terminology become all but impenetrable.[17]

* * *

This proliferation of names suggests that catapults were widely used. Such was indeed the case. Toward the end of the fourth century B.C., Greek siege techniques reached a previously unparalleled level of efficiency. Alexander of Macedon was able to take walled towns with a regularity and rapidity unknown in the Greek world before his time.

[16] Plutarch *Vita Marcelli* xv. 5.

[17] See Charles Oman, *A History of the Art of War in the Middle Ages* (2d ed.; London, 1924), I, 140; II, 45–46. Grose (n. 10 above), II, 302, catalogues engines of war current in the eleventh and twelfth centuries: "Exclusive of the ballista, catapulta, onager, and scorpion, were the mangonel, the trebuchet, the petrary, the robinet, the mategriffon, the bricolle, the bugle or bible, the espringal, the matafunda, the rebaudequin, engine a verge, and the war wolf."

Alexander's success was no simple outgrowth of an improved military technology: the unsurpassed quality of his troops and the vigor and resource with which he pressed home his attacks were doubtless as important as newfangled catapults.[18] Yet brave soldiers and resourceful commanders before Alexander had seldom been able to take Greek cities by storm. Fortifications were far more effective than devices for breaching them; if a city fell, the reason was usually starvation or betrayal. Lacking effective means for neutralizing defenders or penetrating walls, attackers could expect heavy losses. Greek cities were rarely willing to risk such losses among the citizens who composed their armies.[19] Even an attacker's willingness to risk assault, however, guaranteed no success. In the Peloponnesian War, elaborate Spartan attempts to storm the town of Plataea were frustrated by a handful of defenders, who surrendered only after a two-year siege had reduced the town to starvation.[20]

Plataea was by no means unique. In fact, the secondary role of force in Greek siege craft persisted until well into the fourth century. Aeneas Tacticus (fl. *ca.* 350 B.C.) devoted only nine short chapters at the end of his *Poliorketika* to assault techniques.[21] The bulk of his work, a manual for the commanders of besieged cities, dealt with methods to prevent a city's betrayal. His assault techniques differed scarcely at all from those of the Peloponnesian War. Missile engines he mentioned only once,[22] and then so cursorily as to suggest that he had had no personal experience with them, though he was an experienced professional soldier.[23]

Aeneas apparently knew little of Sicily.[24] There siege craft had progressed rapidly, and catapults had been introduced by the beginning of the fourth century. Dionysios of Syracuse, according to the testimony of Diodorus Siculus,[25] gathered artisans from many lands to fabricate

[18] For an excellent brief analysis, see J. F. C. Fuller, *The Generalship of Alexander the Great* (New Brunswick, N.J., 1960), chap. vii ("Alexander's Sieges"), pp. 200–218.

[19] F. E. Adcock, *The Greek and Macedonian Art of War* (Berkeley, Calif., 1957), pp. 58–59.

[20] Thucydides ii. 76–79, iii. 52.

[21] Aeneas Tacticus *Poliorketika* xxxii–xl.

[22] *Ibid.* xxxii. 8.

[23] *Aeneas on Siegecraft*, ed. and trans. L. W. Hunter, rev. S. A. Handford (Oxford, 1927), pp. ix–x, xvii–xxxiv, 222.

[24] *Ibid.*, p. xxxvii.

[25] Diodorus Siculus *Bibliotheke historike* xiv. 47–53.

the siege engines he deployed against his Carthaginian foes, most notably at the siege of Motye in 398. Among these engines were catapults.

What line of development led from the bow-catapults Dionysios used to the torsion catapults used by Alexander remains unknown.[26] In any case, Alexander clearly thought a great deal of his new machines; his army carried the basic components of catapults with it on its most far-flung campaigns.[27] Catapults subsequently became a normal feature of Hellenistic armies and regularly accompanied Roman armies in the field, each legion having an integral complement of artillery.[28] Nor did the use of catapults end with Rome. The catapults described by Procopius,[29] which Belisarius employed against the Goths in the first half of the sixth century, were substantially the same as those described by Ammianus Marcellinus[30] and Vegetius[31] in the last half of the fourth century. None differed in any important way from the catapults that had been perfected by the Hellenistic mechanicians in the third century B.C. (with the possible exception of the *monankon*)—a perfection attested by the decision of Frontinus (*ca.* 40 to *ca.* 104), in the section of his *Stratagems* dealing with the siege and defense of towns, to lay aside "all considerations of works and engines of war, the invention of which has long since reached its limit, and for the improvement of which I see no further hope in the applied arts."[32]

The art of building catapults survived with most of its ancient perfection throughout the Middle Ages at Constantinople.[33] In the West, however, the art languished; techniques persisted but were applied in

26 Where or when torsion was first applied to catapults is impossible to say, though Pliny the Elder claims it was a Phoenician invention (*Naturalis historia* viii. 201). The first certain appearance of torsion catapults was in Alexander's siege of Tyre (333–32). See Tarn (n. 9 above), pp. 105–6, and L. Sprague de Camp, "Master Gunner Apollonios," *Technology and Culture*, II (1961), 240–44.

27 See, e.g., the account of Alexander's Indian campaign (particularly the siege of Aornos where Alexander went to a good deal of trouble to get his catapults into action) in Marc Aurel Stein, *On Alexander's Track to the Indus* (London, 1929), pp. 135–42, 146–47. See also Fuller (n. 18 above), pp. 248–54.

28 Schramm, "Poliorketik" (n. 9 above), pp. 244–45; Kromayer and Veith (eds.) (n. 9 above), pp. 373–75, 442–45.

29 Procopius *History of the Wars* V. xxi. 14–18.

30 Ammianus Marcellinus xxiii. 4.

31 Vegetius *Epitoma rei militaris* iv.

32 Frontinus *The Stratagems* iii. Preface, trans. Charles E. Bennett, ed. Mary B. McElwain (Loeb Classical Library [New York, 1925]), p. 205.

33 Oman (n. 17 above), I, 136, 139; Ferdinand Lot, *L'Art militaire et les armées au Moyen Âge en Europe et dans le Proche Orient* (Paris, 1946), I, 49.

ever cruder form. Nonetheless, they did persist. The Viking siege of Paris in 885–86 saw the employment by both sides of virtually every instrument of siege craft known to the classical world, including a variety of catapults. By then, however, defensive techniques had long since caught up with assault methods; that protracted siege failed.[34]

Torsion catapults had important shortcomings. They were slow and cumbersome–disadvantages that increased as the catapult became more powerful. This mattered little in the leisurely pace of siege operations.[35] Problems inherent in the use of twisted skeins were more serious: the tension of skeins varied with changes in atmospheric humidity; the skeins deteriorated from repeated twisting and untwisting, since they had to be loosened after use, then tightened before being used again. Philon proposed several remedies for these problems, none of which was adopted, if we may accept as evidence the failure of later writers to mention them.[36]

Despite their shortcomings, however, catapults worked well enough to remain in continuous use until they were largely supplanted by the trebuchet, which first appeared around the beginning of the twelfth century. Although the basic design was variously modified, in its simplest form the trebuchet had a single arm, pivoted near one end and moving in a vertical plane. From the shorter end were suspended heavy counterweights, to the longer end was attached a sling to hold the missile. When the longer end was pulled down, either by muscle or mechanically, then suddenly released, the heavy counterpoise swung it through a rapid arc, hurling the missile with considerable force in a high trajectory. The trebuchet was, in fact, a much more powerful machine than any earlier missile-thrower, and far simpler in design.[37] Torsion catapults had nevertheless persisted in effective use for a millennium and a half.

* * *

What did catapults contribute to the highly efficient siege craft of Alexander and his successors? Clearly catapults were not powerful

[34] Oman (n. 17 above), I, 140–48.

[35] Because catapults were heavy, bulky, and slow, they were seldom used in the field. But, as a regular part of Hellenistic and Roman armies, they were sometimes so employed. See Schramm, "Poliorketik" (n. 9 above), pp. 238, 242–43, and Tarn (n. 9 above), pp. 119–20.

[36] See Drachmann, *Mechanical Technology of Greek and Roman Antiquity* (n. 8 above), pp. 189–90.

[37] See Lynn White, jr., *Medieval Technology and Social Change* (New York, 1962), pp. 102–3, 165.

enough to batter down the walls of a well-fortified town. But they could, and did, provide the covering fire that enabled attackers to approach the walls, where other techniques could be brought into play in order to breach or surmount them.[38] Vivid testimony to the effectiveness of catapults is the account of the Roman siege of Jotapata (A.D. 67) given by Josephus, who commanded the Jewish forces defending the city. So overwhelming was the bombardment of the walls by lances and stones that no defender dared mount the ramparts; the force of the engines

> was such that a single projectile ran through a row of men, and the momentum of the stones hurled by the engine carried away battlements and knocked off corners of towers. There is in fact no body of men so strong that it cannot be laid low to the last rank by the impact of these huge stones. . . . Getting in the line of fire, one of the men standing near Josephus on the rampart had his head knocked off by a stone, his skull being flung like a pebble from a sling more than 600 yards; and when a pregnant woman on leaving her house at daybreak was struck in the belly, the unborn child was carried away 100 yards. . . . Even more terrifying than the siege-guns and their missiles was the rushing sound and the final crash.[39]

The performance claimed for catapults by the ancient authors has been substantiated, for the most part, by modern reconstructions. Among the results Schramm obtained in his definitive investigations: the onager threw a 4-pound stone 300 meters, the same distance reached by a 1-pound stone from the *palintonon;* the *euthytonon* shot an 88-centimeter arrow 370 meters, the arrow penetrating an iron-sheathed shield 3 centimeters thick for half its length.[40] This last result suggests the incident related by Procopius in his account of the Gothic siege of Rome (537–38):

> And at the Salarian Gate a Goth of goodly stature and a capable warrior, wearing a corselet and having a helmet on his head, a man who was of no mean station in the Gothic nation, refused to remain in the ranks of his comrades, but stood by a tree and kept shooting many missiles at the parapet. But this man by some chance was hit by a missile from an engine which was on a tower at his left. And

[38] For a brief description of some of these other techniques, see Hall (n. 9 above), 702–3, 715–17.

[39] Josephus *The Jewish War* iii. 240–48, trans. G. A. Williamson (Baltimore, 1959), pp. 189–91.

[40] Schramm, "Poliorketik" (n. 9 above), p. 239.

> passing through the corselet and the body of the man, the missile sank more than half its length into the tree, and pinning him to the spot where it entered the tree, it suspended him there a corpse.[41]

The success and persistence of catapults suggest that they filled an important need. Defensive techniques, particularly those based on elaborate permanent fortifications, have, in most eras, been significantly more effective than the methods available to attackers. This usual superiority of defender over attacker has repeatedly provided the basis for political fragmentation. The ability of the Greek city-states to maintain their independence depended in part on the imperviousness of their walls to assault, just as the inadequacy of medieval siege craft contributed to the diffusion of political power which characterized medieval Europe. Conversely, improvements in siege craft that promoted the equality or ascendancy of the besieger played a part in the establishment of larger political units. Some weight must be given to the development of torsion catapults in accounting for the augmented regularity with which Alexander and his successors, both Hellenistic and Roman, were able to take walled cities. This new ability, in turn, was a factor in the establishment of kingdom and empire. The subsequent reassertion of defensive supremacy that followed on the decline of the legions played its part in invalidating the imperial claims of the German successor states. In much the same way, the rise of nation-states in early modern times owed something to new techniques and tools of siege craft, like gunpowder, which compromised first the security of fortress walls, then the independence those walls protected.

* * *

We have seen that catapults were a successful, long-lived, and significant innovation. I should now like to turn to the developmental work on which the subsequent career of catapults depended. This work, leading to the perfection of catapult technology, was carried out by the so-called mechanicians who flourished at Alexandria and elsewhere in the Hellenistic world. The science of mechanics was traditionally supposed to have been founded by Archytas of Tarentum (fl. *ca.* 400 B.C.) and Eudoxos of Knidos (*ca.* 408 to *ca.* 355), who used mechanical demonstrations for geometrical problems they could not prove rigorously.[42] This approach was rejected by later geometers, although Archimedes (287–212) used a mechanical method to arrive at geometrical conclu-

[41] Procopius *History of the Wars* V. xxiii. 9–11, trans. H. B. Dewing (Loeb Classical Library [New York, 1919]), III, 221.

[42] Plutarch *Vita Marcelli* xiv. 5.

sions preliminary to rigorous mathematical demonstration.[43] Mechanics parted ways with geometry and, in fact, came to be regarded as one of the military arts.[44] The Hellenistic mechanicians earned their livelihood as military engineers.[45] War industry was lavishly supported by the Hellenistic kings, most notably the Ptolemies of Egypt. Kingdoms and cities competed keenly for the services of the most able engineers, whose rewards, in terms of both money and prestige, were great.[46]

Such support was indispensable for the experimental work which led to the perfection of torsion catapults. Philon asserted that the Alexandrian engineers "were heavily subsidized by kings eager for fame and interested in the arts."[47] Heron suggested another motive in the Introduction to his *Belopoiika*: a highly developed art of catapult construction promoted peace, since the state thus endowed would discourage the designs of potential aggressors. Biton made much the same remark in his Introduction. The incessant warfare of Hellenistic times–Hellenistic kings were "always either preparing for war or actually engaged in it"[48]–suggests no dearth of potential aggressors, whatever implication that may have for the efficacy of discouraging attack by strong armament.

The mechanicians displayed an interest in practical applications of their work that contrasts sharply with the usual attitude of Greek scientists. Although Greek science had significant quantitative and experimental aspects, Greek scientists did not universally accept the importance of observation and experiment in linking theory to experience; nor did they all recognize the value of mathematical analysis in more than a few fields of natural inquiry. Furthermore, Greek scientists were primarily interested in knowledge for its own sake. Even in areas where their discoveries might have been turned to practical account, they seldom troubled to do so.[49] Plutarch's allegation that Archimedes repu-

[43] E. J. Dijksterhuis, *Archimedes*, trans. D. Dikshoorn (*Acta historica scientiarum naturalium et medicinalium*, Vol. XII [Copenhagen, 1956]), pp. 321–22.

[44] Plutarch *Vita Marcelli* xiv. 6.

[45] M. Rostovtzeff, *The Social and Economic History of the Hellenistic World* (Oxford, 1941), II, 1234–35.

[46] *Ibid.*, pp. 1232–33, 1236. See also L. Sprague de Camp, *The Ancient Engineers* (Garden City, N.Y., 1963), and Oliver Lyman Spaulding, *Pen and Sword in Greece and Rome* (Princeton, N.J., 1937).

[47] Philon *Belopoiika* 3; quoted from Morris R. Cohen and I. E. Drabkin, *A Source Book in Greek Science* (Cambridge, Mass., 1958), p. 318.

[48] Rostovtzeff (n. 45 above), I, 152.

[49] Marshall Clagett, *Greek Science in Antiquity* (New York, 1955), pp. 22–25, 28–31; Ludwig Edelstein, "Recent Trends in the Interpretation of Ancient Science,"

diated "as sordid and ignoble the whole trade of engineering, and every sort of art that lends itself to mere use and profit,"[50] is often cited as the epitome of this attitude.

Yet Archimedes was famous for his inventions, especially for the variety of engines of war he contrived to protract the Roman siege of Syracuse.[51] Archimedes, however, stood squarely in the strictly logical, geometrical tradition of Greek science and mathematics. His inventions, numerous though they may have been, were diversions. Archimedean science, like Euclidean geometry, demanded a strict axiomatic method. The limitation of available scientific knowledge did not permit the establishment of a science of mechanics comparable in rigor to, say, Archimedes' achievement of a science of hydrostatics.

The mechanicians worked in a quite different tradition. Underlying the strictly logical, axiomatic mathematics and science created by Eudoxos and his contemporaries in the fourth century and exemplified in the work of Euclid, Archimedes, and Ptolemy was another, more elementary tradition rooted in Babylonian algebraic and numerical procedures. This persistent oriental tradition was little influenced by the development of axiomatic geometry, which it both antedated and survived. The writings of Heron of Alexandria form part of this tradition; his geometry reflects the algebraic or arithmetic approach of Babylonian mathematics, with its emphasis on the application of numerical methods to the solution of practical problems.[52] What distinguished the Hellenistic mechanicians from most other Greek scientists was their effort to combine theory with practical utility. According to Pappus of Alexandria (fl. *ca.* 285 to *ca.* 305), "the mechanicians of Hero's school tell us that the science of mechanics consists of a theoretical and a practical part."[53] This approach enabled the mechanicians to make catapult technology something very like an applied science.

What exactly did the mechanicians contribute to catapult development? They made catapult construction systematic. The key element

in Philip P. Weiner and Aaron Noland (eds.), *Roots of Scientific Thought: A Cultural Perspective* (New York, 1957), pp. 91–101.

[50] Plutarch *Vita Marcelli* xvii. 4. Quoted from Plutarch *The Lives of the Noble Grecians and Romans,* trans. John Dryden, rev. Authur Hugh Clough (New York, n.d.), p. 378.

[51] On Archimedes as an inventor, see Dijksterhuis (n. 43 above), pp. 21–29.

[52] O. Neugebauer, *The Exact Sciences in Antiquity* (2d ed.; New York, 1962), pp. 79–80, 146–47.

[53] Pappus of Alexandria *Collectiones mathematicae* viii. 1; quoted from Cohen and Drabkin (n. 47 above), p. 183.

in catapult performance was the size of the twisted skeins, which, of course, determined the diameter of the hole that received the skeins. All other components of the catapult are proportional to this diameter. The proper proportions between the parts were established experimentally, by systematically altering the sizes of the various parts of the catapult and testing the results. "For it is not possible to arrive at a complete solution of the problems involved merely by reason and by the methods of mechanics; many discoveries can, in fact, be made only as a result of trial."[54]

The proportions of the parts of the catapult were expressed in terms of the diameter of the hole which received the skeins, which was in turn related to the size of the missile the catapult was designed to shoot. For arrow-shooting machines, the diameter was taken as one-ninth the length of the arrow.[55] For stone-throwing machines, the diameter was computed from the weight of the stone: the diameter in digits was equal to the cube root of the weight in drachmas plus one-tenth of this root.[56] The results of these computations are presented in the form of tables.[57] Such tables allowed the technical staffs of Hellenistic armies to construct catapults incorporating desired characteristics.[58]

This procedure parallels the method used by Hellenistic astrologers to compute the positions of celestial bodies for horoscopic purposes. Based on empirical material and proceeding on exclusively numerical grounds, Hellenistic astrology completely lacked any geometrical model.[59] Preserved horoscopes contain no doctrinal elaboration or theoretical speculation but simply the bare results of computation.[60] Hellenistic astronomy, like Hellenistic mathematics, followed two distinctive traditions, one of them an axiomatic, geometrically structured science purely Greek in origin and development, the other drawing on Babylonian practical numerical procedures uninfluenced by the axiomatic method.

[54] Philon *Belopoiika* 3; quoted from Cohen and Drabkin (n. 47 above), p. 319.

[55] Heron *Belopoiika* 32.

[56] This formula is given in substantially the same form in Philon *Belopoiika* 6 and Heron *Belopoiika* 32.

[57] Certain discrepancies in the computations of Philon and Vitruvius are discussed in A. G. Drachmann, "Remarks on the Ancient Catapults," *Actes du VIIe Congrès International d'Histoire des Sciences (4–12 Août 1953)* ("Collection des travaux de l'Académie Internationale d'Histoire des Sciences," No. 8 [Paris, 1953]), pp. 279–82.

[58] Rostovtzeff (n. 45 above), II, 1082–84.

[59] Neugebauer (n. 52 above), pp. 157–58.

[60] *Ibid.*, p. 170.

That mechanician and horoscopic astrologer worked in the same tradition seems clear enough. For both, observational data furnished the basis for computing, on the one hand, the size of catapult components and, on the other, the positions of celestial bodies. The aim of both was strictly practical—building catapults or casting horoscopes—and theory played no part in determining the procedures used. Should such activity be termed science? That depends on what we choose to call science, a problem beyond the scope of this paper. Whatever status it is accorded, however, catapult technology was clearly based on experimental investigation, its results systematized mathematically in the form of computed tables which allowed machines of specified characteristics to be constructed. At the same time, the mechanicians were unable to account for their findings or to provide them with a theoretical framework. Nonetheless, the result of this activity, whether we choose to call it scientific or not, was a catapult technology solidly based and well developed.

A. R. Hall has observed "that the purposeful application of science to the art of war (and, it may be, to any technique or useful art) at any period before the nineteenth century is much less than at first appears from non-professional accounts; the conservative traditions of practical men yielded very slowly to the enthusiasms of inventive amateurs, whether scientists or not."[61] To this generalization the work of the Hellenistic mechanicians on catapults may well stand as an exception. Mathematics and experiment, systematically applied to the improvement of catapult technology, produced machines that were subsequently used widely and effectively. Even more striking than any success the mechanicians achieved were the conditions and attitudes that guided their investigations. Many aspects of Hellenistic civilization seem curiously modern; not the least of these were the state support and utilitarian orientation of much of Hellenistic science. When further development lay within the technical possibilities of the time, as the development of catapults obviously did, the fruitful conjunction of favorable conditions and attitudes with adequate technique produced a highly developed technology.

[61] A. R. Hall, *Ballistics in the Seventeenth Century: A Study in the Relations of Science and War with Reference Principally to England* (Cambridge, 1952), p. vii.

The Technical Act

THE ACT OF INVENTION: CAUSES, CONTEXTS, CONTINUITIES AND CONSEQUENCES

LYNN WHITE, JR.*

The rapidly growing literature on the nature of technological innovation and its relation to other activities is largely rubbish because so few of the relevant concrete facts have thus far been ascertained. It is an inverted pyramid of generalities, the apex of which is very nearly a void. The five plump volumes of *A History of Technology*,[1] edited under the direction of Charles Singer, give the layman a quite false impression of the state of knowledge. They are very useful as a starting point, but they are almost as much a codification of error as of sound information.[2] It is to be feared that the physical weight of these books will be widely interpreted as the weight of authority and that philosophers, sociologists, and others whose personal researches do not lead them into the details of specific technological items may continue to be deceived as to what is known.

Since man is a hypothesizing animal, there is no point in calling for a moratorium on speculation in this area of thought until more firm facts can be accumulated. Indeed, such a moratorium—even if it were possible—would slow down the growth of factual knowledge because hypothesis normally provokes counter-hypotheses, and then all factions adduce facts in evidence, often new facts. The best that we can do at present is to work hard to find the facts and then to think cautiously about the facts which have been found.

In view of our ignorance, then, it would seem wise to discuss the problems of the nature, the motivations, the conditioning circumstances, and the effects of the act of invention far less in terms of generality than in terms of specific instances about which something seems to be known.

1. The beginning of wisdom may be to admit that even when we know some facts in the history of technology, these facts are not always fully intelligible, i. e., capable of " explanation," simply because we lack adequate contextual information. The Chumash Indians of the

*Now deceased, LYNN WHITE, JR., at the time this article was published, was Professor of History at the University of California, Los Angeles, and President of the Society for the History of Technology. His best known book is *Medieval Technology and Social Change* (Oxford, 1962).

coast of Santa Barbara County built plank boats which were unique in the pre-Columbian New World: their activity was such that the Spanish explorers of California named a Chumash village "La Carpintería."[3] A map will show that this tribe had a particular inducement to venture upon the sea: they were enticed by the largest group of off-shore islands along the Pacific Coast south of Canada. But why did the tribes of South Alaska and British Columbia, of Araucanian Chile, or of the highly accidented eastern coast of the United States never respond to their geography by building plank boats? Geography would seem to be only one element in explanation.

Can a plank-built East Asian boat have drifted on the great arc of currents in the North Pacific to the Santa Barbara region? It is entirely possible; but such boats would have been held together by pegs, whereas the Chumash boats were lashed, like the dhows of the Arabian Sea or like the early Norse ships. Diffusion seems improbable.

Since a group can conceive of nothing which is not first conceived by a person, we are left with the hypothesis of a genius: a Chumash Indian who at some unknown date achieved a break-away from log dugout and reed balsa to the plank boat. But the idea of "genius" is itself an ideological artifact of the age of the Renaissance when painters, sculptors, and architects were trying to raise their social status above that of craftsmen.[4] Does the notion of genius "explain" Chumash plank boats? On the contrary, it would seem to be no more than a traditionally acceptable way of labeling the great Chumash innovation as unintelligible. All we can do is to observe the fact of it and hope that eventually we may grasp the meaning of it.

2. A symbol of the rudimentary nature of our thinking about technology, its development, and its human implications, is the fact that while the *Encyclopaedia Britannica* has an elaborate article on "Alphabet," it contains no discussion of its own organizational presupposition, alphabetization. Alphabetization is the basic invention for the classification and recovery of information: it is fully comparable in significance to the Dewey decimal system and to the new electronic devices for these purposes. Modern big business, big government, big scholarship are inconceivable without alphabetization. One hears that the chief reason why the Chinese Communist regime has decided to Romanize Chinese writing is the inefficiency of trying to classify everything from telephone books to tax registers in terms of 214 radicals of ideographs. Yet we are so blind to the nature of our technical equipment that the world of Western scholars, which uses alphabetization constantly, has produced not even the beginning of a history of it.

Fortunately, Dr. Sterling Dow of Harvard University is now engaged in the task. He tells me that the earliest evidence of alphabetization is found in Greek materials of the third century B. C. In other words, there was a thousand-year gap between the invention of the alphabet as a set of phonetic symbols and the realization that these symbols, and their sequence in individual written words, could be divorced from their phonetic function and used for an entirely different purpose: an arbitrary but very useful convention for storage and retrieval of verbal materials. That we have neglected thus completely the effort to understand so fundamental an invention should give us humility whenever we try to think about the larger aspects of technology.

3. Coinage was one of the most significant and rapidly diffused innovations of Late Antiquity. The dating of it has recently become more conservative than formerly: the earliest extant coins were sealed into the foundation of the temple of Artemis at Ephesus c. 600 B. C., and the invention of coins, i. e., lumps of metal the value of which is officially certified, was presumably made in Lydia not more than a decade earlier.[5]

Here we seem to know something, at least until the next archaeological spades turn up new testimony. But what do we know with any certainty about the impact of coinage? We are compelled to tread the slippery path of *post hoc ergo propter hoc*. There was a great acceleration of commerce in the Aegean, and it is hard to escape the conviction that this movement, which is the economic presupposition of the Periclean Age, was lubricated by the invention of coinage.

If we dare to go this far, we may venture further. Why did the atomic theory of the nature of matter appear so suddenly among the philosophers of the Ionian cities? Their notion that all things are composed of different arrangements of identical atoms of some "element," whether water, fire, ether, or something else, was an intellectual novelty of the first order, yet its sources have not been obvious. The psychological roots of atomism would seem to be found in the saying of Heraclitus of Ephesus that "all things may be reduced to fire, and fire to all things, just as all goods may be turned into gold and gold into all goods."[6] He thought that he was just using a metaphor, but the metaphor had been possible for only a century before he used it.

Here we are faced with a problem of critical method. Apples had been dropping from trees for a considerable period before Newton discovered gravity:[7] we must distinguish cause from occasion. But the appearance of coinage is a phenomenon of a different order from the fall of an apple. The unprecedented element in the general life of sixth-century Ionia, the chief stimulus to the prosperity which provided leisure for the atomistic philosophers, was the invention of coinage: the age of barter was ended. Probably no Ionian was conscious of any connection between this unique new technical instrument and the brainstorms of the local intellectuals. But that a causal relationship did exist can scarcely be doubted, even though it cannot be "proved" but only perceived.

4. Fortunately, however, there are instances of technological devices

of which the origins, development, and effects outside the area of technology are quite clear. A case in point is the pennon.[8]

The stirrup is first found in India in the second century B. C. as the big-toe stirrup. For climatic reasons its diffusion to the north was blocked, but it spread wherever India had contact with barefoot aristocracies, from the Philippines and Timor on the east to Ethiopia on the west. The nuclear idea of the stirrup was carried to China on the great Indic culture wave which also spread Buddhism to East Asia, and by the fifth century the shod Chinese were using a foot stirrup.

The stirrup made possible, although it did not require, a new method of fighting with the lance. The unstirrupped rider delivered the blow with the strength of his arm. But stirrups, combined with a saddle equipped with pommel and cantle, welded rider to horse. Now the warrior could lay his lance at rest between his upper arm and body: the blow was delivered not by the arm but by the force of a charging stallion. The stirrup thus substituted horse-power for man-power in battle.

The increase in violence was tremendous. So long as the blow was given by the arm, it was almost impossible to impale one's foe. But in the new style of mounted shock combat, a good hit might put the lance entirely through his body and thus disarm the attacker. This would be dangerous if the victim had friends about. Clearly, a baffle must be provided behind the blade to prevent penetration by the shaft of the lance and thus permit retraction.

Some of the Central Asian peoples attached horse tails behind the blades of lances—this was probably being done by the Bulgars before they invaded Europe. Others nailed a piece of cloth, or pennon, to the shaft behind the blade. When the stirrup reached Western Europe c. 730 A. D., an effort was made to meet the problem by adapting to military purposes the old Roman boar-spear which had a metal crosspiece behind the blade precisely because boars, bears, and leopards had been found to be so ferocious that they would charge up a spear not so equipped.

This was not, however, a satisfactory solution. The new violence of warfare demanded heavier armor. The metal crosspiece of the lance would sometimes get caught in the victim's armor and prevent recovery of the lance. By the early tenth century Europe was using the Central Asian cloth pennon, since even if it got entangled in armor it would rip and enable the victor to retract his weapon.

Until our dismal age of camouflage, fighting men have always decorated their equipment. The pennons on lances quickly took on color and design. A lance was too long to be taken into a tent conveniently, so a knight usually set it upright outside his tent, and if one were looking for him, one looked first for the flutter of his familiar pennon. Knights riding held their lances erect, and since their increasingly massive armor made recognition difficult, each came to be identified by his pennon. It would seem that it was from the pennon that distinctive "connoissances" were transferred to shield and surcoat. And with the crystallization of the feudal structure, these heraldic devices became hereditary, the symbols of status in European society.

In battle, vassals rallied to the pennon of their liege lord. Since the king was, in theory if not always in practice, the culmination of the feudal hierarchy, his pennon took on a particular aura of emotion: it was the focus of secular loyalty. Gradually a distinction was made between the king's two bodies,[9] his person and his "body politic," the state. But a colored cloth on the shaft of a spear remained the primary symbol of allegiance to either body, and so remains even in polities which have abandoned monarchy. The grimly functional rags first nailed to lance shafts by Asian nomads have had a great destiny. But it is no more remarkable than that of the cross, a hideous implement in the Greco-Roman technology of torture, which was to become the chief symbol of the world's most widespread religion.

In tracing the history of the pennon, and of many other technological items, there is a temptation to convey a sense of inevitability. However, a novel technique merely offers opportunity; it does not command. As has been mentioned, the big-toe stirrup reached Ethiopia. It was still in common use there in the nineteenth century, but at the present time Muslim and European influences have replaced it with the foot stirrup. However, travellers tell me that the Ethiopian gentleman, whose horse is equipped with foot stirrups, rides with only his big toes resting in the stirrups.

5. Indeed, in contemplating the history of technology, and its implications for our understanding of ourselves, one is as frequently astonished by blindness to innovation as by the insights of invention. The Hellenistic discovery of the helix was one of the greatest of technological inspirations. Very quickly it was applied not only to gearing but also to the pumping of water by the so-called Archimedes screw.[10] Somewhat later the holding screw appears in both Roman and Germanic metal work.[11] The helix was taken for granted thenceforth in western technology. Yet Joseph Needham of Cambridge University assures me that, despite the great sophistication of the Chinese in most technical matters, no form of helix was known in East Asia before modern times: it reached India but did not pass the Himalayas. Indeed, I have not been able to locate any such device in the Far East before the early seventeenth century when Archimedes screws, presumably introduced by the Portuguese, were used in Japanese mines.[12]

6. Next to the wheel, the crank is probably the most important single element in machine design, yet until the fifteenth century the history of the crank is a dismal record of inadequate vision of its potentialities.[13] It first appears in China under the Han dynasty, applied to rotary fans for winnowing hulled rice, but its later applications in the Far East were not conspicuous. In the West the crank seems

to have developed independently and to have emerged from the hand quern. The earliest querns were fairly heavy, with a handle, or handles, inserted laterally in the upper stone, and the motion was reciprocating. Gradually the stones grew lighter and thinner, so that it was harder to insert the peg-handle horizontally: its angle creeps upward until eventually it stands vertically on top. All the querns found at the Saalburg had horizontal handles, and it is increasingly clear that the vertical peg is post-Roman.

Seated before a quern with a single vertical handle, a person of the twentieth century would give it a continuous rotary motion. It is far from clear that one of the very early Middle Ages would have done so. Crank motion was a kinetic invention more difficult than we can easily conceive. Yet at some point before the time of Louis the Pious the sense of the appropriate motion changed; for out of the rotary quern came a new machine, the rotary grindstone, which (as the Latin term for it, *mola fabri*, shows) is the upper stone of a quern turned on edge and adapted to sharpening. Thus, in Europe at least, crank motion was invented before the crank, and the crank does not appear before the early ninth century. As for the Near East, I find not even the simplest application of the crank until al-Jazarī's book on automata of 1206 A. D.

Once the simple crank was available, its development into the compound crank and connecting rod might have been expected quite quickly. Yet there is no sign of a compound crank until 1335, when the Italian physician of the Queen of France, Guido da Vigevano, in a set of astonishing technological sketches, which Rupert Hall has promised to edit,[14] illustrates three of them.[15] By the fourteenth century Europe was using crankshafts with two simple cranks, one at each end; indeed, this device was known in Cambodia in the thirteenth century. Guido was interested in the problem of self-moving vehicles: paddlewheel boats and fighting towers propelled by windmills or from the inside. For such constricted situations as the inside of a boat or a tower it apparently occurred to him to consolidate the two cranks at the ends of the crankshaft into a compound crank in its middle. It was an inspiration of the first order, yet nothing came of it. Evidently the Queen's physician, despite his technological interests, was socially too far removed from workmen to influence the actual technology of his time. The compound crank's effective appearance was delayed for another three generations. In the 1420's some Flemish carpenter or shipwright invented the bit-and-brace with its compound crank. By c. 1430 a German engineer was applying double compound cranks and connecting rods to machine design: a technological event as significant as the Hellenistic invention of gearing. The idea spread like wildfire, and European applied mechanics was revolutionized.

How can we understand the lateness of the discovery, whether in China or Europe, of even the simple crank, and then the long delay in its wide application and elaboration? Continuous rotary motion is typical of inorganic matter, whereas reciprocating motion is the sole movement found in living things. The crank connects these two kinds of motion; therefore we who are organic find that crank motion does not come easily to us. The great physicist and philosopher Ernst Mach noticed that infants find crank motion hard to learn.[16] Despite the rotary grindstone, even today razors are whetted rather than ground: we find rotary motion a bar to the greatest sensitivity. Perhaps as early as the tenth century the hurdy-gurdy was played with a cranked resined wheel vibrating the strings. But by the thirteenth century the hurdy-gurdy was ceasing to be an instrument for serious music. It yielded to the reciprocating fiddle bow, an introduction of the tenth century which became the foundation of modern European musical development. To use a crank, our tendons and muscles must relate themselves to the motion of galaxies and electrons. From this inhuman adventure our race long recoiled.

7. A sequence originally connected with the crank may serve to illustrate another type of problem in the act of technological innovation: the fact that a simple idea transferred out of its first context may have a vast expansion. The earliest appearance of the crank, as has been mentioned, is found on a Han-dynasty rotary fan to winnow husked rice.[17] The identical apparatus appears in the eighteenth century in the Palatinate,[18] in upper Austria and the Siebenbürgen,[19] and in Sweden.[20] I have not seen the exact channel of this diffusion traced, but it is clearly part of the general Jesuit-inspired *Chinoiserie* of Europe in that age. Similarly, I strongly suspect, but cannot demonstrate, that all subsequent rotary blowers, whether in furnaces, dehydrators, wind tunnels, air conditioning systems, or the simple electric fan, are descended from this Han machine which seems, in China itself, to have produced no progeny.

8. Doubtless when scholarship in the history of technology becomes firmer, another curious device will illustrate the same point. To judge by its wide distribution,[21] the fire piston is an old invention in Malaya. Dr. Thomas Kuhn of the University of California at Berkeley, who has made careful studies of the history of our knowledge of adiabatic heat, assures me that when the fire piston appeared in late eighteenth-century Europe not only for laboratory demonstrations but as a commercial product to light fires, there is no hint in the purely scientific publications that its inspiration was Malayan. But the scientists, curiously, also make no mention of the commercial fire pistons then available. So many Europeans, especially Portuguese and Netherlanders, had been trading, fighting, ruling, and evangelizing in the East Indies for so long a time before the fire piston is found in Europe, that it is hard to believe that the Malayan fire piston was not observed and reported. The realization of its potential in Europe was considerable, culminating in the diesel engine.

9. Why are such nuclear ideas sometimes not exploited in new and wider applications? What sorts of barriers prevent their diffusion? Why, at times, does what appeared to be a successful technological item fall into disuse? The history of the faggoted forging method of producing sword blades [22] may assist our thinking about such questions.

In late Roman times, north of the Alps, Celtic, Slavic, and Germanic metallurgists began to produce swords with laminations produced by welding together bundles of rods of different qualities of iron and steel, hammering the resulting strip thin, folding it over, welding it all together again, and so on. In this way a fairly long blade was produced which had the cutting qualities of steel but the toughness of iron. Although such swords were used at times by barbarian auxiliaries in the Roman army, the Roman legions never adopted them. Yet as soon as the Western Empire crumbled, the short Roman stabbing sword vanished and the laminated slashing blade alone held the field of battle. Can this conservatism in military equipment have been one reason for the failure of the Empire to stop the Germanic invasions? The Germans had adopted the new type of blade with enthusiasm, and by Carolingian times were manufacturing it in quantities in the Rhineland for export to Scandinavia and to Islam where it was much prized. Yet, although such blades were produced marginally as late as the twelfth century, for practical purposes they ceased to be used in Europe in the tenth century. Does the disappearance of such sophisticated swords indicate a decline in medieval metallurgical methods?

We should be cautious in crediting the failure of the Romans to adopt the laminated blade to pure stupidity. The legions seem normally to have fought in very close formation, shield to shield. In such a situation, only a stabbing sword could be effective. The Germans at times used a " shield wall " formation, but it was probably a bit more open than the Roman and permitted use of a slashing sword. If the Romans had accepted the new weapon, their entire drill and discipline would have been subject to revision. Unfortunately, we lack studies of the development of Byzantine weapons sufficiently detailed to let us judge whether, or to what extent, the vigorously surviving Eastern Roman Empire adapted itself to the new military technology.

The famous named swords of Germanic myth, early medieval epic and Wagnerian opera were laminated blades. They were produced by the vast patience and skill of smiths who themselves became legendary. Why did they cease to be made in any number after the tenth century? The answer is found in the rapid increase in the weight of European armor as a result of the consistent Frankish elaboration of the type of mounted shock combat made possible by the stirrup. After the turn of the millenium a sword in Europe had to be very nearly a club with sharp edges: the best of the earlier blades was ineffective against such defenses. The faggoted method of forging blades survived and reached its technical culmination in Japan [23] where, thanks possibly to the fact that archery remained socially appropriate to an aristocrat, mounted shock combat was less emphasized than in Europe and armor remained lighter.

10. Let us now turn to a different problem connected with the act of invention. How do methods develop by the transfer of ideas from one device to another? The origins of the cannon ball and the cannon may prove instructive.[24]

Hellenistic and Roman artillery was activated by the torsion of cords. This was reasonably satisfactory for summer campaigns in the Mediterranean basin, but north of the Alps and in other damper climates the cords tended to lose their resilience. In 1004 A. D. a radically different type of artillery appeared in China with the name *huo p'ao.* It consisted of a large sling-beam pivoted on a frame and actuated by men pulling in unison on ropes attached to the short end of the beam away from the sling. It first appears outside China in a Spanish Christian illumination of the early twelfth century, and from this one might assume diffusion through Islam. But its second appearance is in the northern Crusader army attacking Lisbon in 1147 where a battery of them were operated by shifts of one hundred men for each. It would seem that the Muslim defenders were quite unfamiliar with the new engine of destruction and soon capitulated. This invention, therefore, appears to have reached the West from China not through Islam but directly across Central Asia. Such a path of diffusion is the more credible because by the end of the same century the magnetic needle likewise arrived in the West by the northern route, not as an instrument of navigation but as a means of ascertaining the meridian, and Western Islam got the compass from Italy.[25] When the new artillery arrived in the West it had lost its name. Because of structural analogy, it took on a new name borrowed from a medieval instrument of torture, the ducking stool or *trebuchetum.*

Whatever its merits, the disadvantages of the *huo p'ao* were the amount of man-power required to operate it and the fact that since the gang pulling the ropes would never pull with exactly the same speed and force, missiles could not be aimed with great accuracy. The problem was solved by substituting a huge counterweight at the short end of the sling-beam for the ropes pulled by men. With this device a change in the weight of the caisson of stones or earth, or else a shift of the weight's position in relation to the pivot, would modify the range of the projectile and then keep it uniform, permitting concentration of fire on one spot in the fortifications to be breeched. Between 1187 and 1192 an Arabic treatise written in Syria for Saladin mentions not only Arab, Turkish, and Frankish forms of the primitive trebuchet, but also credits to Iran the invention of the trebuchet with swinging caisson. This ascription, however, must be in error; for from c. 1220 onward oriental sources frequently call this engine *magribī,* i. e., "Western." Moreover, while the counterweight artillery has not yet been documented for Europe before 1199, it quickly displaced the older forms of artillery in the West, whereas this new and more effective type of siege machinery became dominant in the Mameluke army only in the second half of the thirteenth century. Thus the trebuchet with counterweights would appear to be a European improvement on the *huo p'ao*. Europe's debt to China was repaid in 1272 when, if we may believe Marco Polo, he and a German technician, helped by a Nestorian Christian, delighted the Great Khan by building trebuchets which speedily reduced a besieged city.

But the very fact that the power of a trebuchet could be so nicely regulated impelled Western military engineers to seek even greater exactitude in artillery attack. They quickly saw that until the weight of projectiles and their friction with the air could be kept uniform, artillery aim would still be variable. As a result, as early as 1244 stones for trebuchets were being cut in the royal arsenals of England calibrated to exact specifications established by an engineer: in other words, the cannon ball before the cannon.

The germinal idea of the cannon is found in the metal tubes from which, at least by the late ninth century, the Byzantines had been shooting Greek fire. It may be that even that early they were also shooting rockets of Greek fire, propelled by the expansion of gases, from bazooka-like metal tubes. When, shortly before 673, the Greek-speaking Syrian refugee engineer Callinicus invented Greek fire, he started the technicians not only of Byzantium but also of Islam, China, and eventually the West in search of ever more combustible mixtures. As chemical methods improved, the saltpeter often used in these compounds became purer, and combustion tended toward explosion. In the thirteenth century one finds, from the Yellow Sea to the Atlantic, incendiary bombs, rockets, firecrackers, and fireballs shot from tubes like Roman candles. The flame and roar of all this has made it marvellously difficult to ascertain just when gunpowder artillery, shooting hard missiles from metal tubes, appeared. The first secure evidence is a famous English illumination of 1327 showing a vase-shaped cannon discharging a giant arrow. Moreover, our next certain reference to a gun, a " pot de fer à traire garros de feu " at Rouen in 1338, shows how long it took for technicians to realize that the metal tube, gunpowder, and the calibrated trebuchet missile could be combined. However, iron shot appear at Lucca in 1341; in 1346 in England there were two calibres of lead shot; and balls appear at Toulouse in 1347.

The earliest evidence of cannon in China is extant examples of 1356, 1357, and 1377. It is not necessary to assume the miracle of an almost simultaneous independent Chinese invention of the cannon: enough Europeans were wandering the Yuan realm to have carried it eastward. And it is very strange that the Chinese did not develop the cannon further, or develop hand guns on its analogy. Neither India nor Japan knew cannon until the sixteenth century when they arrived from

Europe. As for Islam, despite several claims to the contrary, the first certain use of gunpowder artillery by Muslims comes from Cairo in 1366 and Alexandria in 1376; by 1389 it was common in both Egypt and Syria. Thus there was roughly a forty-year lag in Islam's adoption of the European cannon.

Gunpowder artillery, then, was a complex invention which synthesized and elaborated elements drawn from diverse and sometimes distant sources. Its impact upon Europe was equally complex. Its influences upon other areas of technology such as fortification, metallurgy, and the chemical industries are axiomatic, although they demand much more exact analysis than they have received. The increased expense of war affected tax structures and governmental methods; the new mode of fighting helped to modify social and political relationships. All this has been self-evident for so long a time that perhaps we should begin to ask ourselves whether the obvious is also the true.

For example, it has often been maintained that a large part of the new physics of the seventeenth century sprang from concern with military ballistics. Yet there was continuity between the thought of Galileo or Newton and the fundamental challenge to the Aristotelian theory of impetus which appeared in Franciscus de Marchia's lectures at the University of Paris in the winter of 1319-20,[26] seven years before our first evidence of gunpowder artillery. Moreover, the physicists both of the fourteenth and of the seventeenth centuries were to some extent building upon the criticisms of Aristotle's theory of motion propounded by Philoponus of Alexandria in the age of Justinian, a time when I can detect no new technological stimulus to physical speculation. While most scientists have been aware of current technological problems, and have often talked in terms of them, both science and technology seem to have enjoyed a certain autonomy in their development.

It may well be that continued examination will show that many of the political, economic, and social as well as intellectual developments in Europe which have traditionally been credited to gunpowder artillery were in fact taking place for quite different reasons. But we know of one instance in which the introduction of firearms revolutionized an entire society: Japan.[27]

Metallurgical skills were remarkably high in Japan when, in 1543, the Portuguese brought both small arms and cannon to Kyushu. Japanese craftsmen quickly learned from the gunsmiths of European ships how to produce such weapons, and within two or three years were turning them out in great quantity. Military tactics and castle construction were rapidly revised. Nobunaga and his successor, Hideyoshi, seized the new technology of warfare and utilized it to unify all Japan under the shogunate. In Japan, in contrast to Europe, there is no ambiguity about the consequences of the arrival of firearms.

But from this fact we must be careful not to argue that the European situation is equally clear if only we would see it so.

11. In examining the origins of gunpowder artillery, we have seen that its roots are multiple, but that all of them (save the European name *trebuchet*) lie in the soil of military technology. It would appear that each area of technology has a certain self-contained quality: borrowings across craft lines are not as frequent as might be expected. Yet they do occur, if exceptionally. A case in point is the fusee.

In the early fifteenth century clock makers tried to develop a portable mechanical timepiece by substituting a spring drive for the weight which powered stationary clocks. But this involved entirely new problems of power control. The weight on a clock exerted equal force at all times, whereas a spring exerts less force in proportion as it uncoils. A new escapement was therefore needed which would exactly compensate for this gradual diminution of power in the drive.

Two solutions were found, the stackfreed and the fusee, the latter being the more satisfactory. Indeed, a leading historian of horology has said of the fusee: "Perhaps no problem in mechanics has ever been solved so simply and so perfectly." [28] The date of its first appearance is much in debate, but we have a diagram of it from 1477.[29] The fusee equalizes the changing force of the mainspring by means of a brake of gut or fine chain which is gradually wound spirally around a conical axle, the force of the brake being dependent upon the leverage of the radius of the cone at any given point and moment. It is a device of great mechanical elegance. Yet the idea did not originate with the clock makers: they borrowed it from the military engineers. In Konrad Keyser's monumental, but still unpublished, treatise on the technology of warfare, *Bellifortis*, completed c. 1405, we find such a conical axle in an apparatus for spanning a heavy crossbow.[30] With very medieval humor, this machine was called "the virgin," presumably because it offered least resistance when the bow was slack and most when it was taut.

* * * * *

In terms of eleven specific technological acts, or sequences of acts, we have been pondering an abstraction, the act of technological innovation. It is quite possible that there is no such thing to ponder. The analysis of the nature of creativity is one of the chief intellectual commitments of our age. Just as the old unitary concept of "intelligence" is giving way to the notion that the individual's mental capacity consists of a large cluster of various and varying factors mutually affecting each other, so "creativity" may well be a lot of things and not one thing.

Thirteenth century Europe invented the sonnet as a poetic form and the functional button [31] as a means of making civilized life more nearly possible in boreal climes. Since most of us are educated in terms of traditional humanistic presuppositions, we value the sonnet but think that a button is just a button. It is doubtful whether the chilly northerner who invented the button could have invented the sonnet then being produced by his contemporaries in Sicily. It is equally doubtful whether the type of talent required to invent the rhythmic and phonic relationships of the sonnet-pattern is the type of talent needed to perceive the spatial relationships of button and buttonhole. For the button is not obvious until one has seen it, and perhaps not even then. The Chinese never adopted it: they got no further than to adapt the tie-cords of their costumes into elaborate loops to fit over cord-twisted knobs. When the Portuguese brought the button to Japan, the Japanese were delighted with it and took over not only the object itself but also its Portuguese name. Humanistic values, which have been cultivated historically by very specialized groups in quite exceptional circumstances, do not encompass sufficiently the observable human values. The billion or more mothers who, since the thirteenth century, have buttoned their children snugly against winter weather might perceive as much of spirituality in the button as in the sonnet and feel more personal gratitude to the inventor of the former than of the latter. And the historian, concerned not only with art forms but with population, public health, and what S. C. Gilfillan long ago identified as "the coldward course" of culture,[32] must not slight either of these very different manifestations of what would seem to be very different types of creativity.

There is, indeed, no reason to believe that technological creativity is unitary. The unknown Syrian who, in the first century B. C., first blew glass was doing something vastly different from his contemporary who was building the first water-powered mill. For all we now know, the kinds of ability required for these two great innovations are as different as those of Picasso and Einstein would seem to be.

The new school of physical anthropologists who maintain that *Homo* is *sapiens* because he is *faber*, that his biological differentiation from the other primates is best understood in relation to tool making, are doubtless exaggerating a provocative thesis. *Homo* is also *ludens*, *orans*, and much else. But if technology is defined as the systematic modification of the physical environment for human ends, it follows that a more exact understanding of technological innovation is essential to our self-knowledge.

REFERENCES

[1] (Oxford, 1954-58).

[2] Cf. the symposium in *Technology and Culture*, I (1960), 299-414.

[3] E. G. Gudde, *California Place Names*, 2nd ed. (Berkeley and Los Angeles, 1960), 52; A. L. Kroeber, "Elements of Culture in Native California," in *The California Indians*, ed. R. F. Heizer and M. A. Whipple (Berkeley and Los Angeles, 1951), 12-13.

[4] E. Zilsel, *Die Entstehung des Geniebegriffes* (Tübingen, 1926).

[5] E. S. G. Robinson, "The Date of the Earliest Coins," *Numismatic Chronicle*, 6th ser., XVI (1956), 4, 8, arbitrarily dates the first coinage c. 640-630 B. C. allowing "the Herodotean interval of a generation" for its diffusion from Lydia to the Ionian cities. But, considering the speed with which coinage appears even in India and China, such an interval is improbable.

D. Kagan, "Pheidon's Aeginetan Coinage," *Transactions and Proceedings of the American Philological Association*, XCI (1960), 121-136, tries to date the first coinage at Aegina before c. 625 B. C. when, he believes, Pheidon died; but the argument is tenuous. The tradition that Pheidon issued a coinage is late, and may well be no more than another example of the Greek tendency to invent culture-heroes. The date of Pheidon's death is uncertain: the belief that he died c. 625 rests solely on the fact that he is not mentioned by Strabo in connection with the war of c. 625-600 B. C.; but if Pheidon, then a very old man, was killed in a revolt of 620 (cf. Kagan's note 21) his participation in this long war would have been so brief and ineffective that Strabo's silence is intelligible.

[6] H. Diels, *Fragmente der Vorsokratiker*, 6th ed. (Berlin, 1951), 171 (B. 90).

[7] The story of the apple is authentic: Newton himself told William Stukeley that when "the notion of gravitation came into his mind [it] was occasion'd by the fall of an apple, as he sat in a contemplative mood"; cf. I. B. Cohen, "Newton in the Light of Recent Scholarship," *Isis*, LI (1960), 490.

[8] The materials on pennons, and other baffles behind the blade of a lance, are found in L. White, jr., *Medieval Technology and Social Change*, (Oxford, 1962), 8, 33, 147, 157.

[9] See the classic work of Ernst Kantorowicz, *The King's Two Bodies*, (Princeton, 1957).

[10] W. Treue, *Kulturgeschichte der Schraube*, (Munich, 1955), 39-43, 57, 109.

[11] F. M. Feldhaus, *Die Technik der Vorzeit, der Geschichtlichen Zeit und der Naturvölker*, (Leipzig, 1914), 984-987.

[12] E. Treptow, "Der älteste Bergbau und seiner Hilfsmittel," *Beiträge zur Geschichte der Technik und Industrie*, VIII (1918), 181, fig. 48; C. N. Bromehead, "Ancient Mining Processes as Illustrated by a Japanese Scroll," *Antiquity*, XVI (1942), 194, 196, 207.

[13] For a detailed history of the crank, cf. White, *op. cit.*, 103-115.

[14] A. R. Hall, "The Military Inventions of Guido da Vigevano," *Actes du VIIIe Congrès International d'Histoire des Sciences*, (Florence, 1958), 966-969.

[15] Bibliothèque Nationale, MS latin 11015, fols. 49r, 51v, 52v. Singer, *op. cit.*, II, figs. 594 and 659, illustrates the first and third of these, but with wrong indications of folio numbers.

[16] H. T. Horwitz, "Uber die Entwicklung der Fahigkeit zum Antreib des Kurbelmechanismus," *Geschichtsblätter fur Technik und Industrie*, XI (1927), 30-31.

[17] White, *op. cit.*, 104 and fig. 4. For what may be a slightly earlier specimen, now in the Seattle Art Museum, see the catalogue of the exhibition *Arts of the Han Dynasty* (New York, 1961), No. 11, of the Chinese Art Society of America.

[18] I am so informed by Dr. Paul Leser of the Hartford Theological Foundation.

[19] L. Makkai, in *Agrártörténeti Szemle*, I (1957), 42.

[20] P. Leser, "Plow Complex; Culture Change and Cultural Stability," in *Man and Cultures: Selected Papers of the Fifth International Congress of Anthropological and Ethnological Sciences*, ed. A. F. C. Wallace (Philadelphia, 1960), 295.

[21] H. Balfour, "The Fire Piston," in *Anthropological Essays Presented to E. B. Tylor*, (Oxford, 1907), 17-49.

[22] É. Salin, *La Civilisation Mérovingienne*, III (Paris, 1957), 6, 55-115.

[23] C. S. Smith, "A Metallographic Examination of Some Japanese Sword Blades," *Quaderno II del Centro per la Storia della Metallurgia*, (1957), 42-68.

[24] White, *op. cit.*, 96-103, 165.

[25] *Ibid.*, 132.

[26] A. Maier, *Zwei Grundprobleme der scholastischen Naturphilosophie*, 2nd ed. (Rome, 1951), 165, n. 11.

[27] D. M. Brown, "The Impact of Firearms on Japanese Warfare, 1543-98," *Far Eastern Quarterly*, VII (1948), 236-253.

[28] G. Baillie, *Watches*, (London, 1929), 85.

[29] Singer, *op. cit.*, III, fig. 392.

[30] Göttingen University Library, Cod. phil. 63, fol. 76ᵛ; cf. F. M. Feldhaus, "Uber den Ursprung vom Federzug und Schnecke," *Deutsche Uhrmacher-Zeitung*, LIV (1930), 720-723.

[31] Some buttons were used in antiquity for ornament, but apparently not for warmth. The first functional buttons are found c. 1235 on the "Adamspforte" of Bamberg Cathedral, and in 1239 on a closely related relief at Bassenheim; cf. E. Panofsky, *Deutsche Plastik des 11. bis 13. Jahrhundert*, (Munich 1924), pl. 74; H. Schnitsler, "Ein unbekanntes Reiterrelief aus dem Kreise des Naumburger Meisters," *Zeitschrift des Deutschen Vereins fur Kunstwissenschaft*, I (1935), 413, fig. 13.

[32] In *The Political Science Quarterly*, XXXV (1920), 393-410.

Air Pollution and Fuel Crises in Preindustrial London, 1250–1650

WILLIAM H. TE BRAKE

Throughout much of the great mass of literature generated by the present environmental crisis there runs a persistent misconception: that environmental problems result from modern industrialization and thus are no older than the Industrial Revolution of the 18th and 19th centuries. Such a view implies that there were no serious environmental problems before industrialization, that, could we eliminate certain offending industries or develop the proper technology to control them, our present ecological ills would be cured. More properly, however, the basic problems of disposing of wastes and finding adequate sources of food, water, and fuel, though certainly aggravated by modern industrialization, are as old as civilization itself. I intend to show in this paper that the occurrence of air pollution in London before the Industrial Revolution was symptomatic of one of these basic environmental problems—the exhaustion of a society's preferred source of fuel and the subsequent difficulty of finding an adequate substitute—and, further, that it was intimately connected to certain demographic and economic developments within that society.

* * *

Air pollution was already a very serious nuisance in London by the middle of the 17th century. John Evelyn, a fellow of the Royal Society of London but perhaps better known as a diarist, wrote in 1661: "It was one day, as I was Walking in Your MAJESTIES Palace at WHITE-HALL, . . . that a presumptuous Smoake . . . did so invade the Court" that ". . . Men could hardly discern one another for the Clowd, and none could support, without manifest Inconveniency." This smoke, he explained, came from "one or two Tunnels" (smokestacks) nearby, "indangering as well the Health [of the king and his subjects] as it sullies the Glory of this . . . Imperial Seat." "And what is all this, but that Hellish and dismall Cloud of SEA-COALE," an "impure and thick Mist, accompanied with a fuliginous and filthy vapour, which renders them obnoxious to a thousand inconveniences, cor-

Mr. Te Brake, professor of history at the University of Maine, teaches and researches European medieval and environmental history.

rupting the *Lungs,* and disordering the entire habit of their Bodies," causing "*Catharrs, Phthisicks, Coughs* and *Consumptions* [to] rage more in this one City, than in the whole Earth besides."[1]

John Graunt, a fellow member of the Royal Society, noted four years later that "little more than one of 50 dies in the Country, whereas in *London* it seems manifest that about one in 32 dies, over and above what dies of the *Plague*." The reasons for this, he concluded, were that London was too crowded and sea coal was too often burned. Whereas before 1600 the death rate for the city was no higher than that of the countryside when little sea coal was burned, by 1665 London was "more *unhealthful*" because "*Sea-Coals* . . . are now universally used." Many people "cannot at all endure the smoak of *London,* not only for its unpleasantness, but for the suffocations which it causes."[2]

These statements by Evelyn and Graunt clearly define the problem. For both, the source of the air pollution was sea coal, a very soft, sulfurous, low-grade coal that when burned emitted a "continual cloud of choking, foul-smelling smoke . . . , leaving behind a heavy deposit of thick black soot on the clothing and faces of all attending."[3] In contrast, the traditional and preferred fuel, firewood (and especially its derivative, charcoal), gave off relatively small quantities of smoke and fumes.

[1]John Evelyn, *Fumifugium: or The Inconveniencie of the Aer and Smoak of London Dissipated. Together with some Remedies humbly Proposed by J. E. Esq; to His Sacred Majestie, and to the Parliament now Assembled* (London, 1661), from the dedicatory epistle and p. 5. The origin of the term "sea coal" is shrouded in mystery. It may first have been used to refer to the coal originating along the Northumberland coast, because it was usually shipped by sea to its destination. It seems more likely, however, that sea coal was first discovered washed up on the beaches of northern England, since the coal seams there extend below the sea. Because it resembled charcoal in appearance it was probably called "sea charcoal" first and "sea coal" later; see John U. Nef, *The Rise of the British Coal Industry* (London, 1932), vol. 2, appendix P, "Note on the Origin of the Word 'Sea Coal' "; see also Raymond Smith, *Sea-Coal for London: History of the Coal Factors in the London Market* (London, 1961), pp. 2, 7, and appendix A; L. F. Salzman, *English Industries of the Middle Ages,* 2d ed. (Oxford, 1923), pp. 2–3; and Howard N. Eavenson, *Coal through the Ages,* 2d ed. (New York, 1939), p. 9.

[2]John Graunt, *Natural and Political Observations Mentioned in a following Index and made upon the Bills of Mortality . . . with reference to the Government, Religion, Trade, Growth, Air, Diseases and the several Changes of the said City,* 5th rev. ed. of 1676, repr. in William Petty, *The Economic Writings of Sir William Petty,* ed. Charles Henry Hull (Cambridge, 1899), 2:393–94. Graunt's name is closely associated with the origins of demographic research in Great Britain; see, e.g., D. V. Glass, "John Graunt and His Natural and Political Observations," *Proceedings of the Royal Society,* ser. B, 159 (1963): 1–32.

[3]Nef, 1:130. Actually, coal containing sulfur eventually produces sulfuric acid. Burning emits oxides of sulfur, some of which—sulfur trioxide (SO_3) especially—readily react with water to form the acid H_2SO_4. Thus, sea coal not only produced choking clouds of foul-smelling smoke and soot but also a very powerful and irritating acid which entered the eyes and breathing passages.

Contrary to Graunt's opinion, however, air pollution was not unknown before the 17th century. Nor were Evelyn and Graunt the first to locate its cause in the use of a particular type of fuel. The smoke of sea coal presumably drove Queen Eleanor from Nottingham Castle at the time of the feast of Saint Michael in 1257.[4] One must look to London, however, to find evidence suggesting that air pollution at this time could be more than simply an isolated problem.[5]

The smoke of sea coal fires was a general nuisance in London by the last quarter of the 13th century. A royal commission appointed in 1285 to inquire into the operation of certain lime kilns found "that whereas formerly the lime used to be burnt with wood, it is now burnt with sea-coal." Consequently, "the air is infected and corrupted to the peril of those frequenting . . . and dwelling in those parts."[6] The lime burners, however, were persistent; a second commission of inquiry was appointed for the same reason in 1288 "on complaint by many inhabitants that they are annoyed by lime kilns."[7] In 1298 a group of London smiths voluntarily decided "that none [of their trade] should work at night on account of the unhealthiness of [sea] coal and damage to their neighbors."[8]

The commissions of inquiry and the voluntary efforts of the smiths

[4]"Propter fumum carbonum maris nullo modo potuit demorari," in *Rerum Britannicarum Medii AEvi Scriptores* (Rolls Series), *Annales Monastici*, no. 36, 3:203–4.

[5]Archaeological investigations have disclosed that coal occasionally was used as a fuel in Roman Britain. Coal cinders have been found amid the remains of some Roman towns and villas, especially near outcropping coal seams in Northumberland as well as in the Forest of Dean, in southwestern England. There is no reason to assume, however, that it amounted to anything more than local use at that time. The Domesday Book, compiled in 1086 by the new Norman administration, was an attempt to catalog everything of economic value in England. It contains no mention of coal or coal mining, although lead and iron mining did figure in it. By 1200, however, the monks of Holyrood and Newbattle abbeys were mining coal along the Firth of Forth at Carriden as well as at Linlithgow; see, for example, Eavenson, pp. 5, 8; as well as Cyril E. Hart, *Royal Forest: A History of Dean's Woods as Producers of Timber* (Oxford, 1966), p. 5; Nef, 1:2; and Salzman, p. 1. When or under what conditions sea coal was first introduced into London is not known. Perhaps it was used as ballast in ships returning from Newcastle. By 1228, however, it was common enough for a street to be named "Sea-Coal Lane," and in 1257 definite mention was made of imports of sea coal into London. The London sea coal trade was extensive enough by the late 13th century to warrant the appointment of a coal meter, an official charged with regulating its import and sale; see H. T. Riley, ed. and trans., *Memorials of London and London Life in the XIIIth, XIVth, and XVth Centuries* (London, 1868), p. xvi, n. 7; *Calendar of Early Mayor's Court Rolls* (1298–1307), p. 30; Salzman, pp. 3–4; Smith, p. 2; J. B. Blake, "The Medieval Coal Trade of Northeast England," *Northern History* 2 (1957): 1–2; and M. M. Postan, *Medieval Economy and Society: An Economic History of Britain, 1100–1500* (Berkeley and Los Angeles, 1972), p. 198.

[6]*Calendar of Patent Rolls,* Edward I (1281–92), p. 207.

[7]Ibid., p. 296.

[8]*Calendar of Early Mayor's Court Rolls* (1298–1307), p. 34.

were not sufficient to combat London's air pollution problem. Edward I, therefore, issued a royal proclamation in 1307 prohibiting the use of sea coals in kilns,

> as the King learns from the complaint of prelates and magnates of his realm, who frequently come to London for the benefit of the commonwealth by his order, and from the complaint of his citizens and all his people dwelling there and in Southwark that the workmen in the city and town aforesaid and in their confines now burn them [kilns] and construct them of sea-coal instead of brushwood and charcoal, from the use of which sea-coal an intolerable smell diffuses itself throughout the neighboring places and the air is greatly infected, to the annoyance of the magnates, citizens and others there dwelling and to the injury of their bodily health.[9]

Two weeks later another royal commission, authorized "to punish offenders by grievous ransoms," attempted to determine why the royal proclamation was not being observed.[10]

For the next two centuries or more, air pollution seems to have been much less of a problem in London. After a complaint in 1371 that the smelting operations of certain plumbers in Wodhawe in the parish of Saint Clements were causing annoying smoke and fumes,[11] which sounds as if sea coal may have been the cause, it is difficult to find further mention of the nuisance until well into the 16th century.[12] This silence may simply reflect a gap in the records, since

[9] *Calendar of Close Rolls,* Edward I (1302–7), p. 537.

[10] *Calendar of Patent Rolls,* Edward I (1301–7), p. 549.

[11] *Calendar of Letter Books of the City of London,* ed. R. R. Sharpe, *Book G* (London, 1905), pp. 283–84.

[12] A document of 1467 from the town of Beverley, some 150 miles due north of London, contains a reference to damaged fruit trees and other inconveniences caused by the stench of brick kilns. In the future, such kilns would be allowed no closer to Beverley than they were at that time: "Item salubriter ordinatum est quod proper fetorem et aeris intemperiem ad destruccionem fructuum arborum, aliaque incommoda, que ex inde provenire poterit, nullus edificare presumat decetero aliquod thorale pro cremacione tegularum infra predictam villam Beverlaci, aut propius eandem villam quam throalia tegularum edificata existunt in presenti, sub pena centum solidorum applicandorum et solvendorum usui Communitatis ville antedicte" (in Arthur F. Leach, ed., *Beverley Town Documents,* Publications of the Seldon Society, vol. 14 [London, 1900], p. 58). The actual wording of this document, however, suggests that this was a sort of zoning ordinance against brick kilns in general, whether fired by firewood, charcoal, or sea coal. Even firewood and charcoal would emit some smoke, especially, perhaps, in the brick-baking process. In any case, there is nothing here to suggest that the town of Beverley was suffering from an air pollution problem at all comparable to that caused by the burning of sea coal in London at the beginning of the 14th century.

sea coal continued to be imported into London.[13] But for reasons that will be discussed in detail below, it is perhaps best to view the lack of complaints about air pollution in London in the late 14th, 15th, and early 16th centuries as actually reflecting a diminution of the problem.

By the second half of the 16th century, air pollution had once again assumed serious proportions in London. Queen Elizabeth was so "greved and annoyed with the taste and smoke of sea cooles" in 1578 that the Company of Brewers promised to use only wood in their brewing operations in the future.[14] In 1623, the House of Lords passed an act forbidding the use of sea coal in brewhouses "within one mile of any house in which His Majesty's Court or the Court of the Prince of Wales shall be usually held," but the Commons dropped it at the end of the session.[15] Hereafter, however, the complaints become very frequent and begin to assume a much more sweeping character. Thus, a 1627 petition aimed at "the farmers of the alum works, on account of the loathsome vapour from [their] works to the great annoyance of the inhabitants within a mile compass," maintained that sea coal smoke was responsible for "tainting the pastures, and poisoning the very fish in the Thames."[16]

If the complaints of air pollution were becoming more and more common, the city of London was also becoming more and more dependent on sea coal for fuel. In the winter of 1637–38, Charles I attempted to sell the monopoly of the coal trade to the city of Newcastle, a measure bitterly opposed by London shipowners. The ensuing quarrel temporarily interrupted the supply of sea coal to London, causing considerable discomfort in the city.[17] Six years later, Richard Gesling reported that, whereas previously "some fine Nosed City *Dames*" used to complain about "the smell of this *Cities Seacoale Smoke,*" they now cry "would to *God* we had *Seacoale.* O the want of Fire undoes us! O the sweete *Seacoale* fire we used to have, how we want them now, no fire to your *Seacoale!*"[18] London was one of the

[13]See, e.g., *Calendar of Patent Rolls,* Edward III (1348–50), p. 50. In 1388 alone, nine persons were issued licenses to ship sea coal from Newcastle to London (*Calendar of Patent Rolls,* Richard II [1385–89], pp. 400, 407, 410).

[14]*Calendar of State Papers, Domestic Series,* Elizabeth I (1578), p. 612.

[15]The Royal Commission on Historical Manuscripts, *Third Report* (London, 1872), p. 28.

[16]*Calendar of State Papers, Domestic Series,* Charles I (1627–28), pp. 269–70.

[17]Ibid., Charles I (1637), p. 295.

[18]Richard Gesling, *Artificiall Fire, or Coale for Rich and Pore* (1644). The relevant page is reproduced in Nef, vol. 1, facing p. 247.

world's leading industrial cities by the reign of Charles I.[19] Its innumerable factories and workshops and tens of thousands of domestic hearths demanded immense quantities of fuel. Sea coal and air pollution had become facts of life for London's inhabitants.

* * *

All of the complaints listed above expressed an obvious distaste for sea coal. It was clearly identified as the source of air pollution. Why then was sea coal burned? Did it have certain characteristics demanded by new industries that could not be matched by wood fuels?

For the technological requirements of pre–Industrial Revolution England, both wood and coal fuels produced sufficient quantities of heat energy. Lime burners and smiths were the first to use sea coal in their fires. By the 16th century, the artisans of many other trades were also beginning to make the fuel substitution: brewers, brick and tile makers, salt makers, dyers, and malt dryers, to name a few.[20] But all of these trades had been practiced for centuries with wood or charcoal. Thus, the changeover was not made to meet the higher energy requirements of new or developing technologies.

On the other hand, sea coal contained such impurities that, in addition to letting off clouds of smoke and fumes, it could cause harmful side effects in the industrial processes themselves. Iron smiths preferred to use charcoal or wood in their forges because the high sulfur content of sea coal made the iron overly brittle. From time to time, brewers would attempt to brew with sea coal fires, though the drinking public generally insisted that beer or ale thus brewed was tainted by the smoke.[21] In the 17th century, artisans and amateur scientists began to experiment with "charring sea coale" so that it could be used in smelting processes as well. Such attempts at producing coke were not really successful, however, until the 18th century.[22]

Sea coal cannot have been an ideal fuel for domestic heating either. During the 12th and 13th centuries major technological innovations in home heating resulted in the widespread diffusion of the fireplace, flue, and chimney throughout all classes of society. In combination,

[19] John U. Nef, "A Comparison of Industrial Growth in France and England from 1540 to 1640," *Journal of Political Economy* 44 (1936): 663.

[20] Nef, *Rise of the British Coal Industry,* 1:11–12, 201–23.

[21] Ibid., 1:11–13, 201–23; Blake, pp. 4–5; and G. T. Salusbury, *Street Life in Medieval England,* 2d ed. (London, 1948), p. 116.

[22] For a description of one of these attempts, see John Evelyn, *The Diary of John Evelyn,* ed. E. S. de Beer (Oxford, 1955), 3:180–81, entry for July 10, 1656. For a recent discussion of 18th-century advances in the use of coal for iron production, see Charles K. Hyde, "Technological Change in the British Wrought Iron Industry, 1750–1815: A Reinterpretation," *Economic History Review,* 2d ser., 27 (1974): 190–206.

these elements provided more usable heat and better control of smoke and fumes in dwellings than did the earlier, central hearth, which allowed smoke and fumes to swirl around the room before exiting through a hole in the roof. But even so, the burning of poor or uncured firewood could still fill rooms equipped with fireplace, flue, and chimney with a considerable amount of smoke.[23] If sea coal were burned, presumably even less satisfactory results would be expected. Thus, people like the 17th-century English diarist Samuel Pepys kept a supply of charcoal, when such could be obtained, to be used instead of sea coal for heating the dining room and bedroom.[24]

If sea coal was no better than wood fuels as far as the amount of heat produced was concerned, if it actually produced undesirable side effects in certain industrial processes, and if it compared unfavorably to firewood or charcoal for domestic purposes, why was sea coal in fact substituted for wood fuel in the late 13th and early 14th centuries and again from the late 16th century on? As the 18th-century British economist Adam Smith suggested, the substitution of sea coal for wood fuels was a matter of economics. Writing in 1776, he too concluded that "coals are a less agreeable fewel than wood: they are said to be less wholesome." Therefore, he observed, "the expence of coals . . . at the place where they are consumed must generally be somewhat less than that of wood."[25] Indeed, what he said can perhaps be applied to an earlier period as well.

Though it is impossible to construct satisfactory price indices for wood products and sea coal during the 13th and 14th centuries,[26] I intend to show that in London, at least, there was a genuine shortage of wood fuels at that time which would have led to an increase in prices. Many Londoners, therefore, began to turn to relatively inexpensive sea coal as an alternative source of fuel despite its other dis-

[23]LeRoy J. Dresbeck. "The Chimney and Fireplace: A Study in Technological Development Primarily in England during the Middle Ages" (Ph.D. diss., University of California, Los Angeles, 1971), esp. chaps. 3, 6. For the effects of poor fuel in the new heating systems, see p. 222.

[24]Samuel Pepys, *The Diary of Samuel Pepys,* ed. Robert Latham and William Matthews, vol. 1 (Berkeley and Los Angeles, 1971), pp. 14, 405, 409.

[25]Adam Smith, *An Enquiry into the Nature and Causes of the Wealth of Nations,* ed. Edwin Cannan (New York, 1937), bk. 1, chap. 11, pt. 2, p. 165. Asta Moller, "Coal-Mining in the Seventeenth Century," *Transactions of the Royal Historical Society,* 4th ser., 8 (1925): 81, observes that during the 17th century, "where no immediate scarcity of fuel, either domestic or industrial use, was felt, the inducement to engage in [coal] mining was only slight."

[26]Postan, p. 231. Hart, pp. 22–28, 60–65, 318–20, compiled some interesting details concerning costs and prices of timber and wood fuels in the Forest of Dean for the 13th and 14th centuries. Unfortunately, they cannot be properly compared with sea coal prices at London, nor are they complete enough to indicate any long-term price trends.

advantages. When supplies of wood fuels once more became adequate in the late 14th, 15th, and early 16th centuries, firewood and charcoal again replaced sea coal in many of London's forges, furnaces, and kilns, thus reducing the frequency of complaints about air pollution. A second fuel crisis began in the second half of the 16th century, and once again sea coal became a less expensive alternative.

Because the necessary price indices are lacking, the alternating phases of shortage and adequate supplies of wood fuels that I have postulated can be indicated only by an examination of some general features of late medieval life and a number of long-term demographic and economic trends common to all of western Europe. Stated most simply and mechanically, land, or rather the use of land for agricultural purposes, was the basis of economic life. Under normal conditions, woodland and other waste (i.e., land not under cultivation) played a very important role in peasant ecology as, among many other things, a constant source of timber and wood fuels. Significant and sustained population increase, however, upset the balance. Demands for timber and wood fuels grew proportionately with population increase, while much waste, including woodland, was reclaimed and cultivated to meet increased demands for food. If these demands were severe enough, woodland dwindled to the point where local fuel shortages developed and alternative sources of fuel were sought. If, on the other hand, population suddenly decreased substantially, fuel and timber demands immediately decreased as well, and some previously cultivated land was allowed to revert to brush and eventually woods capable of supplying most fuel needs.[27]

* * *

[27]On medieval peasant ecology and economy, see Georges Duby, *Rural Economy and Country Life in the Medieval West,* trans. Cynthia Postan (London, 1968), *The Early Growth of the European Economy: Warriors and Peasants from the Seventh to the Twelfth Century,* trans. Howard B. Clarke (London, 1974), and "Medieval Agriculture, 900–1500," in *The Fontana Economic History of Europe,* ed. Carlo M. Cipola, vol. 1, *The Middle Ages* (London, 1972), pp. 175–220; Roger Grand and Raymond Delatouche, *L'agriculture au moyen âge de la fin de l'Empire romain au XVIe siècle* (Paris, 1950); B. H. Slicher van Bath, *The Agrarian History of Western Europe, A.D. 500–1850,* trans. Olive Ordish (London, 1963); Guy Fourquin, *Le paysan d'Occident au moyen âge* (Paris, 1972); Marc Bloch, *French Rural History: An Essay on Its Basic Characteristics,* trans. Janet Sondheimer (Berkeley and Los Angeles, 1966); *The Cambridge Economic History of Europe,* vol. 1, *The Agrarian Life of the Middle Ages,* 2d ed., ed. M. M. Postan (Cambridge, 1966); Lynn White, jr., *Medieval Technology and Social Change* (Oxford, 1962); and others. For the process involved in the expansion and contraction of arable lands during the Middle Ages, it may be useful to keep in mind the explanation of Ester Boserup, *The Conditions of Agricultural Growth: The Economics of Agrarian Change under Population Pressure* (Chicago, 1965), as well as her more recent "Environment, population et technologie dans les sociétés primitives,"

At the time of the Norman Conquest, in 1066, England had extensive woodlands. Though Anglo-Saxon and Scandinavian peasants had been busy for centuries converting much of them into pasture and arable land, the Domesday survey of 1086 gave detailed accounts of how much remained. For the county of Middlesex, excluding London, it listed several villages with enough woodland to provide acorns and beech mast for as many as 2,000 swine, with many more villages possessing lesser amounts.[28] William Fitz Stephen, in his 12th-century *Description of London* (ca. 1174), maintained that within a very short distance of London there was "a great forest with wooded glades and lairs of wild beasts, deer both red and fallow, wild boars and bulls."[29]

Though England's woodlands were still extensive at the end of the 11th and into the 12th century, one must not suppose, therefore, that they were unused or vacant. Rather, they were the scene of much activity. Since the earliest times, the woods were an important source of supplemental food, both animal and vegetable. Most household and agricultural tools and implements were constructed, at least in part, of wood. Most dwellings and all ships were constructed of timber. For centuries swine had been allowed to forage on acorns and beech mast; this right of pannage often meant the difference between starvation and subsistence. Much medieval industry was located near camps of charcoal burners, who, for a ton of charcoal—the essential fuel for smelting, glassmaking, and many other industries—leveled

---

Annales: Economies, sociétés, civilisations 29 (1974): 538–52. Briefly, she maintains that change in population density is the major factor in determining agricultural change. As population increases in a given area, the use of the land is intensified (or the fallow is shortened), producing food for more people but also requiring more labor. Thus, if population declines, with a reduction in food requirements and with fewer hands to perform the labor, the intensiveness of land use is relaxed (or the fallow is lengthened).

[28]H. C. Darby, "The Clearing of the Woodland in Europe," in *Man's Role in Changing the Face of the Earth*, ed. William L. Thomas et al. (Chicago, 1956), pp. 191–92 and fig. 55, and "Domesday Woodland," *Economic History Review*, 2d ser., 3 (1950): 22; William Page, ed., *The Victoria History of the County of Middlesex* (London, 1911), 2:223; and N. Neilson, "Early English Woodland and Waste," *Journal of Economic History* 2 (1942): 56.

[29]William Fitz Stephen, *A Description of London*, trans. H. E. Butler, published in *Norman London: An Essay*, ed. F. M. Stenon et al., Historical Association Leaflets, nos. 93, 94 (London, 1934), p. 27. It is quite possible, however, that Fitz Stephen was using the term "forest" in its legal sense, to describe an area reserved by the Crown or nobility for the chase and, therefore, off limits to most people in search of firewood; yet his statement does give us an idea of woodland in close proximity to London—perhaps it was some of the woodland described by the Domesday Book a century before. On the other hand, G. J. Turner, ed., *Select Pleas of the Forest*, Publications of the Seldon Society, vol. 13 (London, 1901), p. cviii, maintains that there were no forests or forest law in the counties of Kent and Middlesex, except for a warren at Staines, Middlesex, already disaforested in 1227 (i.e., its legal status was changed from forest to ordinary land).

huge stretches of timber.[30] In almost every way imaginable, woodland was essential to preindustrial European society. In Germany, Heriger, bishop of Mainz (913–27), when told of a false prophet "who with many good reasons had advanced the idea that Hell was completely surrounded by a dense forest," is supposed to have replied, "I would like to send my swineherd there with my lean pigs to pasture."[31] One suspects that, say, the bishop of Winchester would have reacted in a similar manner.

Because of the central role that woodland played in the medieval economy, an elaborate system of rights and usages based on customary law codified and regulated its exploitation. One should remember that, in addition to obtaining the Magna Carta from King John, the English barons also placed a check, in the form of the Forest Charter, on royal attempts to put additional land in the royal game preserves. This Forest Charter, in Doris Stenton's opinion, was perhaps as important as the Magna Carta itself and became one of the foundations of the medieval English social scene.[32] The complex body of forest law, rights, and usages was designed to guarantee access for a number of groups to England's woodlands. At the same time, it was intended, for various reasons, to guard against their destruction, for if the woodlands disappeared, they would take with them the very life of the medieval economy.[33]

In spite of such efforts to preserve the woodlands, there were compelling reasons for destroying them as well. Peasants often had an interest in extending their arable lands by assarting, that is, by cutting down trees and digging out their roots. Until well into modern times, the margin of security against starvation was extremely thin. The failure of a single year's harvest often brought on genuine famine

[30]Three lime kilns consumed 26 acres of timber in 1229 (*Calender of Close Rolls,* Henry III [1227–31], p. 268). In the 16th century, two iron forges and their charcoal works consumed 2.7 million cubic feet of timber in two years; see Salzman, p. 39; and H. R. Schubert, *History of the British Iron and Steel Industry from c. 450 B.C. to A.D. 1775* (London, 1957), pp. 112–14. For an extensive list of the uses of woodland in the Middle Ages, see Clarence Glacken, *Traces on the Rhodian Shore: Nature and Culture in Western Thought from Ancient Times to the End of the Eighteenth Century* (Berkeley and Los Angeles, 1967), pp. 320–22; as well as the discussions of Duby, *Rural Economy and Country Life in the Medieval West,* pp. 143–44; Grand and Delatouche, pp. 410–44; Slicher van Bath, pp. 72–73; Salzman, throughout; Neilson, pp. 54–62; and Cyril E. Hart, *The Verderers and Forest Laws of Dean* (Newton Abbot, 1971), p. 19, and *Royal Forest,* p. xix.

[31]Quoted in Glacken, p. 321.

[32]Doris Stenton, *English Society in the Early Middle Ages,* 4th ed., Pelican History of England, vol. 3 (Harmondsworth, 1971), pp. 100–122, esp. p. 109. For mostly Continental examples, see Glacken's discussion of medieval forest laws in France and Germany, pp. 322–40.

[33]Glacken, p. 336. See also Grand and Delatouche, p. 433; and Neilson, pp. 54–62.

conditions.[34] Therefore, efforts were constantly made to increase the supply of foodstuffs. This was done in essentially two ways: by improving the yield per acre through the use of better seeds, different types of crops, fertilizer, and a more efficient organization of fallow and cultivation; or by increasing the acreage under the plough. The latter method, the expansion of arable land at the expense of woodland and other waste, was the one most used until the end of the Middle Ages; between 1000 and 1300 it was sufficient to prevent serious food shortages in most of Europe.[35]

The period from the 11th through the 13th centuries was a significant era of environmental change in western Europe, the great age of reclamation.[36] Almost everywhere woodland and marsh yielded to arable farmland. Though the patterns and chronology of reclamation varied from place to place, certain common factors were involved. The key element was the growth of population; plots of agricultural land had been divided and subdivided for generations to the point where they were no longer large enough to provide the food needed by the people who depended on them for subsistence.[37] The

[34]F. Curschmann, *Hungersnöte im Mittelalter: Ein Beitrag zur deutschen Wirtschaftsgeschichte des 8. bis 13. Jahrhunderts,* Leipziger Studien aus dem Gebiet der Geschichte, vol. 6, pt. 2 (Leipzig, 1900); Hans van Werveke, "De middeleeuwse hongersnood," *Mededelingen, Koninklijke Vlaamse Academie voor Wetenschappen, Letteren, en Schone Kunsten,* klasse der letteren, vol. 29, no. 3 (1969), entire issue, and "La famine de l'an 1316 en Flandre et dans les régions voisines," *Revue du Nord* 41 (1959): 5–14; Henry S. Lucas, "The Great European Famine of 1315, 1316, and 1317," *Speculum* 5 (1930): 343–77; Leopold Genicot, "Crisis: From the Middle Ages to Modern Times," in *The Cambridge Economic History of Europe,* vol. 1, *The Agrarian Life of the Middle Ages,* 2d ed., ed. M. M. Postan (Cambridge, 1966), pp. 672–74; Ian Kershaw, "The Great Famine and Agrarian Crisis in England, 1315–1322," *Past and Present,* no. 59 (1973), pp. 3–50; and Odo Rigaldus, *The Register of Eudes of Rouen,* ed. Sydney M. Brown (New York and London, 1964), pp. 253, 306–7, 410–11, 502–3.

[35]Duby, "Medieval Agriculture, 900–1500," p. 199; and B. H. Slicher van Bath, "The Rise of Intensive Husbandry in the Low Countries," in *Britain and the Netherlands,* vol. 1, ed. J. S. Bromley and E. H. Kossmann (London, 1959), pp. 130–53.

[36]Glacken, p. 291; and Grand and Delatouche, pp. 237–46.

[37]For medieval population changes in general, see Josiah C. Russell, "Late Ancient and Medieval Population," *Transactions of the American Philosophical Society,* n.s., vol. 48, pt. 3 (1958), and "Population in Europe, 500–1500," in *The Fontana Economic History of Europe,* ed. Carlo M. Cipolla, vol. 1, *The Middle Ages* (London, 1972), pp. 25–70. See also Duby, *Rural Economy and Country Life in the Medieval West,* p. 67; F. L. Ganshof and A. E. Verhulst, "Medieval Agrarian Society in Its Prime: France, the Low Countries, and Western Germany," in *The Cambridge Economic History of Europe,* vol. 1, *The Agrarian Life of the Middle Ages,* 2d ed. ed. M. M. Postan (Cambridge, 1966), p. 294; M. M. Postan, "Medieval Agrarian Society in Its Prime: England," in ibid., pp. 563–65; J. Z. Titow, "Some Evidence of the Thirteenth Century Population Increase," *Economic History Review,* 2d ser., 14 (1941): 218–24; Norman J. G. Pounds, "Overpopulation in France and the Low Countries in the Later Middle Ages," *Journal of Social History* 3 (1970):

simultaneous growth of nonagricultural centers of merchants and craftsmen, as at Paris and London and in Flanders, demanded agricultural surpluses and tended to inflate prices of foodstuffs.[38] As a result, a snowballing process began: as the number of people increased and agricultural prices remained high, new lands were wrested from the waste; as population continued to increase, more cropland was added. This process proceeded virturally without interruption until, by the beginning of the 14th century, few lands remained to be cleared.[39] At the same time, the marshlands of northwestern Europe were drained and occupied for much the same reason.[40]

It is not possible to give any set of accurate figures or percentages of the amount of land brought into cultivation during this period, but, while it lasted, the growth was significant.[41] In the mountain areas, especially the Alps and the Pyrenees, the edge of human settlement crept to ever higher elevations.[42] The huge wooded areas of France, Germany, and England shrank dramatically. The most striking evidence of this growth is found in the vast numbers of new towns

---

225–47; and Slicher van Bath, *Agrarian History of Western Europe,* p. 73. For a general discussion of medieval population trends and some problems associated with their study, see Carlo Cipolla, Jean Dhont, M. M. Postan, and Philippe Wolff, "Anthropologie et démographie: moyen âge" (a collective report given to the Ninth International Congress of Historical Sciences at Paris, August 28–September 3, 1950), *Rapports* (Paris, 1950), 1:55–80.

[38]See, for example, Slicher van Bath, *Agrarian History of Western Europe,* pp. 116, 132.

[39]Ibid., pp. 134–36; and Genicot, pp. 669–70.

[40]Concerning the reclamation of some English marshes, see H. E. Hallam, *Settlement and Society: A Study of the Early Agrarian History of South Lincolnshire,* Cambridge Studies in Economic History (Cambridge, 1965). For coastal Flanders, see A. Verhulst, "Historische geografie van de Vlaamse kustvlakte tot omstreeks 1200," *Bijdragen voor de geschiedenis der Nederlanden* 14 (1959–60): 1–37. Peat bog reclamation in the county of Holland and the bishopric of Utrecht is carefully studied by H. van der Linden, *De Cope: bijdrage tot de rechtsgeschiedenis van de openlegging der Hollands-Utrechtse laagvlakte* (Assen, 1955). For an environmental history approach to marshland reclamation in a part of the county of Holland, see William H. TeBrake, "The Making of a Humanized Landscape in the Dutch Rijnland, 1000–1500" (Ph.D. diss., University of Texas, Austin, 1975).

[41]Ganshof and Verhulst, p. 295; Postan, "Medieval Agrarian Society in Its Prime," pp. 548–52; Richard Koebner, "The Settlement and Colonization of Europe," in *The Cambridge Economic History of Europe,* vol. 1, *The Agrarian Life of the Middle Ages,* 2d ed., ed. M. M. Postan (Cambridge, 1966), pp. 43–91; and Glacken, pp. 291, 328. H. Draye (*Landelijke cultuurvormen en kolonisatie geschiedenis van B. Huppertz: "Räume und Schichten bäuerlicher Kulturformen in Deutschland,"* Toponymica: bijdragen en bouwstoffen uitgegeven door de Vlaamsche Toponymische Vereeniging te Leuven [Louvain and Brussels, 1941], p. 60) maintains that two-thirds of the presently cultivated land in the originally wooded areas of Belgium were reclaimed in the 10th through 13th centuries.

[42]Slicher van Bath, *Agrarian History of Western Europe,* pp. 133.

founded during this period.[43] Place names containing elements referring to woods, clearings, assartings, and the like (as well as marshes, dikes, and drainage canals), are common throughout all of western Europe, and most date from the 11th through the 13th centuries.[44]

The patterns of medieval growth and expansion so prevalent in western Europe in general were present in England in particular. It is estimated that England's population grew from 1.1 million at the time of the Domesday survey (1086) to about 3.7 by the early 14th century.[45] Meanwhile, London grew from about 20,000 inhabitants in 1200 to at least 40,000 by 1340.[46] And, as was the case for Europe as a whole, ploughland made heavy inroads into the woodland. An Anglo-Saxon poet once described the ploughman as the "grey enemy of the wood."[47] This is also the picture that one must keep in mind for the 11th through the 13th centuries. The ever increasing food needs of the burgeoning population eventually made lands so scarce that even soil of vastly inferior quality was cultivated—areas "where no or virtually no grain was grown in any other period of English history." These were lands which a society would cultivate only in times of "real land hunger."[48]

[43]See, for example, Maurice Beresford, *New Towns of the Middle Ages: Town Plantation in England, Wales, and Gascony* (New York and Washington, D.C., 1967), p. 366, table 13.6. Beresford concludes that, of the total of 368 new creations between 1066 and the end of the Middle Ages, all but 51 occurred between 1066 and 1300. See also R. P. Bekinsale, "Urbanization in England to A.D. 1420," in R. P. Bekinsale and J. M. Houston, *Urbanization and Its Problems* (Oxford, 1968), p. 18.

[44]Slicher van Bath, *Agrarian History of Western Europe,* p. 133; Darby, "Clearing of the Woodland in Europe," pp. 191–95; Duby, *Rural Economy and Country Life in the Medieval West,* p. 80; M. Gysseling, *Toponymisch woordenboek van België, Nederland, Luxemburg, Noord-Frankrijk en West-Duitsland (vóór 1226),* 2 vols. (Brussels, 1962), passim; and D. P. Blok, "Plaatsnamen in Westfriesland," *Philologia Frisica anno 1966,* no. 319 (1968), pp. 11–19.

[45]Josiah C. Russell, *British Medieval Population* (Albuquerque, N.M., 1948), pp. 246–60, and "Late Ancient and Medieval Population," p. 105. After the demographic decline of the second half of the 14th century, England's population reached 3.7 million again only shortly before 1600, a time for which we have better evidence of a fuel crisis; see Julian Cornwall, "English Population in the Early Sixteenth Century," *Economic History Review,* 2d ser., 23 (1970–71): 44, table 5.

[46]Russell, *British Medieval Population,* p. 287, estimated London's population at 30,000 early in the 13th century and not much below 60,000 by 1340. For a criticism of Russell's figures, see Gwyn A. Williams, *Medieval London: From Commune to Capital* (London, 1963), pp. 315–17. See also Russell, "Late Ancient and Medieval Population," p. 47 (table 48), p. 61 (table 64), p. 68, and p. 69 (table 71).

[47]Darby, "Clearing of the Woodland in Europe," p. 191.

[48]Postan, "Medieval Agrarian Society in Its Prime," pp. 551–54, 565. Also see J. B. Harley, "Population Trends and Agricultural Developments from the Warwickshire Hundred Rolls of 1279," *Economic History Review,* 2d ser., 11 (1958–59): 11–12; Maurice

It should be remembered, however, that there were still substantial amounts of woodland in England forming the nuclei of royal and noble forests and warrens which could be exploited only with special charters and grants. The fines for trespassing were so severe that for the ordinary man it was disastrous to be caught simply lopping off the branch of a tree. Essentially, therefore, forests and other preserves of this sort were not viable sources of fuel, since their reason for existence was to provide cover for game, not firewood or charcoal.[49]

By the middle of the 13th century, the effects of increasing population, requiring more and more wood products and the extension of arable lands at the expense of waste, began to be strongly felt. Consequently, one can find an occasional complaint about the destruction of England's woodlands during the 13th century. In 1232, Henry III was told that the timbers he had ordered for his building project at Westminster could not be found in the forests of Windsor or Cornbury.[50] In 1255, he learned that the forges in the Forest of Dean were "harmful to the forest because the destruction of the forest exceeds the issues of the forges,"[51] meaning that the fuel consumed was worth more than the iron produced. Later in the same year the king rescinded an order to sell off part of his timber in Northhampton because "in process of time one trunk will fetch as much as three or four now,"[52] indicating that the king, or at least his councillors, were aware of a growing shortage of good timber. In 1258 and 1259 Henry III again expressed concern about the destruction of his forests and ordered that sales of timber be stopped.[53]

That a demand for timber and wood fuels existed is shown by the fact that, during the 1260s, a certain Peter de Neville managed to dispose of 7,000 oaks and other trees from the king's park of Ridlington at a value of 350 pounds and caused further inestimable damage to the underwood and branchwood by illegal sales or gifts.[54] And on

---

Beresford, *The Lost Villages of England* (London, 1954), p. 201; W. B. Hoskins, *The Making of the English Landscape* (Harmondsworth, 1970), chap. 3; Titow, pp. 222–23; and Pounds, p. 241.

[49]Margaret L. Bazeley, "The Extent of the English Forest in the Thirteenth Century," *Transactions of the Royal Historical Society,* 4th ser., 4 (1921): 140–72; Stenton, pp. 100–122; and Hart, *The Verderers and Forest Laws of Dean,* pp. 20–24, and *Royal Forest,* pp. 7–8, 13–14.

[50]*Calendar of Close Rolls,* Henry III (1231–58), pp. 41, 85. By the late 12th and the early 13th centuries, imports of Baltic and Scandinavian timber were already quite common at ports such as Lynn and Boston; see E. Carus-Wilson, "The Medieval Trade of the Ports of the Wash," *Medieval Archaeology* 6–7 (1962–63): 191.

[51]*Calendar of Patent Rolls,* Henry III (1247–58), p. 432.

[52]Ibid., p. 436.

[53]*Calendar of Close Rolls,* Henry III (1256–59), p. 345.

[54]See the Rutland Eyre of 1269 in Turner, ed., *Select Pleas of the Forest,* pp. 44–45.

February 12, 1290, Edward I ordered that exports of timber and wood fuels through the Cinque Ports (Hastings, Romney, Hythe, Dover, Sandwich, Winchelsea, and Rye) be suspended:

> To Stephen de Penecestre, warden of the Cinque Ports. Order to cause proclamation to be made throughout his bailiwick prohibiting all persons, under pain of loss of their goods and chattels, from taking out of the realm or causing to be taken out any timber, brushwood or charcoal without the king's license, as the king learns that many men cause timber, brushwood and charcoal to be taken from Sussex and Kent to divers places by sea, whereby inestimable damage may arise to the king and the men of those parts when they need such things.[55]

If shortages of timber and wood fuels existed at all, they were felt most keenly, of course, where the demand was the greatest, where population was most dense. In this respect no part of England could compete with London with its 40,000 inhabitants or more by 1340.[56] The woodlands within easy reach of London were the first to go. Costs of transportation made the price of firewood prohibitive if it had to be carried overland for any great distance.[57] On the other hand, water transport was much more economical.[58] As firewood had to be transported over ever greater distances to reach London, the prices rose steadily. Eventually Londoners began to look for less expensive alternatives. What they found was sea coal from Newcastle, easily transported by water to London and therefore much less costly.[59]

* * *

[55]*Calendar of Close Rolls,* Edward I (1288–96), p. 70.

[56]See above, n. 46.

[57]In the middle of the 13th century, the costs of transporting timber overland from the Forest of Dean to Gloucester or Bristol (about 15–25 miles) resulted in at least a doubling, at times even a quadrupling, of the retail price of the commodity (Hart, *Royal Forest,* p. 27). In the 16th and 17th centuries, the price of sea coal doubled for every 10 miles it was transported by land carriage (Nef, *Rise of the British Coal Industry,* 1:102).

[58]During the 17th century, e.g., the price of Newcastle coal at Stockton, carried 50 miles by water and loaded several times, competed successfully with coal from the Durham pits, 15–20 miles overland from Stockton; see Nef, *Rise of the British Coal Industry,* 1:102, as well as R. Smith, p. 5.

[59]L. F. Salzman, *Building in England down to 1540: A Documentary History* (Oxford, 1952), p. 150. The point here is not that England experienced a general wood crisis but, rather, that there were shortages and price increases in particular localities. If England's 3.7 million inhabitants had been evenly distributed over the countryside, presumably there would have been no problems. But the fact that at least 40,000 lived in London alone put a severe strain on those areas that could supply wood fuels to the city without excessive overland transport—essentially the lower Thames valley and immediately adjacent lands. Thus, if there had been plentiful supplies of wood fuels at Leicester, for example, they would nevertheless have been too far away overland to be

The peak of the expansion of arable land at the expense of woodlands and other waste in western Europe was reached by the last quarter of the 13th century, except for some scattered areas on the eastern frontier where population was less dense. In the most densely settled areas of France, western Germany, the southern Low Countries, and England, the extent of agricultural land began to decrease by the first years of the 14th century.[60]

The first areas to be abandoned were marginal lands capable of providing a few crops of cereals but unsuitable for continued exploitation. They became exhausted very quickly with the depletion of their meager supplies of nutrients.[61] And because much common land and pasture had been turned into cropland during the 12th and 13th centuries, there was less livestock per acre of cultivated land, severely curtailing the amount of manure for fertilizer.[62] Some of the lighter, marginal soils in the Low Countries, parts of Germany, and some areas of south Wales were so seriously eroded after being

---

easily accessible to London and thus have any beneficial effect on the fuel supplies and prices there. See G. Hammersley, "The Crown Woods and Their Exploitation in the Sixteenth and Seventeenth Centuries," *Bulletin of the Institute of Historical Research* 30 (1957): 156–59, who considers the accessibility factor in relation to local shortages and rising prices of wood products in the 16th and 17th centuries. Reference should be made here to the pioneer work of J. H. von Thünen, *Der isolierte Staat in Beziehung auf Landwirtschaft und Nationalökonomie* (Rostock, 1826), and the large body of literature on the subject of location that has appeared since then. Von Thünen found that, around a given central city, different types of land use were arranged in more or less concentric circles based on "the Economic Rent accruing to each type of land use at various distances from the central city" (Michael Chisholm, *Rural Settlement and Land Use: An Essay on Location,* 2d ed. [London, 1968], p. 28). In preindustrial Europe, wood products were in constant demand for construction and fuel; because they were bulky and incurred high transportation costs, they were normally located in close proximity to the market—sandwiched between a band of horticulture and dairying very close to the city and a band of intensive cereal cultivation a little farther away (see ibid., p. 27, fig. 4, and p. 28). Under the abnormal conditions pertaining to the London area, with its unprecedented population densities in the late 13th and early 14th centuries, the band of sylviculture near the city was eliminated by the dual pressures of increased demands for wood products and the ability of cereal cultivation to outcompete (because of rising food prices; see Slicher van Bath, *Agrarian History of Western Europe,* pp. 116, 132) sylviculture in such close proximity to the city.

[60]Alan R. H. Baker, "Evidence in the 'Nonarum Inquisitiones' of Contracting Arable Lands in England during the Early Fourteenth Century," *Economic History Review,* 2d ser., 19 (1966): 518–32; M. M. Postan, "Some Economic Evidence of Declining Population in the Later Middle Ages," *Economic History Review,* 2d ser., 2 (1950): 245, and "Medieval Agrarian Society in Its Prime," p. 557; Slicher van Bath, *Agrarian History of Western Europe,* pp. 133, 142; and Wilhelm Abel, *Die Wüstungen des ausgehenden Mittelalters,* 2d ed. (Stuttgart, 1955), pp. 74–75.

[61]Slicher van Bath, *Agrarian History of Western Europe,* p. 87; and Duby, *Rural Economy and Country Life,* p. 89.

[62]Slicher van Bath, *Agrarian History of Western Europe,* p. 89.

cleared that it is impossible to grow anything on them even today.[63] That lands of inferior quality were being cultivated is shown by the fact that the average yield per quantity of seed generally declined in Western Europe between the years 1230 and 1350.[64] Assarting, the extension of agricultural land at the expense of woodland, had arrived at the point of diminishing returns.

While the rate of expansion of cultivation leveled off, or even reversed, population continued to increase. Plots of arable land became so fragmented that many were no longer large enough to support those who depended on them for subsistence. On the manor of Taunton, in Somerset, the average acreage per person decreased from 3.3 in 1248 to 2.5 or less by 1311.[65] If it is true that an average yield of three acres of good farmland was required at that time to support one person,[66] then this manor was overpopulated by the early 14th century. In such a situation, the margin of security against widespread starvation became most fragile. Add to this the widespread use of lands of low fertility, and the situation becomes critical. Even under poor growing conditions, land of good quality will produce something. But dependence in part on the harvests of inferior marginal croplands meant that harvests would at best be barely adequate in the face of the pressure of overpopulation.[67]

The worst shortages of food occurred in 1315–17. Because of an unusually wet spring and summer in 1315, there was a very poor harvest in most of Western Europe that year. Food stores were completely inadequate, and what few reserves remained from the previous year were soon selling at highly inflated prices. By the spring of 1316, many of those unable to pay the high prices were dying of starvation. Though the harvest in 1316 was somewhat better, the starvation continued; various diseases spread through a badly weakened population, taking a heavy death toll well into 1317.[68] Mal-

[63]Ibid., pp. 142, 162. Duby, "Medieval Agriculture, 950–1500," maintains that much land was nearing exhaustion because "the food needs of the ever-increasing population had necessitated an abusive exploitation of the land" (p. 198).

[64]Slicher van Bath, *Agrarian History of Western Europe,* p. 136; and Duby, "Medieval Agriculture, 950–1500," p. 198.

[65]Titow, pp. 222–23.

[66]Pounds, p. 241.

[67]Genicot, p. 669; and Slicher van Bath, *Agrarian History of Western Europe,* pp. 134–36. In addition, see the comments of Duby, "Medieval Agriculture, 950–1500," p. 200.

[68]Kershaw, pp. 3–50; Curschmann, pp. 209–17; Lucas (who maintains, p. 352, that the average price for a quarter of wheat in England in 1313 was 5 shillings; by June 24, 1315, it was reportedly selling for 40 shillings per quarter); van Werveke, "La famine de l'an 1316 en Flandre"; Josiah C. Russell, "Effects of Pestilence and Plague, 1315–1385," *Comparative Studies in Society and History* 8 (1966): 466; and H. van Werveke, "Bronnenmateriaal uit de Brugse Stadsrekeningen betreffende de hongersnood

nutrition, aggravated by continued food shortages, became a common, recurring phenomenon.[69] Population growth slowed, and a period of stagnation followed.[70] The medieval boom period was over.

The Black Death of 1348–51 transformed this stagnation into a collapse that affected virtually all of Europe. A population already severely weakened by malnutrition had little chance in combating the virulence of the plague. Approximately one out of every four people died in Europe during this initial outbreak,[71] with some areas relatively spared and others much harder hit. Those areas with the densest populations often suffered the most.[72] Repeated attacks of the plague continued to force down the population tallies.[73] At times it would linger in an area; at other times it would lie dormant in one place and flare up in another—generally local in scope but very

van 1316," *Bulletin de la Commission royale d'histoire, Académie royale de Belgique* 125 (1959): 431-510.

[69]Van Werveke, "La famine de l'an 1316 en Flandre," pp. 5–14; Slicher van Bath, *Agrarian History of Western Europe,* pp. 84, 87–88, 90; and Kershaw, pp. 3–50.

[70]Pounds, p. 239; Norman J. G. Pounds and Charles C. Roome, "Population Density in Fifteenth Century France and the Low Countries," *Annals of the Association of American Geographers* 61 (1971): 117; and Genicot, p. 667. See also Russell, "Effects of Pestilence and Plague," and the reply by Sylvia L. Thrupp, "Plague Effects in Medieval Europe," *Comparative Studies in Society and History* 8 (1966): 474–83.

[71]Karl Helleiner, "Europas Bevölkerung und Wirtschaft im späteren Mittelalter," *Mitteilungen des Instituts für Oesterreichische Geschichtsforschung* 62 (1954): 254–60, and "The Population of Europe from the Black Death to the Eve of the Vital Revolution," in *The Cambridge Economic History of Europe,* vol. 4, *The Economy of Expanding Europe in the Sixteenth and Seventeenth Centuries,* ed. E. E. Rich and C. H. Wilson (Cambridge, 1967), p. 9.

[72]Hans van Werveke, "De Zwarte Dood in de Zuidelijke Nederlanden (1349–1351)," *Mededeelingen, Koninklijke Vlaamse Academie voor Wetenschappen, Letteren en Schone Kunsten,* klasse der letteren, vol. 12, no. 3 (1952), entire issue. On the other hand, a broad band in the Low Countries (parts of modern-day Belgium and the Netherlands, especially along the North Sea coast) was hardly struck at all. It is quite possible that the inhabitants of this area, by regularly supplementing their diets with fish containing valuable protein, were in better health and therefore less susceptible to the plague (see ibid., pp. 20–23; Pounds and Roome, p. 117; and Slicher van Bath, *Agrarian History of Western Europe,* pp. 84, 89–90). Van Werveke's exceptions for parts of the Low Countries have not gone completely unchallenged, see, e.g., P. Rogghé, "De Zwarte Dood in de Zuidelijke Nederlanden," *Revue belge de philologie et d'histoire* 30 (1952): 834–37, and Hans van Werveke's reply, "Nogmaals: De Zwarte Dood in de Zuidelijke Nederlanden," *Bijdragen voor de geschiedenis der Nederlanden* 8 (1954): 251–58.

[73]Even the Low Countries succumbed in 1400–1401 (van Werveke, "De Zwarte Dood in de Zuidelijke Nederlanden," p. 23). T. S. Jansma says that in 1369 the plague struck Leiden in the county of Holland ("Un document historico-démographique concernant le plat pays de Hollande [environ 1370]," in *Miscellanea medievalia in memoriam Jan Frederik Niermeyer* [Groningen, 1967], p. 335).

deadly all the same. Often, also, the plague would be aggravated by famine and dearth.[74]

In England, outbreaks of the plague in 1348–51, 1360–61, 1369, and 1374 reduced population totals from the preplague level of over 3.7 million to 2.25 million in 1374. The bottom was reached sometime between 1400 and 1430 at around 2.1 million, less than 60 percent of the preplague figure.[75] The population of London suffered as severely.[76]

Because of these losses in population, cultivated lands were abandoned in many areas of western Europe. It is estimated, for example, that one out of every four settlements established in Germany in the period of medieval expansion was depopulated during the Later Middle Ages.[77] There was a considerable contraction of the arable in England as well. Those lands least suitable for raising cereals, the marginal lands, were the first to be abandoned.[78] Fields, farms, and in some cases whole villages were deserted beginning in the first years of the 14th century and accelerating after the mid-century pestilence.[79]

Much of the land taken out of cultivation became pasture for increasing numbers of cattle and especially sheep.[80] But some also was

[74]Helleiner, "The Population of Europe from the Black Death to the Eve of the Vital Revolution," pp. 10–11; and Pounds and Roome, pp. 116–17.

[75]Russell, *British Medieval Population,* pp. 263, 269, and "Late Ancient and Medieval Population," p. 121. Thrupp says that there was probably no recovery until the 1470s or 1480s. Still others say that there was no real recovery until the 16th century (see Ian Blanchard, "Population Change, Enclosure, and the Early Tudor Economy," *Economic History Review,* 2d ser., 23 [1970]: 428, 435, 437).

[76]Russell, *British Medieval Population,* pp. 287, 297, concludes that London's population totals fell from nearly 60,000 to somewhere around 35,000 by 1377; in contrast, see Williams, pp. 315–17. For a description of the conditions for mortality in medieval London, see Sylvia L. Thrupp, *The Merchant Class of Medieval London (1300–1500)* (Chicago, 1948), pp. 201–2.

[77]Wilhelm Abel, "Verdorfung und Gutsbildung in Deutschland zu Beginn der Neuzeit," in *Morphogenesis of the Agrarian Cultural Landscape* (papers of the Vadstena Symposium at the Nineteenth International Geographical Congress, August 14–20, 1960), published in *Geografiska Annaler* 43 (1961): 1–3. See also Abel, *Die Wüstungen des ausgehenden Mittelalters,* throughout; Helleiner, "Europas Bevölkerung und Wirtschaft in späteren Mittelalter," pp. 262–64; and Duby, "Medieval Agriculture, 950–1500," p. 201.

[78]Baker, "Evidence in the 'Nonarum Inquisitiones' of Contracting Arable Lands"; Beresford, *The Lost Villages of England,* pp. 200–1; Postan, "Some Economic Evidence of Declining Population in the Later Middle Ages," pp. 245–46; Neilson, p. 61; and Duby, "Medieval Agriculture, 950–1500," p. 201.

[79]Hoskins, pp. 117–23; Postan, "Some Economic Evidence of Declining Population in the Later Middle Ages," pp. 227, 238–40; Beresford, *The Lost Villages of England,* p. 202; and Kershaw, pp. 29–36.

[80]Beresford, *The Lost Villages of England,* pp. 200–202.

allowed to grow up in brushwood and eventually timber. With the concurrent decrease in demands for building materials and fuel, the woodlands were given a chance to regenerate. It is difficult to find direct evidence for this, but it is indicated indirectly in some cases. As noted above, exports of timber and wood fuels through the Cinque Ports were banned in 1290. By 1357, however, they had resumed in one of these ports, Winchelsea, at least. On July 7, 1357, Edward III intervened in a quarrel of competing groups of merchants over which had the right to export firewood through the port. The same document noted that some shipowners, English as well as foreign, were in the habit of loading their ships with wool or hides and covering them with firewood, thus avoiding payment of the customs on the goods so concealed;[81] in other words, they were using firewood as a cover for smuggling operations. In the Forest of Dean in 1369, it was reported that there were "200 acres of underwood [used for firewood and charcoal] worth nothing a year for want of buyers."[82] More convincing evidence comes from the complaints about air pollution from the 16th and 17th centuries. Almost all of these complaints contain implied references to a previous age when firewood and charcoal, not sea coal, were burned—reference to a recent, not a distant, past, perhaps within the lifetimes of many.[83] The lapse in complaints concerning air pollution in London corresponded perfectly to this period, from the late 14th to the late 16th century, when England's woodlands were once again able to supply most of London's fuel needs.

* * *

After being severely reduced by famines and plagues during the entire 14th century, the English population made only a very slow recovery in the 15th and early 16th centuries. From a low of about 2.1 million between 1400 and 1430, the numbers slowly began to increase to about 2.3 million by 1525 and nearly 2.8 million by the middle of the 16th century. Then by 1600, there was a sharp increase to a level equaling that of the early 14th century—over 3.7 million.[84]

Again, increased demands for wood products and new woodland

[81]*Calendar of Patent Rolls,* Edward III (1354–58), pp. 578–79.

[82]Hart, *Royal Forest,* pp. 58, 64. The eyre rolls of the Forest of Dean for 1270, in contrast, had complained about the wholesale destruction of both tall wood and underwood in the Forest (ibid., p. 48).

[83]See, e.g., Graunt (p. 394), who maintained in 1665 that the use of sea coal was quite uncommon in London before 1600. Graunt's opinion, in particular, is valuable here. Because he was always a very careful researcher, one is inclined to believe him and not write off such a statement as an antiquarian's longing for a proverbial golden age.

[84]See Cornwall, p. 44, table 5, as well as the items cited in n. 75 above.

clearances for agricultural purposes began to take their toll on England's woodlands. By the second half of the 16th century, a second period of fuel and timber shortages began.[85] A draft of an act "prohibiting coal export" (ca. 1595) stated: "Whereas as by reason of the great scarcity, dearth, and decay of woods" in England, "the use of coals commonly called sea coal or stone coal is of late years greatly augmented not only for fuel but also to serve divers tradesmen and artificers of this our realm."[86] A few years earlier Parliament had passed an act "for the bringing in clapboard from the parts beyond the seas, and the restraining of transporting of winecasks, for the sparing and preserving of timber within the realm."[87] By 1662 the English timber supply was so depleted that the commissioners of the Royal Navy, concerned that the miserable condition of England's woods was jeopardizing the future of the Navy, asked the fledgling Royal Society for its assistance in finding ways to remedy the situation.[88]

This second fuel and timber crisis manifested itself in a serious inflation that can be traced in the sources. The prices of firewood increased by 780 percent between 1540 and 1640, compared with a contemporaneous 291 percent general inflation for most other commodities.[89] The increase in the retail price of sea coal, meanwhile, corresponded closely to the general inflation of 291 percent.[90] Once again Londoners began to look around them for less expensive fuels,

[85]Nef, "Comparison of Industrial Growth," pp. 173, 402, says that England's woodlands were suffering severely already by the reign of Henry VIII and were virtually depleted by the second decade of the reign of James II.

[86]Paul L. Hughes and James F. Larkin, eds., *Tudor Royal Proclamations* (New Haven, Conn., and London, 1969), 2:154.

[87]*Statutes of the Realm,* 35 Elizabeth I, cap. xi.

[88]Thomas Birch, *The History of the Royal Society of London for improving of Natural Knowledge from its first rise* (London, 1756), 1:110–20; Evelyn, 3:340; and Robert K. Merton, *Science, Technology, and Society in Seventeenth-Century England* (New York, 1970), pp. 178, 244–45. Evelyn's masterpiece, *Silva; or a Discourse of Forest-Trees . . .* (1664), was written in response to the commissioners' request. It is a careful, minute consideration of the many types of trees that grew in England, with many hints about planting and raising them. Among other things, he encouraged widespread plantings and the banishment of industries that were particularly destructive to the woods (especially iron making) to the New World colonies; see esp. secs. 1 and 2 of the introduction and chap. 30 to the end. I read the third edition of 1679.

[89]See the table in Nef, *Rise of the British Coal Industry,* 1:158. It is important to keep in mind once again that these shortages were most keenly felt in London and that places outside the London hinterland may well have been spared. Again, it was the cost of overland transport that made the difference. Hammersley, esp. pp. 156–59, rightly criticizes Nef for assuming a wood crisis and rise in prices in all parts of England, allowing instead only local crises much as I have tried to indicate.

[90]Nef, *Rise of the British Coal Industry,* 20:402–5, appendix E.

and once again they found sea coal from Newcastle.[91] This time, however, the change was permanent.

* * *

My purpose has not been to undermine the arguments of those who blame modern industrialization for the deterioration in the quality of life on our planet. To do so would ignore the extent to which the pressures on our natural environment have been aggravated during the last two centuries. Rather, I have attempted to show that the present environmental crisis is much more complex than is often admitted, that it is the result of centuries of struggle by Western man to achieve dominance over his natural world, and that there is no magic date which marked the transition from a simple, pristine world to our complex, polluted world.

This implies an alternate view of the course that Western civilization has taken in the past 1,000 years. Because many of the great achievements of the Middle Ages were temporarily aborted by the effects of overpopulation, famine, and plague during the 14th and 15th centuries, historians have had great difficulty in disposing of the notion that the entire Middle Ages were nothing more than a low water mark between two highs: classical antiquity and the modern era. If, on the other hand, one thinks of the last 1,000 years of history as part of Western man's rise to ecological dominance, then the Middle Ages no longer remain such an insignificant interlude. In fact, the medieval period of land reclamation constituted one of the great turning points in the process, in spite of temporary setbacks. It was during this period that a number of new, powerful weapons were first added to Western man's arsenal, weapons that were essential to the more recent scientific, technological, and industrial revolutions. Lynn White, jr., has found that only in the Latin West were the activities that change man's relationship to his environment—manual labor and invention—considered real acts of piety. He concludes that, with the blessing of the church, medieval man invented invention.[92] Clarence Glacken has shown that the key to the great period of medieval European landscape change lay in man's growing awareness of his ability to alter and control nature, to create something new, and to

[91]This new demand for sea coal was a great stimulant to the Newcastle mining and shipping industry. Newcastle came to be "worth more to the King in customs and coals than all the revenue of Scotland by far" (*Calendar of State Papers, Domestic Series,* Charles I [1640–41], p. 50).

[92]Lynn White, jr., "Cultural Climates and Technological Advance in the Middle Ages," *Viator: Medieval and Renaissance Studies* 2 (1971): 171–201.

destroy. He concludes that these attitudes, largely unknown to classical civilization, have been with Western society ever since.[93] It is difficult to conceive of today's technological achievements and their almost complete control of the natural world without these medieval underpinnings.

Just as today's environmental crisis is the unwanted side effect of man's alteration and exploitation of his natural environment, so it was in the preindustrial era. Early air pollution in London was the unwanted result of population increase and the subsequent clearing of England's woodlands. Only scale and degree are different today. Contemporary man possesses a much greater capacity for environmental manipulation and exploitation than did his medieval ancestors, but the ecological disturbances he causes are also proportionately more severe. Perhaps it will be instructive for us to keep in mind that the one great environmental crisis caused by human abuse before our own was resolved by a population reduction of 40 percent during the 14th century.

[93]Glacken, pp. 288–351, esp. pp. 330–51.

The Replacement of the Longbow by Firearms in the English Army

THOMAS ESPER

In a letter to Charles Lee dated February 11, 1776, Benjamin Franklin made a curious suggestion:[1]

> But I still wish, with you, that pikes could be introduced and I would add bows and arrows. These were good weapons, not wisely laid aside;
>
> 1st. Because a man may shoot as truly with a bow as with a common musket.
>
> 2dly. He can discharge four arrows in the time of charging and discharging one bullet.
>
> 3dly. His object is not taken from his view by the smoke of his own side.
>
> 4thly. A flight of arrows, seen coming upon them, terrifies and disturbs the enemies' attention to their business.
>
> 5thly. An arrow striking in any part of a man puts him *hors du combat* till it is extracted.
>
> 6thly. Bows and arrows are more easily provided everywhere than muskets and ammunition.

In 1792 Lee, an officer in the British army, recommended the longbow as a more effective weapon than the flintlock musket.[2]

Considering the reasons given for the longbow's superiority over firearms in the late eighteenth century, one wonders why that weapon was replaced two hundred years earlier by much inferior firearms. Scholarly works on the history of the English army or on firearms have not adequately explained this curious development.

Charles Cruickshank simply stated that "by the end of the [sixteenth] century the bow, which had been for so long the national weapon, had

THOMAS ESPER, Research Fellow at the Russian Research Center of Harvard University, has translated and edited *The Land and Government of Muscovy* by Heinrich von Staden (1587). Thomas Esper is now retired after twenty-five years of teaching history and currently resides in Cleveland Heights, Ohio.

[1] Benjamin Franklin, *Works* (Boston, 1839), VIII, 169–70.

[2] Charles Ffoulkes, *Arms & Armament* (London, 1945), p. 54. Ffoulkes did not realize that Lee was quoting Franklin.

almost completely disappeared, and the new 'weapons of fire' were firmly established in its place."[3] William Carman understood the deficiencies of sixteenth-century firearms and the cheapness and effectiveness of archery; but he gave no more clear an explanation for the longbow's replacement than did other scholars.[4] Charles Oman was very much aware of the incongruity of this development. Early in this century he remarked that the range of the musket was quite inferior to that of the longbow and that the bow's rate of fire was five times greater than that of the musket. "Indeed," he concluded, "it is not easy to make out the reasons why it superseded the bow in the end of the reign of Elizabeth, till one has gone through the series of controversial pamphlets which were written by the advocates of the rival weapons between 1590 and 1600."[5] In a later work, however, Oman failed to explain clearly the reason for the replacement. He did note that the number of archers in England had decreased; but he did not develop that point. Instead he passed on to a consideration of the late sixteenth-century controversy of archery versus firearms in which he presented the various arguments for and against the two weapons and then simply accepted the arguments of the advocates of firearms.[6]

The controversy of the 1590's over England's abandonment of military archery resembles in many ways the publicistic efforts of some of our Pentagon officials upon the "phasing out" of a particular weapons system. Pamphlets were written by advocates of the bow and by the supporters of firearms laying out in great detail the various advantages of each weapon.

* * *

The firearms around which the dispute centered were the harquebus and the musket. The harquebus was lighter, smaller, and less effective than the musket, but it was commonly used in the sixteenth century because it was also cheaper. The musket was used more and more toward the end of the century until it became the dominant handgun of the infantry. Although various forms of firing mechanisms were known and used in that century, the matchlock (touching a smoldering string to a pan of powder) was the most generally used because it was far less expensive and more reliable.

Since the harquebus and, to a certain extent, the musket were already

[3] C. G. Cruickshank, *Elizabeth's Army* (London, 1946), p. 60.

[4] W. Y. Carman, *A History of Firearms* (London, 1955), p. 100.

[5] "Columns and Line in the Peninsular War," *Proceedings of the British Academy, 1909–1910*, pp. 321–42.

[6] *A History of the Art of War in the Sixteenth Century* (New York, 1937), pp. 380–84.

commonly used in the English army, the debate was primarily concerned, on the one hand, with arguments favoring the complete abandonment of the longbow and, on the other, the defense of the bow as fully superior to firearms. Furthermore, it was urged that the longbow be revived as a primary projectile weapon of the infantry. Also, as today, the traditional weapon was defended as one which had been responsible for past victories and the abolition of which would also cause widespread unemployment. Cockle mentioned that one propagandist probably wrote "at the instance of the bowyers and fletchers, whose trade was declining as fire-arms were improving."[7]

The controversy over abandoning military archery did not arise suddenly, nor did the argument fully subside once the bow was discarded in 1595. The decisions taken during the 1590's to replace the bow and bill (halberd) with the pike and harquebus or musket had been in preparation for decades. For a large part of the sixteenth century the English were at peace, and the military engagements that did occur were conducted along rather traditional lines. Henry VIII, of course, took pride in his artillery, but he was also a patron of archery. Handguns, therefore, were only gradually introduced, while the longbow remained the primary projectile weapon.[8] Finally, during the last quarter of the century, England, confronted with a new military challenge, sought definite reforms in military organization which were demanded by soldiers who had seen service in the Netherlands.

Soldiers who served in the Low Countries were greatly impressed by what they termed the "new discipline." Elizabeth's army was never a standing force, and its organization remained inferior to that of the Spanish. The poorly organized English forces sent to fight the Spaniards soon proved unequal to the superior training, armament, and staff organization of the Iberians. Only after long service, during which the weak and unfit fell (or ran) out, did the troops of Robert Dudley, Earl of Leicester, acquire the professionalism and discipline required to face their foes successfully; and they were trained in the new system of pike and musket drill that had been developed on the continent.[9] Leices-

[7] Maurice J. D. Cockle, *A Bibliography of English Military Books up to 1642 and of Contemporary Foreign Works* (London, 1900), p. 55. The work Cockle referred to is R. S., *A Briefe Treatise To Proove the Necessitie and Excellence of the Use of Archerie* (London, 1596).

[8] The crossbow was inferior to the longbow in that its rate of fire was greatly less than the latter's. At the Battle of Crécy (1326) the Genoese crossbowmen were decimated by the English as they stood defenseless, cranking up their cumbersome machines after they had "shot their bolts."

[9] Oman, *Art of War*, p. 376.

ter was instrumental in introducing the "new discipline" to the English army. He realized that troops, to be effective, must not be poorly trained levies, mustered for one campaign and released to civilian life immediately thereafter, never really acquiring sufficient familiarity with arms or military discipline. Long service in the Dutch War of Independence convinced Leicester that such extended duty and improved organization were absolutely necessary for an effective fighting force. He took pride in the troops he had trained, and it was with particular chagrin that he had to report that "[500] Englishmen of our oldest Flemish training ran flatly and shamefully away" (at Grave, 1586).[10]

One of the English officers in the Low Countries, Sir Roger Williams, saw service on both sides, excusing his service with the Spaniards by claiming that at the time England and Spain were not at war. His experience converted him to an enthusiastic supporter of the "new discipline." Williams described and recommended the military organization of the Spaniards in *A Briefe Discourse on Warre*, published in 1590. He was very up to date in his views, preferring the musket to the harquebus, and preferring both to the bow:[11]

> In our ancient wars, our enemies used crossbows, and such shoots; few, or any at all had the use of long bows as we had. Wherefore none could compare with us for shot: but God forbid we should try our bows with their muskets and calivers [harquebuses], without the like shot to answer them. I do not doubt but all, honorable and others, notwithstanding some will contrary it, although they never saw the true trial of those weapons belonging either to horse or foot; alleging antiquity without other reasons, saying, we carried arms before they were born. . . . True it is, long experience requires age, age without experience requires small discipline. Therefore we are deceived, to judge men expert because they carried arms 40 years, and never in action 3 years, during their lives counting all together.

In a section entitled "To prove Bow-men the worst shot used in these days," Williams enumerated in some detail the inferiority of the longbow. He maintained that five hundred musketeers were better than three times that many bowmen; and that, of five thousand archers, only one thousand could shoot with sufficient force to be effective. He also complained that after a period of three or four months' service only one-tenth of all bowmen were worthwhile. Bowmen were good

[10] John Bruce, ed., *Correspondence of Robert Dudley, Earl of Leicester* (London, 1844), p. 244.

[11] London, 1590, pp. 47–48. I have modernized the spelling in this and the following quotations.

against horsemen, he admitted, but only if they were well defended with trenches and pikes. Even this was of little advantage, however, since few horsemen would charge a defended foot position unless they saw an opening. Williams was convinced that archery was ineffective against armored horsemen, although it could kill horses and thus put the rider out of the fight. Horses supported by firearms, however, especially muskets, terrified bowmen because the latter knew no defense against such a combination. In addition, "few or none [bowmen] do any great hurt 12 or 14 score [yards] off."[12] Williams also made the curious statement that "munition that belongs unto bow men, are not so commonly found in all places, especially arrows, as powder is unto other shot."[13] He must have meant that firearms and powder were common on the Continent and fletchers scarce. All this gathers greater importance when we realize that William's opinions were in accord with Leicester's and those of most of England's military leaders.

Another champion of the "new discipline" was Humfrey Barwick, who, like Williams, was an old and well-experienced soldier. He wrote his little work in 1594 and claimed to have read the work of Williams and of Sir John Smythe (which will be considered below). Barwick discussed the matter of archery versus firearms in much greater detail than did Williams. He first maintained that archery was responsible for more bloodshed than firearms, and for a peculiar reason:[14]

> When battles and great encounters chanced to fall out, by reason that the shot were no more offensive, then by the use of long bows and crossbows: the enemies did then commonly join both with long and short weapons . . . whereby the fight continued unto the last end of the one party, but in these days where the weapons of fire have been rightly used, it has been scarcely seen that either pike or halberd has come to join at any time before the one party did turn their faces, by reason of the terrible force of the great and small shot.

The claim of having bloodier battles with archery, while not supported by history, reflects Barwick's conviction that arrows could not stop advancing infantry. Oman noted that the battle at Flodden Field (1513), which was viewed by some contemporaries to be a victory by archery, was actually won by bills, since the Scottish infantry were so

[12] *Ibid.*, p. 46. Williams put the effective range of muskets at 600 yards, which is certainly an exaggeration (*ibid.*, pp. 40–41).

[13] *Ibid.*, p. 47.

[14] *A Briefe Discourse concerning the Force and Effect of All Manuall Weapons of Fire and the Disability of the Long Bowe or Archery in Respect of Others of Greater Force* (London, 1594), Introduction.

well armored that the arrows of the English did not penetrate.[15] Barwick likewise reported that a Frenchman explained to him that when he saw arrows coming he would lower his head so that his helmet would protect his face and let the arrows strike his body armor without effect. While citing many examples in which arrows put few men *hors du combat,* Barwick insisted that to every enemy killed with an arrow, one hundred were killed with firearms. One reason he gave for this is that the arrowheads were rusty and therefore they did not penetrate; another was that the bowmen did not pull their bows to their full extent, thus discharging their arrows at a low velocity. Furthermore, by asserting that weather rendered bows ineffective, Barwick disputed the contention that the bow was serviceable in all weather. Rather, he felt, rain dissolved the glue in the horns of the bow and ruined them, while the good soldier with a harquebus keeps his powder dry. One could also say the same of the bowmen, who, indeed, regularly covered his weapon to protect it from the wet. This latter practice has been suggested as one reason for the English victory at Crécy (1346).[16]

With regard to range, accuracy, and rate of fire, Barwick claimed superiority for the harquebus. He did not, however, consider the long-range capabilities of the longbow, except to scorn the notion that the harquebus had a shorter range. He merely said: "I leave that to the judgment of all such captains and soldiers as have seen the true trial of both the weapons."[17] Barwick only considered the point-blank effectiveness of the two weapons. The harquebus is best because it is the first to fire; the archers must still pull their bows. While Barwick did not claim that the harquebus had an accurate range greater than 8–10 yards, he assumed the same range for the bow and dismissed the greater rapidity of fire the bowman enjoyed by saying that the soldier with a harquebus need only put several balls in his piece to have greater fire power. He did admit that a man with a harquebus could discharge only once every 40 seconds (if he was an expert), while during that same period a bowman could let loose six arrows. Still, Barwick claimed

[15] Oman, *Art of War,* p. 314. In the *Grafton Chronicle* (ed. 1809, II, 275) we read the same complaint concerning Flodden Field: "And they abode the most dangerous shot of arrows, which sore them annoyed, and yet except it hit them in some bare place it did them no hurt."

[16] "Previous to this engagement fell a heavy rain, which is said to have much damaged the bows of the French, or perhaps rather the strings of them. Now our long-bow (when unstrung) may be most conveniently covered, so as to prevent the rain's injuring." Daines Barrington, "Observations on the Practice of Archery in England" (*Archaeologia,* VII [London, 1785]).

[17] Barwick, *op. cit.,* p. 3.

greater accuracy for the harquebus and stated that the bowman shot by guess whereas the harquebusier shot by rule. One must keep in mind that Barwick constantly compared the weapons with a 12-yard range in mind. Contrary to his assertion, which incidently indicates a lack of understanding on aiming an arrow, the good bowman did not just shoot in the general direction of the enemy. He had the capacity, if trained, to shoot at a particular target. "Any qualified archer was expected to shoot a dozen arrows in one minute at a man-sized target two hundred and forty yards away—and *hit* it with all twelve."[18] This statement, while exaggerated, echoes, in essence, the opinion of a number of sixteenth-century writers.

As we have seen, Barwick's arguments were not as cogent as Williams'. His statements on range, accuracy, and fire power were incorrect. He also revealed a certain prejudice against archers and archery. This attitude was common among the supporters of the "new discipline"; Barwick only expressed it with greater force. The following clearly reveals that prejudice or snobbery:[19]

> We have the like estimation of the long bow, as the Irish have of their darts, the Dansker of their hatchets, and as the Scotch men have had of their speares: all of which are more meeter for savage people or poor potentates, who are not able to maintain others of greater force, than puissant princes. The Scots and Irish for the defense of their countries do use bows, and so do the Burgundians and Walloons in the time of the wars, guard their caves, churches & small piles: the countrymen for the safety of their goods, but the soldiers in pay do never use them.

* * *

The longbow had its supporters too, and some of them published pamphlets defending their favorite weapon. Chief of the archery advocates was Sir John Smythe, whose booklet appeared in 1590.[20] Smythe criticized those who proposed the "suppression" of the longbow as lacking in military experience. Their experience, he complained, was mostly in the civil wars in the Low Countries or in France—wars which were disorderly and unusual in their lack of discipline. In England the long years of peace had led to a decay of military exercise and science. In regard to firearms, Smythe enumerated the various imperfections of those weapons and their inferiority when compared to the longbow: Firearms are not accurate except at close range. A musket ball will

[18] Edwin Tunis, *Weapons* (Cleveland, 1954), p. 62.

[19] Barwick, *op. cit.*, p. 2.

[20] *Certain Discourses concerning the Formes and Effects of Divers Sorts of Weapons, etc.* (London, 1590).

carry 400 yards but it is only effective at 200 or 240 yards. (Smythe reported that Spanish officers prohibited firing at targets over 10 or 12 score yards away.) Harquebuses, which are less bulky, have an even shorter range. (Smythe said that they were generally discharged at 8–12 paces, while muskets were only accurate at 15 or 20 yards.) The weapons themselves become dangerous to their users after seven or eight discharges. They heat up and explode powder prematurely. Also, powder must be of good quality and always be kept dry. Firearms also have a much lower rate of fire, archers being able to shoot four to five arrows before the harquebusier is ready to fire once. With regard to the weather:[21]

> If in the time of any battle, great encounter, or skirmish, the weather does happen to rain, hail, or snow, the aforementioned weapons of fire can work no effect, because the same does not only wet the powder in their pans and touch holes, but also does wet the match, put out, or at least damp the fire, or does mar the powder in their flasks and touchboxes. . . . Whereas contrarywise, neither hail, rain, nor snow, can let or hinder the archers from shooting, and working great effects with their arrows.

Smythe felt that archery was being neglected and he suggested that the already existing legislation commanding its practice be enforced.

Another pamphleteer was a certain "R. S.," whose work appeared in 1596.[22] The author wrote "to deplore . . . the discontinuance, yea (almost) the utter extirpation [of archery], within this realm: and withal, our own miserable estates, who with many other poor artificers that have had their maintenance thereby are (in great number) brought to utter ruin and decay." R. S. offered the usual arguments: cheapness, greater range, and rate of fire, as evidence of the longbow's superiority over firearms.

* * *

It would be worthwhile to test the claims of those disputants in regard to the capabilities of the longbow. One foreign commentator living almost a century after the controversy claimed that arrows were more deadly than bullets because it is extremely difficult to extract the former without tearing the wound.[23] A recent evaluation of the range of the longbow, moreover, agrees with the claims of its sixteenth-century advocates:[24]

21 Smythe, *op. cit.*, pp. 21–22.

22 R. S., *op. cit.*

23 Louis de Gaya, *Traité des armes* [1678] (London, 1911), pp. 48–49.

24 R. Ewart Oakeshott, *The Archaeology of Weapons* (London, 1960), p. 297.

> We may remember that Shakespeare's Old Double, upon whom John of Gaunt betted much money, could shoot an armed arrow 240 yards and a flight arrow 280 or 290, but a good professional archer of Edward III's time would have been able to beat that up to the traditional 410 yards, for Shakespeare's facts were taken from contemporary practice when the art of archery was in decline.

One may also recall that the term "bow shot" was considered to express a distance of 400 yards,[25] and that the expression "don't shoot until you see the whites of their eyes" indicates the accurate range of firearms into the nineteenth century.

The longbow was obviously cheaper than firearms. In the latter part of Elizabeth's reign, when the longbow was replaced, the cost of equipping a company went up by 50 per cent.[26] It is likewise self-evident that arrows could be discharged several times more rapidly than bullets. The advantage that firearms had over archery concerned penetrating power.

Contemporary estimations of the longbow as a poor weapon can be reduced to this point: archery could not stop a charge—arrows did not penetrate. On the other hand, Carew, in his *Survey of Cornwall,* stated that "their pricks . . . would pierce any ordinary armour."[27] A recent writer, describing the type of arrowhead that was used, the pile, said: "This was the war head which would pierce chain mail or kill a horse at two hundred yards; at closer range it could puncture ordinary plate armour."[28] Although there appears to be conflict in the various claims, they are all correct. The force of a shot depended upon the training of the archer, and thus there was considerable variety in the power of shots. Archers as a whole were more poorly trained in the sixteenth century, and especially at the end of it, than they were in preceding centuries; consequently the efficiency of an army of bowmen was reduced. The task remaining is to determine why archery "decayed."

* * *

In the *Holinshed Chronicle* we read the complaint that the French and Germans scoffed at English archers. They raised their backsides high in derision of the bowmen. "But if some of our Englishmen now

[25] Robert C. Clephan, *The Defensive Armour and the Weapons and Engines of War of Medieval Times and of the "Renaissance"* (London, 1900), pp. 180–81.

[26] Cruickshank, *op. cit.,* p. 61.

[27] Cited by Joseph Strutt, *The Sports and Pastimes of the People of England* (London, 1834), p. 67.

[28] Tunis, *op. cit.,* p. 61.

lived, that served King Edward III in his wars with France: the breech of such a varlet had been nailed to his back with an arrow; and another feathered in his bowels, before he should have turned about to see who shot the first."[29] Even the most enthusiastic defenders of archery could not deny that the art had decayed. Roger Ascham, whose *Toxophilus* is a long treatise in praise of the bow, admitted that men bought bows and kept them as the law required, but they did not use them because they did not know how to shoot.[30] Ascham and his contemporaries understood that the training of an archer was more important than the latter's physical strength. "And thus strong men, without use, can do nothing in shooting to any purpose, neither in war nor peace."[31] Trained weak men can outshoot stronger but unpracticed ones.

Long years of regular practice were necessary for the development of good bowmen. This fact was very clearly expressed by Hugh Latimer in 1549:[32]

> In my time my poor father was as diligent to teach me to shoot, as to learn me any other thing; and so I think other men did their children: he taught me how to draw, how to lay my body in my bow, and not to draw from strength of the body: I had my bows bought me, according to my age and strength; as I increased in them, so my bows were made bigger and bigger; for men shall never shoot well, except they be brought up in it.

Latimer praised bygone days when men would go into the fields and practice archery for exercise. "But now," he protested, "we have taken up whoring in towns, instead of shooting in fields."[33] It is evident, therefore, that the "decay of archery" referred to the abandonment of archery as the national pastime of the English people. In earlier generations men levied for the army would be practiced in shooting the longbow, and with those men England won her great victories on the Continent during the Hundred Years' War (1339–1453). If one considers the manner in which armies were raised in Elizabeth's reign, when quite often vagabonds and the most wretched were pressed into serv-

[29] 1807 edition, I, 333.

[30] [1544] (Birmingham, 1868), p. 103.

[31] *Ibid.*, p. 89. Likewise R. S., *op. cit.*: "For experience does teach us, that the strongest men do not always make the strongest shot, which thing proves that drawing strong lies not so much in the strength of men, as in the use of shooting."

[32] *Sermons*, Parker Society, Vol. XXVII (Cambridge, 1844), p. 197. Preached before King Edward VI, April 12, 1549.

[33] *Ibid.*

ice,[34] it is understandable that the soldiers were generally poor archers and their weapons, poorly used, inferior to firearms.

The defenders of archery in the sixteenth century—Ascham, Latimer, the military men—all ascribed superior moral qualities to those who practiced archery, and the opinion common among them was that archery had decayed because of the popularity of "unlawful games." As early as 1365 Edward III, compelled by that fear, ordered that all the able-bodied practice archery on church holidays. Other games were prohibited. It is curious that Edward, fifty years before the great victory at Agincourt, was fearful about the state of archery and that he felt that "now the said art is almost wholly disused."[35] Two hundred years later, when archery was indeed decayed, the practice of "unlawful games" was considered by many to have been the cause of the problem. "So by means of the said exercise [strictly banishing all unlawful games] there is no doubt but our archery will in short time become as strong & effectual as ever it was."[36] This same attitude was expressed by Elizabeth in 1591. The state's interests in the maintenance of archery were clearly indicated:[37]

> Whereas her Majesty is informed that diverse unlawful games are daily used in most places of this realm and that thereby archery is greatly decayed and in a manner altogether laid aside, being an exercise not only of good recreation but otherwise of good use and defense to the realm, and that in all ages in this land has been specially used, and by laws and statutes necessarily provided for. Wherein though at this time, considering the great charge the subjects do sustain in furnishing themselves with musquets, harquebuses and other weapons both of defense and offense that are now more in use, it should seem a hard course to lay that burden on them which the law in this case does strictly impose, nevertheless her Majesty for very good considerations thinks meet and accordingly has willed us to require you in her name forthwith to take special care that such kinds of exercises, games and pasttimes as are prohibited by law, namely bowls, dicing, carding and such like, may be forthwith forbidden . . . but that instead thereof archery may be revived and practised and that kind of ancient weapon whereby our nation in times past has gotten so great honor may

[34] Cruickshank, *op. cit.*, p. 9.

[35] *Calendar of Close Roles.* Edward III, 1364–68, June 12, 1365.

[36] Henry Knyuett, *The Defense of the Realme* [1596] (Oxford, 1906), p. 18. Latimer listed some of the games as "glossing, gulling, and whoring within the house" (p. 196).

[37] *Acts of the Privy Council,* XXI, 174–75 (June 6, 1591).

> be kept in use, and such poor men whose stay of living with their whole families do chiefly depend thereon, as bowyers, fletchers, stringers, arrowhead makers, being many in number throughout the realm, may be maintained and set on work according to their vocations, and her Majesty's said gracious intent and meaning thereby duly executed.

Another contemporary attributed the decay of archery, at least in London, to enclosures. "For by the means of closing in of common grounds, our archers for want of room to shoot abroad, creep into bowling allies, and ordinary dicing houses nearer home."[38] Enclosures, nevertheless, were a lesser cause of the problem.

The decay of archery, long in process, resulted from a number of social and economic changes. As Lynn White has pointed out, "The acceptance or rejection of an invention, or the extent to which its implications are realized if it is accepted, depends quite as much upon the condition of society, and upon the imagination of its leaders, as upon the nature of the technological item itself."[39]

But why the English abandoned archery in favor of other sports is a fit subject of another, lengthy paper. The fact that archery continued as a sport as long as it did, nevertheless, must be attributed more to the moral qualities associated with its practice than to legal injunctions.

In summary, the replacement of the longbow by firearms occurred at a time when the former was still a superior weapon. The longbow, as used in the late sixteenth century, however, was inferior in penetrating power, because the practice of archery had declined. The capabilities of the longbow had been fully exploited in earlier generations, when archery was the national sport of England. If archery could have been restored to its previous condition, the longbow could have remained the primary projectile weapon of the English infantry for generations longer.

[38] John Stow, *The Survey of London* [1598] (London, 1618), p. 162.

[39] Lynn White, *Medieval Technology and Social Change* (Oxford, 1962), p. 28.

On the Social Explanation of Technical Change: The Case of the Portuguese Maritime Expansion

JOHN LAW

It is a truth almost universally acknowledged that a technological artifact whose origins and development are insufficiently understood must be in want of a social or economic explanation. For some, the pursuit of manifest economic or social interests ranks as a central explanatory resource.[1] Others, perhaps more numerous, also make reference to profit and social control but link them to the operation of the capitalist labor process and the way in which they structure technological possibilities.[2] Yet others, while not necessarily disagreeing with such an approach, are more committed to a systems model in which complex social, economic, technical, and political elements are seen to influence one another, and distant developments may create the perception of technological problems.[3] My purpose here is to consider the applicability of the last of these approaches through an analysis of Portuguese transoceanic expansion.

My particular object is to offer a nonreductionist sociological per-

PROFESSOR LAW is with the Centre for Social Theory and Technology, and the Department of Sociology and Social Anthropology, University of Keele. He would like to thank Serge Bauin, Wiebe Bijker, Michel Callon, Rich Feeley, Elihu Gerson, Antoine Hennison, Tom Hughes, Bruno Latour, Jean Lave, Mike Lynch, Donald MacKenzie, Chandra Mukerji, Trevor Pinch, Arie Rip, and Leigh Star, who all read and commented on earlier drafts, and also the University of Keele, l'École Nationale Supérieure des Mines de Paris, la Fondation Fyssen, the CNRS, and the Leverhulme Foundation for support and study leave. Another article by the author on this subject appears in *The Social Construction of Technological Systems*, ed. Wiebe E. Bijker, Thomas P. Hughes, and Trevor J. Pinch (Cambridge, Mass., 1987).

[1]See, e.g., Langdon Winner's outline account of the road-building activities of Robert Moses on Long Island: "Do Artifacts Have Politics?" *Daedelus* 109 (1980): 121–36.

[2]Philip Kraft, *Programmers and Managers: The Routinization of Computer Programming in the United States* (New York, 1977); David F. Noble, *America by Design: Science, Technology, and the Rise of Corporate Capitalism* (New York, 1977).

[3]Thomas P. Hughes, *Networks of Power: Electrification in Western Society, 1880–1930* (Baltimore and London, 1983); Edward W. Constant, "On the Diversity and Co-Evolution of Technological Multiples: Steam Turbines and Pelton Water Wheels," *Social Studies of Science* 8 (1978): 183–210.

spective on technological change. Thus sociologists of science are starting to join historians in the search for social influences, causes, or correlates of technology. They come equipped with a battery of techniques. For the last dozen years sociologists have been arguing, with great success, that scientific knowledge is not fully determined by nature but is also directed by social factors. Sometimes they have argued that the relevant interests are local and professional in character—that scientists buy into their theories and hence become committed to them and unwilling to give them up.[4] On other occasions, they have sought to show that, while professional interests tell a part of the story, background and class interests also exert a guiding influence on scientific knowledge—and that the latter may serve social as well as intellectual roles.[5] However, now that the issues have been mapped out in the sociology of science, they (I include myself among this "they") are starting to apply their methods to the analysis of the origins of technological artifacts.[6] There is no doubt that these methods are powerful and little doubt that they will bear fruit.[7] It is perhaps ungracious to examine the mouths of gift horses too closely, but the purpose of this essay is to suggest that the benefits of drawing on the sociology of science may not be unalloyed: unless caution is exercised, there may be costs as well as benefits.

That this is the case can be seen by looking at debates within the sociology of science, for there is no consensus within that field. Many of these debates revolve around two related issues of great potential for the explanation of technological change.[8] First, there is disagreement

[4]Harry Collins and Trevor J. Pinch, *Frames of Meaning: The Social Construction of Extraordinary Science* (London, 1982).

[5]Donald MacKenzie, "Statistical Theory and Social Interests," *Social Studies of Science* 8 (1978): 35–83.

[6]Trevor J. Pinch and Wiebe E. Bijker, "The Social Construction of Facts and Artefacts: Or How the Sociology of Science and the Sociology of Technology Might Benefit Each Other," *Social Studies of Science* 14 (1984): 399–441; Donald MacKenzie, "Missile Accuracy—a Case Study in the Social Processes of Technological Change," in *The Social Construction of Technological Systems: New Directions in the Sociology and History of Technology*, ed. Wiebe E. Bijker, Thomas P. Hughes, and Trevor J. Pinch (Cambridge, Mass., 1987); Wiebe E. Bijker, "The Social Construction of Bakelite: Towards a Theory of Invention," in Bijker, Hughes, and Pinch, ibid.

[7]The most obvious example of this at present is MacKenzie's study of the development of the guidance systems of missiles (n. 6 above), in which he has fruitfully explored the relations between the systems approach recommended, inter alia, by Hughes, and the notion of social interest developed in the sociology of science.

[8]Since I am going to talk about disagreements, let me note that there are substantial points of *agreement* between many sociologists of scientific knowledge. Thus they are concerned with analysis of the *content* of scientific knowledge which they take to be (at least) underdetermined by nature and (at least) partially constructed by scientists. They

about the degree to which scientific knowledge (or, by extension, technological devices) may be explained *by the operation of social factors alone*. Most sociologists are committed to the idea that social factors are important—it would, after all, be odd if they were not. However, some are committed to the view that it is social factors and social factors *alone* that explain the growth of knowledge.[9] The first question for the historian is thus: Does he or she wish to be committed wholeheartedly to such social reductionism? If the answer to this question is no, then the new sociological enthusiasm must be greeted with caution.

The second set of issues is equally important. Assuming that social factors have some relevance for the explanation of technology, then what *kinds* of social factors should we be looking for? Sociologists of science—even those who are not committed to the first form of reductionism mentioned above—tend to limit both the *type* and *number* of explanatory social factors. In fact, so far as *type* of social factors is concerned, as I have already indicated, they typically find room for only one category: that of social interest. So far as *numbers* of relevant social interests are concerned, reference is often limited, as I have indicated above, to professional or class interests. This, then, is a second form of sociological reductionism, and again one that the historian of technology should treat with caution. Are interests really the only social category worth considering when trying to explain cases of technological development?

There are, of course, good reasons for sociological reductionism. First, studies posed in such terms have often been highly successful in the sociology of science and indeed run parallel to "labor process" studies undertaken by historians of technology.[10] Second, such enthusiastic, albeit one-sided, analyses may deepen our understanding of technological change because they test the power of sociological models and establish their limits. This is surely preferable to a wishy-washy eclecticism. Third, in committing itself to the importance of a limited number of social factors, sociology is simply adopting a strategy that is both widespread and highly successful in natural science—that of explanatory parsimony. In effect, the sociologist is committing himself or herself to a model of society and deriving predictions from it: that, for instance, social class interests often influence the growth of knowl-

---

are also committed—and this is a most important methodological principle—to the *symmetrical* analysis of "true" and "false" knowledge (and their analogues).

[9]See in particular H. M. Collins, "The Seven Sexes: A Study in the Sociology of a Phenomenon, or the Replication of Experiments in Physics," *Sociology* 9 (1975): 205–24; but for a counterview see David C. Bloor, *Knowledge and Social Imagery* (London, 1976).

[10]See, e.g., Larry Lankton, "The Machine *under* the Garden: Rock Drills Arrive at the Lake Superior Copper Mines, 1868–1883," *Technology and Culture* 24 (1983): 1–37.

edge. If the alternative is to abandon any model of technological change and instead to celebrate the complexity and contingency of history, then perhaps sociological reductionism is to be preferred.

This said, I still believe that the historian should be cautious of sociological reductionism. This is because as an explanatory strategy it tends toward the assumption that relatively stable social forces lie *behind* and *outside* the technological and direct the latter. Social factors act as an explanatory backcloth, but (and this is the crucial point) it is a backcloth that is both taken to be relatively unchanging and left substantially unanalyzed. This explanatory strategy has recently been the subject of critical comment in the sociology of science. Though the debate is as yet unconcluded,[11] there are a number of studies that suggest that scientists and technologists not only attempt to manipulate beliefs about the *natural* world but also operate on the *social* world and, in particular, manipulate the interests of others.[12] If, as these studies suggest, the social and the natural are subject to continuous reworking, then the idea that social forces lie behind and explain other events or processes is difficult to sustain, at least in any simple form, and it becomes necessary to work the "social" into the explanation in a rather different way. In this alternative view the latter is seen as lying *within* the system rather than outside it. Accordingly, the focus of analysis shifts from the study of background social variables to scrutiny of the *heterogeneous engineering* attempted by the system-builder as he or she seeks to juxtapose diverse elements and so create a relatively stable

[11]See Steve Woolgar, "Interests and Explanation in the Sociology of Science," *Social Studies of Science* 11 (1981): 295–315; Barry Barnes, "On the 'Hows' and 'Whys' of Cultural Change," *Social Studies of Science* 11 (1981): 481–95.

[12]Thus, Williams, Callon, and I have analyzed interest talk in a biochemical laboratory. See Rob Williams and John Law, "Beyond the Bounds of Credibility," *Fundamenta Scientiae* 1 (1980): 295–315; John Law and Rob Williams, "Putting Facts Together: A Study in Scientific Persuasion," *Social Studies of Science* 12 (1982): 535–58; Michel Callon and John Law, "On Interests and Their Transformation," *Social Studies of Science* 12 (1982): 615–25; John Law, "Laboratories and Texts," in *Mapping the Dynamics of Science and Technology: The Sociology of Science in the Real World*, ed. Michel Callon, John Law, and Arie Rip (London, 1986), pp. 35–50. Callon has shown how both social and technical features were manipulated in a struggle between EDF and Renault about the future of the electric vehicle, in "Struggles and Negotiations to Define What Is Problematic and What Is Not: The Sociologic of Translation," in *The Social Process of Scientific Investigation*, ed. Karin D. Knorr, Roger Krohn, and Richard Whitley (Boston and Dordrecht, 1980), pp. 197–219. Bruno Latour has argued persuasively that it is impossible to understand Pasteur unless he is seen as engaging in large-scale social engineering, in "Give Me a Laboratory and I Will Raise the World," in *Science Observed: Perspectives on the Social Study of Science*, ed. Karin D. Knorr-Cetina and Michael Mulkay (London and Beverly Hills, Calif., 1983), pp. 141–70; and Latour, *Les Microbes, guerre et paix, suivi de irréductions* (Paris, 1984).

network of components. That is, just as laboratory scientists may be seen as entrepreneurs combining everything from rats and machines to grant-giving bodies and academic journals in a situation where other actors may be trying to undo their work, so the technologist has to be seen as attempting to build a world where bits and pieces, social, natural, physical, or economic, are interrelated and *keep each other in place* in a hostile and dissociating world.[13]

Although this view sounds somewhat radical, I suggest that it is tacitly or explicitly shared by many historians of technology and even by a few sociologists of science. The notions of system-building devised by Thomas Hughes and coevolution used by Edward Constant both represent attempts to grapple with the interrelatedness of heterogeneous elements and to handle the finding that the social as well as the technical is being constructed.[14] The analysis of traditions of "technological testability" developed by Constant can be seen as a study of the way in which a wide range of actors comes to a locally enforceable agreement that certain social/technical relations are appropriate and may not (for the time being) be reconstructed.[15] Finally, the treatment by Trevor Pinch and Wiebe Bijker of the notion of social group and the role of advertising[16] seems to imply a recognition that the analysis of technological change involves the study of social and not only technical engineering. Historians and some sociologists are groping toward a method for studying the construction of and interaction between heterogeneous structures. It is time to place this concern fair and square on the agenda.

My argument, then, is that reductionist modes of sociological explanation may have relatively little to offer the social study of technology and that historians should be correspondingly cautious of them. Explanations in terms of "social interests" or other social factors, while sometimes useful, tend to conceal many of the most interesting features of technological change. This is because the latter are best seen as a function of heterogeneous engineering in which systems containing both the social and the technological are constructed in a context of conflict with other actors, both natural and social. The addition of a social deus ex machina offers little explanatory assistance under such

[13]Law (n. 12 above); Callon (n. 12 above).

[14]Thomas P. Hughes, "The Electrification of America: The System Builders," *Technology and Culture* 20 (1979): 124–61; Hughes, *Networks of Power* (n. 3 above); Constant (n. 3 above).

[15]Edward W. Constant, "Scientific Theory and Technological Testability: Science, Dynamometers, and Water Turbines in the 19th Century," *Technology and Culture* 24 (1983): 183–98.

[16]Pinch and Bijker (n. 6 above).

circumstances. If the struggle to construct "systems" is to be properly understood, we will thus need a method for analyzing such heterogeneous engineering. The present article, by using secondary material about the Portuguese maritime expansion, seeks to contribute to such an approach and to use it in order to reinterpret the concepts of "system," "adaptation," and "technological testability."

The Struggle between Cape Bojador and the Galley

In 1291 Ugolino and Vadino Vivaldi set sail from Genoa in two galleys, passed through the Pillars of Hercules "ad partes Indiae per mare oceanum," and vanished, never to be seen by any European again.[17] In 1497 Vasco da Gama sailed from the Tagus in Lisbon. He too was headed for the Indies by way of the ocean, but unlike the case of the brothers Vivaldi we know what became of his expedition. On May 20, 1498, he anchored in the Calicut Road off the Malabar coast of southwest India. He entered into (ultimately unsuccessful) negotiations with the Samorin of Calicut about trading in spice. So unsuccessful were these negotiations that on his second expedition in 1502 da Gama's now heavily armed fleet bombarded the town of Calicut in an effort to force the Samorin into submission.[18] The Portuguese spice trade had begun, and Portugal's sailors were in a position to impose a monopoly by force. Muslims in general, and Arabs in particular, were squeezed out, and the lucrative trade in spice was channeled round the cape to Lisbon and Antwerp rather than through the Persian Gulf and the Red Sea to Alexandria, to Syria, and ultimately to Venice.

The historical importance of the Portuguese control of the Indian Ocean and the passage via the cape hardly needs to be emphasized. When taken with the near-simultaneous discovery by Christopher Columbus of America it signals a change in the balance of power between Europe and the rest of the world whose consequences we are still assimilating today. It also represents a shift in power between the Mediterranean, and particularly the city-states of Genoa and Venice, and Atlantic Europe. Though historians have, in general, been more concerned with the motives or background causes of this expansion than with the methods used by the Portuguese,[19] the whole episode can also be looked at from the standpoint of system-building or heter-

[17]Bailey W. Diffie and George D. Winius, *Foundations of the Portuguese Empire, 1415–1580* (Minneapolis, 1977); Pierre Chaunu, *European Expansion in the Later Middle Ages*, trans. Katharine Bertram (Amsterdam, 1979).

[18]J. H. Parry, *The Age of Reconnaissance* (London, 1963).

[19]For exceptions, see Chaunu (n. 17 above); Carlo M. Cipolla, *Guns and Sails in the Early Phase of European Expansion, 1400–1700* (London, 1965); Parry (n. 18 above).

ogeneous engineering. Sometimes the opponents were people and sometimes they were objects. Let me start, then, by talking of galleys.

The galley was primarily a war vessel. It was light and maneuverable, a method for converting the power of between 150 and 200 men into efficient forward motion. In order to reduce water resistance it was long and thin—typically, at least in Venice, about 125 feet in length and 22 feet wide, including outriggers.[20] The hull was lightly sparred and the planks laid carvel fashion, edge to edge, to minimize water resistance. It was also low. The oarsmen pulled, usually three to a bench each with his oar, on between twenty-five and thirty oars on each side. The vessel also carried at least one (and frequently more than one)[21] mast, stepped well forward, which carried a lateen sail. This assisted the oarsmen, though it was never more than an auxiliary source of power. This ship was steered by means of two rudders projecting from each side of the stern—and after 1300 by a single rudder attached to the sternpost, an idea introduced from northern Europe.[22] The stern was slightly raised, and it was here that the helmsmen and the commander of the vessel stood. By contrast, the bow was low and pointed, being designed to ram other ships.

Now let me state the obvious: the galley is an *emergent phenomenon.* It has attributes possessed by none of its individual components. The galley builders associated wood and men, pitch and sailcloth, and they built an array that floated and could be propelled and guided. It was able to associate wind and manpower to make its way to distant places. It became a "galley" that allowed the merchant or the master to depart from Venice, to arrive at Alexandria, to trade, to make a profit, and so to fill his palace with fine art.

The galley is, of course, a technological object. Let me, then, define technology as a family of methods for associating and channeling other entities, both human and nonhuman. It is a method, one method, for the conduct of heterogeneous engineering, for the construction of a system of related bits and pieces with emergent properties in a hostile or indifferent environment. When I say this I do not mean that the methods are somehow different from the forces that they channel. Technology does not act as a kind of traffic policeman that is distinct in nature from the traffic it directs. It itself is nothing other than a set of channeled forces or associated entities. Thus there is always the danger that the associated entities that constitute a piece of technology will be

[20]Frederic C. Lane, *Venetian Ships and Shipbuilders of the Renaissance* (Baltimore, 1934).

[21]Björn Landström, *Sailing Ships in Words and Pictures from Papyrus Boats to Full-Riggers* (London, 1978).

[22]Frederic C. Lane, *Venice, a Maritime Republic* (Baltimore, 1973).

dissociated in the face of a stronger and hostile system. Let us, therefore, consider the limitations of the galley.

As a war machine in the relatively sheltered waters of the Mediterranean it was a great success. As a cargo-carrying vessel it had, however, its drawbacks. Its carrying capacity was extremely limited. The features that made it a good man-of-war—the fact that it was slim and low and carried a large crew that might repel boarders—were an impediment to the carriage of cargo. Lane reckons that 50 tons was about the limit, though Denoix mentions the figure of 100.[23] Furthermore, the *endurance* of the galley was restricted by the size of its crew. It could not pass far from the sight of land and the possibility of water and provisions. Although the Venetians and the Genoese used galleys to transport valuable cargoes where reliability was called for, they were replaced in this role by the "great galleys" after about 1320.[24]

It must have been in such vessels that the brothers Vivaldi left Genoa in 1291 for what they thought would be a ten-year trip to the Indies.[25] Perhaps their galleys were larger than normal, precursors of the great galley. Perhaps they had higher freeboards. But their endurance would have been limited and their seaworthiness doubtful—one can imagine all too well the consequences of running into a storm off the Saharan coast. And, if indeed the Vivaldis were attempting to row down the west coast of Africa, then they would have had to pass what may be regarded as a point of no return—Cape Bojador or the Cape of Fear. One 15th-century description speaks of this in the following terms: "A wise pilot will pass Bojador eight leagues out at sea. . . . Because Cape Bojador is most dangerous, as a reef or rock juts out in the sea more than four or five leagues, several ships have already been lost. This cape is very low and covered with sand; . . . in ten fathoms you cannot see the land because it is so low."[26]

Chaunu summarizes the problem in the following way: "At twenty-seven degrees north, Cape Bojador is already in the Sahara, so there could be no support from the coast. The Cape is 800 kilometres from the River Sous; the round trip of 1,600 kilometres was just within reach of a galley, but it was impossible to go any further without sources of fresh water, except by sail. In addition there were the difficulties . . . [of] the strong current from the Canaries, persistent mists, the depths

[23]Ibid., p. 122; L. Denoix, "Caractéristiques des navires de l'époque des grandes découvertes," in *Actes du Cinquième Colloque International d'Histoire Maritime: Les Aspects internationaux de la découverte océanique aux XVe et XVIe siècles*, ed. Michel Mollat and Paul Adam (Paris, 1966), p. 142.

[24]Lane (n. 22 above), pp. 122, 126.

[25]Diffie and Winius (n. 17 above), pp. 24–25.

[26]Duarte Pacheco Pereira, quoted by Diffie and Winius (n. 17 above), p. 69.

of the sea bed, and above all the impossibility of coming back by the same route close hauled."[27]

How brave, then, were the Vivaldi brothers and their men when they sailed their galleys past the Pillars of Hercules and out of recorded history! We do not know in what form disaster finally struck. What we can say, however, is that the galleys, emergent objects constituted by heterogeneous engineers, were dissociated into their component parts. The technological object was dissolved in the face of a stronger adversary, one better able to associate elements than the Italian system-builders. It was a trial of strength, in which part of the physical world had the final say. Accordingly, it is a paradigmatic case of the fundamental problem faced by system-builders: how to juxtapose and relate heterogeneous elements together such that they stay in place and are not dissociated by other actors in the environment in the course of inevitable struggles, whether these be social or physical or some mix of the two. And it also suggests why we must be ready to handle that heterogeneity in all its complexity, rather than adding the social as an explanatory afterthought. For a system—here, the galley—associates everything from men to the wind. It depends precisely on a combination of social and technical engineering in an environment filled with indifferent or overtly hostile physical and social actors.

The Portuguese versus Cape Bojador: The Lines of Force

In the struggle between the Atlantic and the galley, the Atlantic was the winner. We might say that the forces associated by the Europeans were not strong enough to dissociate those that constituted the Atlantic. The heterogeneous engineers of Europe needed to associate and channel more and different forces if they were to dissociate such a formidable opponent and put its component parts in their place. So for over a hundred years Cape Bojador remained the point of no return. Where were the new allies to come from? How might they be associated with the European enterprise?

Three types of technological innovation were important. The first of these took the form of a revolution in the design of the sailing ship in the 14th and early 15th centuries. The details of this revolution remain obscure, circumstantial, and in any case beyond the scope of this essay, but the end result was a square-rigged seagoing vessel that had much greater endurance and seaworthiness than its predecessors, one that was able to convert winds from many directions into forward motion. There were no oarsmen, so manpower was reduced, and it was thus

[27] Chaunu (n. 17 above), p. 118.

possible to carry sufficient stores to undertake a considerable passage without foraging. This, then, was the first step in the construction of a set of allied entities capable of putting the North Atlantic in its place. The second was the fact that the magnetic compass became generally available in Christian Europe in the late 12th century. I shall consider methods of navigation in a later section, but here it should be noted that the initial importance of this innovation was that it allowed a reasonably consistent heading to be sailed in the absence of clear skies. Combined with dead reckoning and a *portolan* chart, it took some of the guesswork out of long-distance navigation, and in particular it meant that the sailor did not need to hug the coast to have some idea of his location. This, then, was the second decisive step toward a change in the balance of forces. When new ships combined newly channeled winds with new methods of navigation and consequent knowledge of position, the ground was prepared for a possible change in the balance of power.

What was the decisive third step? To answer this question it is necessary to know a little about the currents and winds between Portugal and the Canaries. It is relatively easy to sail out from Lisbon or the Algarve in a southwesterly direction along the Atlantic coast of Africa. The ship is carried along by the Canaries current and is also carried before the northeast trade winds. These are particularly strong in summer. So far, then, the forces of wind and current assist the project of the sailor. It is, however, more difficult to make the return journey for precisely these same reasons. In a ship good at beating to windward it is no doubt possible to make some northeasterly headway. But this requires frequent tacking, something that was difficult in the square-rigged ships of the day that could not, in any case, sail very close to the wind. Although the wind blows from the southwest for a period in the winter—thus making the return journey easy[28]—at some unrecorded point sailors decided to try and put the adverse winds and currents to good use by beating out to seaward, away from the Moroccan coast. For it turns out that, so long as one has an appropriate vessel, some means of determining a heading, and an appropriate dose of courage, it is much easier to return to Lisbon or the Algarve this way than by the coast. The vessel sails, on a northwesterly heading, close-hauled against the northeasterly trades. It is gradually able to take a more northerly course as the trades are left behind until the westerlies and the North Atlantic drift (current) are encountered, when it becomes possible to head east in the direction of Iberia.[29] It was the invention of this circle, called by the Portuguese the *volta*, that marks the decisive third step.

[28]Diffie and Winius (n. 17 above), pp. 61, 136.
[29]Chaunu (n. 17 above), pp. 111–15.

Ships were no longer forced to stay close to the coast. Cape Bojador, the classic point of no return, was no longer the obstacle it had previously been. The masters could sail beyond it and hope to return by using the *volta* in just the way they had previously made use of it to return from, say, the Canaries.

The *volta* can thus be seen as a geographical expression of a struggle between heterogeneous bits and pieces assembled by the Portuguese system-builders and their adversaries, the winds, the currents, and the capes. It traces, on a map, the solution available to the Portuguese. It depicts what, *with the forces they had available*, they were able to impose on the dissociating forces of the ocean. It shows us in a graphic manner how they were able to convert the currents and the winds from opponents into allies and associate them with their ships and navigational techniques in an acceptable and usable manner.

Now we begin to see the advantages of a "system" metaphor. It underlines heterogeneity and interrelatedness while revealing the *struggles* that shape a network of heterogeneous and mutually sustaining elements. The system-builder tries to associate his or her elements in what is hoped will be a durable array. He or she tries to dissociate hostile systems and reassemble their components in a manner that contributes to what is being built. But the particular form of association or dissociation depends on the state of forces. Some of these are obdurate—the currents and winds were beyond tampering with, such was their strength. Some of them can be manipulated but only with difficulty. Here, for instance, the square-rigged ship and navigational practices, while not immutable, were difficult to influence. Others, however, may be more easily altered. In this case, the course sailed by the vessels on their return journey was a matter for discretion, *and had become so as a result of the advances in shipbuilding and navigation* of the previous 150 years briefly discussed above. Here there was, in the most literal sense, new room for maneuver. The course was no longer rigidly overdetermined for the system-builder. Accordingly, the *volta* may be seen as tracing the state of forces and measuring their relative strengths in a very literal way. It represents the state of shipbuilding, the state of navigation, the state of seamanship, and their collision with the forces of nature. The *volta* is the extra increment of force that allows a new system to be fashioned, for the course was suddenly the most malleable element in the conflict between the Portuguese desire to return to Lisbon and the natural forces of the Atlantic.

The Caravel and the African Littoral: Adaptation

Africa, as the Portuguese were to discover, does not reduce to Cape Bojador. The capacity to get round the cape and then return to

European waters is all very well, but there was more coastline to explore. South of the cape this becomes even more inhospitable—empty desert. There are no rivers, few if any inhabitants. You search for hundred of miles and find nothing. And then in 1441 you reach Cabo Branco, find people, and barter a few goods—leather, oil, and sealskins. You also conduct a slave hunt. This isn't gold, but at least it is proof that there are human beings beyond the Maghreb. Two years later you reach Arguin. Here there are both gold and people to be turned into slaves. You establish a trading post. Then, in 1444, ten years after passing Cape Bojador, you find the Senegal River and sail into its estuary. You find yourself at last in direct contact with black Africa.

For most of this tricky exploration the Portuguese made use of caravels. Though the origins of this type of vessel are shrouded in mystery,[30] its 15th-century features are well known. Weighing less than 100 tons and being perhaps 70–80 feet from stem to stern, the caravel was unusual in being a long sailing ship, having a length-to-breadth ratio of between 3.3 and 3.8 to one, as opposed to the 2 to 2.5 to one ratio of the older cog.[31] It was carvel-built (smooth-sided), quite light and fine in lines, and drew little water, having a flat bottom and little freeboard.[32] It had only one deck and, indeed, was sometimes even open or only half-decked. There was no forecastle, and the superstructure of the poop was modest, at best containing one room.[33] A virtue of the caravel, which was increasingly utilized as the 15th century wore on, was that it could be rerigged very quickly to suit different conditions. Thus Columbus's caravel, the *Niña*, was changed in a matter of days from lateen to square rig in the Canaries during the outward stage of his first voyage in 1492.[34] Parry suggests that such events were entirely routine.[35] However, during the middle of the 15th century, and certainly in the early voyages of discovery, the caravel normally appears to have been lateen-rigged on all its masts.

[30]Landström (n. 21 above), p. 100; Chaunu (n. 17 above), p. 243; Parry (n. 18 above), p. 65; Richard W. Unger, *The Ship in the Medieval Economy, 600–1600* (London, 1980), pp. 212–15.

[31]Parry (n. 18 above), p. 65; on length-to-breadth ratio, see Paul Gille, "Navires lourds et navires rapides avant et après les caravelles," *Actes du Cinquième Colloque* (n. 23 above), pp. 175–76. Diffie and Winius (n. 17 above), p. 118, suggest a length-to-breadth ratio of 3 to 1.

[32]Parry (n. 18 above), p. 65; Denoix (n. 23 above), p. 143; Landström (n. 21 above), p. 100.

[33]Parry (n. 18 above), p. 65.

[34]Chaunu (n. 17 above), p. 160; Bartolemé de las Casas, "Digest of Columbus's Log-Book on His First Voyage," in *The Four Voyages of Christopher Columbus*, ed. J. M. Cohen (Harmondsworth, 1969), p. 40; Parry (n. 18 above), p. 66.

[35]Parry (n. 18 above), pp. 65–66.

We might say that the caravel was "well adapted" to the context of offshore exploration. Thus we might note (as have many historians, Denoix for instance)[36] that for such a task one needs a vessel that will not blunder into reefs, is light and handy, draws little water, sails well against the wind, and does not require a large crew. All of these attributes were true for the caravel, which was indeed "well adapted" to its task. But what are we really saying when we say this?

The answer to this question is to be found in the notion of network that was introduced in the last section. The system-builder, it was argued, seeks to create a network of heterogeneous and mutually sustaining elements. He or she seeks to dissociate hostile forces and associate these with the enterprise by transforming them. The crucial point, however, is that the structure of this network reflects both the power and nature of the forces available and those with which it collides. To say, then, that an artifact is well adapted to its environment is to say that it forms a part of a system or network that is able to assimilate (or turn away) potentially hostile external forces. It is, consequently, to note that the network in question is relatively stable. Again, to say of an artifact like the caravel that it is adaptable is to note that a network of associated heterogeneous elements has been generated that is stable because it is able to resist the dissociating efforts of a wide variety of potentially hostile forces and to use at least some of these by transforming them and associating them with the project. And this, of course, is precisely the beauty of the caravel in the 15th-century context in which it was used by the Portuguese. Properly manned and provisioned, it was able to convert whatever the West African littoral threw at it into controlled motion and controlled return. It was a network of people, spars, planks, and canvas that was able to convert a wide range of circumstances into exploration without falling apart in any of the numerous ways open to vessels when things start to go wrong. Like the *volta*, then, the caravel reflected the forces around it. It was "well adapted" because it maintained stable relations among its component parts by associating everything it encountered with that network as it moved around.[37]

[36]Denoix (n. 23 above), p. 142.

[37]Elsewhere I have argued that successful scientific papers may be seen as arraying elements that have the effect of converting a wide range of readers from skeptics into believers by working on and adapting their "interests." The structure of papers may be seen as a "funnel of interests" that moves the reader from a general interest (say, in cancer) to a specific concern with the subject of the paper (the pinocytic effects of a particular polymer) by presenting a link between a general problem and a specific topic. See, inter alia, John Law, "The Heterogeneity of Texts," in *Mapping the Dynamics of Science and Technology* (n. 12 above), pp. 67–83. The caravel may be analyzed in analogous terms as an array of elements that converts a wide range of outside forces into elements

Navigation and the Raising of the Sun: The Importance of Metrication

Between 1440 and 1490 the Portuguese explored most of the West African coast. As they moved further south and used increasingly large *voltas*, problems of navigation began to become more acute. How, then, did they know where they were going? In the 14th century, like their Mediterranean neighbors, they had used the *portolan* chart, the rutter (a written guide), and the magnetic compass.[38] The principle was simple. One determined the desired heading on the chart by looking for an appropriate rhumb line from a convenient wind rose. If there was no appropriate wind rose, then dividers were used to discover a rhumb line parallel to the desired course. The master then used the magnetic compass to sail that course and dead reckoning to determine how far he was along it. If he was obliged to tack, his progess along or deviation from his course was calculated by means of a set of *marteloio* tables.[39] He was assisted by the rutter, which described places of importance or danger, and distances between points of arrival and departure along important routes.

Despite the sophistication of *portolan* chart and compass navigation, the method had various drawbacks. One was the danger of getting lost by sailing a compass course for a given reckoned distance. The difficulty arose partly because there was no way of being absolutely certain how much distance the ship had traveled. Worse, however, than inaccuracies in distance were those in direction. Here the problem was one of guessing the influence of currents and adverse winds—and of correctly calculating from the *marteloio* tables the direction (and distance) covered if it proved necessary to tack. These difficulties were real in the Mediterranean, but in general surmountable since most courses set were quite short and within sight, at times at least, of land. This was not true for Atlantic navigation. If one set a compass course for, say, Madeira, then one either arrived there or one didn't. And if one failed, there was no way of calculating where one was in relation to it.

These problems greatly concerned the Portuguese, and in the 1480s they developed a practical method for the astronomical determination of latitude on board ship. The general idea was that if the *altura* or height above the horizon of the sun or a star (normally the polestar) could be determined and then compared with the known *altura* of the port of destination, then the ship could sail north or south until it

---

that contribute to the project in hand. The general description of such a process would be a "funnel of translation." On the notion of translation, see Callon (n. 12 above).

[38] Diffie and Winius (n. 17 above), pp. 131–32.

[39] E. R. G. Taylor, *The Haven-Finding Art: A History of Navigation from Odysseus to Captain Cook* (London, 1956), pp. 112, 117–20; Diffie and Winius (n. 17 above), p. 130.

reached that latitude and then sail, as appropriate, east or west in the certainty of finding its destination point.

The measurement of *altura* was possible using either the quadrant or the astrolabe. Both of these were standard university instruments of astronomy and astrology. They were used, in combination with the ephemerides and tables of latitude and longitude, to calculate the local time of astronomical events and thus of propitious and unpropitious hours.[40] Accordingly, they carried a great deal of information that was both unnecessary to the calculation of latitude and simply incomprehensible to the layman. However, on the back of the astrolabe there was an alidade, which was a rule on a swivel with two pinhole sights. The observer held the instrument upright by a swivel suspension ring, peered along the alidade, and measured the *altura* of the star by reading off the position of the alidade on a scale marked on the rim. Horizontal was 0° and vertical 90°. The quadrant was an instrument with similar functions. It was in the form of a quarter-circle, and the star sight was taken along one of the "radii." The artificial horizon was provided by a plumb line suspended from the center of the "circle" and measured with a scale along the circumference.[41] Like the astrolabe, in its university and astrological version it also carried information about the movements of planets, seasons, and hours. Both of these instruments, shorn of all but their essentials for the measurement of *altura*, were used by Portuguese explorers, though it seems that the somewhat simpler quadrant was the first to be used by navigators.[42]

By themselves these instruments were, of course, powerless. The mere fact of sighting a heavenly body through the pinholes of an alidade had nothing, per se, to do with navigation. That sighting, or the reading that corresponded to it, had to undergo a number of complex transformations before it could be converted into a latitude. The construction of a network of artifacts and skills for converting the stars from irrelevant bits of light in the sky into formidable allies in the struggle to master the Atlantic is an example of heterogeneous engineering—of associating diverse elements in such a way that they constitute a relatively stable set of interrelated entities which reflect both the struggle between different forces and the aims of the system-builder.

I have already mentioned the simplification of the quadrant and astrolabe. This can be treated as the first step in the process. The second stage involved what may be treated as social engineering—the

[40]Taylor (n. 39 above), pp. 151–52.
[41]Ibid., pp. 158–59.
[42]Ibid., p. 159.

construction of a network of practices which, associated with the instruments themselves, would lead to the necessary transformations of sun- and starlight.[43] This social engineering itself came in three stages. First, in the early 1480s King John II convened a "scientific commission" to find improved methods for measuring the *altura*. This was made up of four experts: the royal physician, Master Rodrigo; the royal chaplain, Bishop Ortiz; Martin Behaim, the geographer; and finally José Vizinho, who had been a disciple of the astronomer Abraham Zacuto of Salamanca.[44] The convocation of a "scientific commission" for the purpose of converting esoteric scientific knowledge into a set of widely applicable practices is already remarkable. Even more noteworthy was the fact that these men, and probably in particular Vizinho, were able to effect that transformation by producing a set of rules for the calculation of the latitude by semieducated mariners. These rules, which form the second part of this experiment in social engineering, took the form of the *Regimento do Astrolabio e do Quadrante*, which were probably available from the late 1480s, at least in handwritten form. They may be read as instructions about how to turn the vessel and its instruments into an observatory—how, in other words, to create a stable if heterogeneous association of elements that had the property of converting measurements of the *altura* into determinations of the latitude. Let me describe a small part of the "rule of the sun."

The observer is told that he must take the height of the sun "and this must be at midday when the Sun is at its greatest elevation." He is told to keep a note of the altitude—probably on a slate. He is then required to look up the table for the month and day, and undertake the appropriate calculation. For example, one of the rules states: "Take out the declination, and if the Sun is in a northern Sign, and if the shadow is falling to your north, then subtract the altitude that you found from ninety, and add the declination. The sum will be the number of degrees you are north of the equinoctial. . . . and if you find an altitude of 90° know that you are distant from the line as many degrees as the Sun has of declination, no more and no less."[45] Similar

[43]In what follows I have been highly selective with respect to material in order to highlight what I take to be the essentials of the process and to avoid getting bogged down in detail. For similar reasons, I have also taken the liberty of reorganizing the chronology of events by talking of the establishment of the latitudes of important points on the coast after discussing the *Regimento*.

[44]Chaunu (n. 17 above), p. 257; Taylor (n. 39 above), p. 162; Guy Beaujouan, "Science livresque et art nautique au XVe siècle," *Actes du Cinquième Colloque* (n. 23 above), p. 74; D. W. Waters, *Science and the Techniques of Navigation in the Renaissance* (Greenwich, 1980), pp. 9–10.

[45]Taylor (n. 39 above), p. 165.

rules are spelled out for the other cases, and exceptions, which must have been likely to confuse the navigator, are also detailed.

If this gave a latitude, it was still necessary to convert the latter into practical instructions for sailing. These took the form of another set of rules, those for "raising a degree." Since the new navigation depended on finding the appropriate latitude and then sailing along it, it became important to know, for any given compass course or "wind," how far it was necessary to sail to cover 1° of latitude and, also, how far the vessel had moved to the east or west. "Know," starts this section of the *Regimento*, "that the degree of north-south is seventeen and a half leagues, and sixty minutes make a degree." The text then explains, for eight quarters, the distance to be sailed to raise a degree and the amount of easting and westing.

Even this, however, was not enough. To adopt the new method of sailing, a third piece of heterogeneous engineering was called for: it was necessary to know the latitudes of important coastal features and in particular the major ports and capes. It was, in other words, *necessary to generate a metric within which the observations might be given north-south meaning* and the observatory of the ship accordingly located. The measurement of important coastal latitudes again involved a major organizational effort. It involved sending out competent observers armed with large wooden astrolabes on the vessels of exploration and having them report back to Lisbon. It further required that known latitudes be available to mariners, and indeed, a further section of the *Regimento* listed these. It seems likely (though there is historical controversy here) that Henry the "Navigator," before his death in 1460, was asking his mariners to go ashore and measure the altitude of the polar star at places of importance. By 1473 the astronomers in Lisbon had a list of latitudes which reached the equator,[46] a list that was extended as the century wore on.

The new method of navigation proved difficult for most mariners. Only the most up-to-date attempted its practice, and there is evidence that Columbus, among others, understood it only imperfectly. Though details remain unclear, it appears that in the early 16th century—and probably earlier—classes on navigation were taught to pilots at Lisbon.[47] Here is further evidence, if this is needed, of the propensity of the Portuguese to undertake social engineering. Such instruction was not, however, invariably successful. There were complaints in the 16th century that many pilots were inexpert. It seems, then, that in the attempt to create a stable network of elements for the conversion of

[46]Ibid., p. 159.
[47]Diffie and Winius (n. 17 above), p. 142.

stars into measurements of the latitude—in the attempt, in other words, to convert ships into observatories—the mariners constituted the weakest link. The stars were always there, as were the oceans. They could not be budged. Again, once the instruments and the inscriptions were in place, these proved to be fairly durable. But instruments, inscriptions, and stars were not enough. Part of the association of elements to convert stars into latitudes lay in the practices of men, and it was these that were most prone to distortion. Interestingly, then, it proved easier to practice social engineering in the form of associating experts than it did in the form of associating skills in the daily conduct of pilots. It was difficult—though not ultimately impossible—to create a new and essential social group, that of the astronomical navigator.

There was one other element vital for the Portuguese system—the construction of a metric against which the readings of the nautical observatory might be compared. For, as I have noted above, it would not have been possible to convert a latitude into a plan for action in the absence of a list of shorebound reference points. Here again, as I have indicated, the heterogeneous engineering skills of the Portuguese were evident. But there is a further point to be made. So far I have tacitly made the assumption that when a technological system works this fact is obvious. If one arrives at one's port of destination (or for that matter runs aground on the reefs of Cape Bojador), the success (or failure) of the enterprise is readily apparent to all. We might say that, in the ultimate analysis, it was the capacity of the Portuguese to *return* to their point of departure that marked success. The success of astronomical navigation was that it contributed to that return. Yet however much final success depended on the capacity to return, decision making on the voyage would not have been possible without a scale of reference. The success of any course sailed could only be measured, in the interim, against an entirely man-made metric, a metric that depended on inscriptions and the capacity to interpret those inscriptions. We have here, then, the construction of a background against which to measure success—something akin if not identical to the technological testing tradition described by Constant in the context of water turbine engineering.[48] The history of navigation can, I believe, be understood as the construction of progressively more general systems of metrication against which the adequacy of particular courses and navigational decisions may be measured.

[48]Constant (n. 15 above).

The Muslim and the Gun: System-Building and Dissociation

On July 8, 1497, Vasco da Gama's fleet weighed anchor in the Tagus at Lisbon and set sail. His four tiny vessels carried 170 men and twenty cannon. They also carried merchandise. Though this expedition is normally described as one of exploration, it is clear that this was exploration of a very particular kind. The Portuguese were not sailing into the unknown. They were sailing with very definite goals in mind: they were seeking out Christians and spices.[49] Furthermore, they knew, or thought they knew, where to look and how to get there. The journey was long and difficult—I shall not detail it here. Nevertheless it was, within limits, successful. Two years later two of the original four vessels returned to Lisbon. The cape route to India had been opened and spices brought back.

In part the Portuguese difficulties arose from the hostility of Muslim traders.[50] Such merchants organized and controlled the Indian Ocean section of the spice trade. They bought spices in the Calicut bazaars and shipped these, either via the Persian Gulf or the Red Sea, to Arabian ports for transshipment to the Mediterranean and Venice. The arrival of the Portuguese on the Malabar coast at Calicut was not welcomed by the traders. Not only was there religious hostility between the two groups, but the Muslims rightly saw the Portuguese as potential commercial competitors. Negotiations went badly between the Portuguese and the Hindu ruler of Calicut, the Samorin. There were a number of reasons for this—the Portuguese were disappointed to discover that the Hindus, although not Muslim, were not Christian either, and the Samorin was angered by the paucity of the gifts that the Portuguese had brought him.[51] The major reason, however, appears to have been the hostility of the Muslim traders on whom the Portuguese were obliged to depend for translation. These spread the rumor that the Portuguese were simply marauders and were able to have da Gama and his small landing party imprisoned. After a series of adventures and lucky chances, all the Portuguese escaped and, indeed, despite a boycott by the Muslim merchants, managed a certain amount of trade direct with Hindu merchants.[52] Hence the relative success of the expedition.

[49]Boies Penrose, *Travel and Discovery in the Renaissance, 1420–1620* (Cambridge, 1952), pp. 50 ff; Parry (n. 18 above), pp. 140 ff; Vitorino Magalhaes-Godinho, *L'Économie de l'empire portugais aux XVe et XVIe siècles* (Paris, 1969); Diffie and Winius (n. 17 above), p. 177.

[50]Magalhaes-Godinho (n. 49 above), p. 558.

[51]Penrose (n. 49 above), p. 54; Magalhaes-Godinho (n. 49 above), p. 559.

[52]Diffie and Winius (n. 17 above), pp. 182–83.

It was clear from this point on (as had previously been suspected) that it would be necessary to exercise force in the Indian Ocean. Da Gama's first expedition had carried guns, but more would be needed if the hostility of the Muslims was to be mastered. In fact, the Portuguese had come to this conclusion even before da Gama's return. A much larger and more heavily armed second expedition under the command of Cabral had already set out, consisting of thirteen vessels and between 1,000 and 1,500 men. Cabral's orders were clear. He had to install an agent to buy spices in Calicut and was instructed to display force where this was necessary although to refrain from conquest.[53] This time the Portuguese approached the Samorin with due deference. On arrival, they saluted him with guns, and sent him fine gifts of silver, tapestries, carpets, and velvet. In return they were granted the right to set up a factory. Once again, however, things went wrong. Muslims, aware of the danger that the Portuguese represented to their position, fomented discontent, and in a riot, only twenty of fifty Portuguese ashore succeeded in regaining their vessels. Cabral put to sea, destroyed a number of Muslim vessels, and bombarded the town of Calicut in response. In so doing he, as Diffie and Winius put it, "literally blasted the Samorin into the enemy camp," though Parry suggests that the amount of real damage inflicted must have been small.[54]

In 1502, da Gama used even more force. His fleet was powerful—twenty heavily armed vessels—and all went well until they got to Calicut. Even though the Samorin, doubtless all too aware of the firepower of the fleet, was willing to negotiate, discussions broke down. Though he was prepared to give up those Muslims primarily responsible for the massacre of Cabral's men, he was not willing to expel all Muslims and close his port to certain categories of valued customers. In the middle of the negotiations da Gama decided that it was time to exert force: ". . . the admiral suddenly gave the order to hoist anchor, swung his ships in close to shore, and opened fire, hurling stone and metal cannonballs into the city's streets for an entire day before sailing off towards Cochin. In addition to this inexplicable and barbarous act which killed indiscriminately, he committed one even more appalling: he butchered or burned alive several hundred innocent fishermen."[55] If this incident serves to illustrate contemporary European attitudes to non-Christians, it also cast the die for Portuguese control of the Indian Ocean over the next few years. It would have to be primarily by force; there was no room for both Muslim and Portuguese commerce. Fur-

[53]Magalhaes-Godinho (n. 49 above), p. 561.
[54]Parry (n. 18 above), p. 122.
[55]Diffie and Winius (n. 17 above), p. 224.

thermore, the Samorin was now a sworn enemy, and the Portuguese were obliged to set up their trading posts in adjacent kingdoms.

At sea the Portuguese were, at least in the short run, able to exert the necessary military power to choke Muslim maritime trade. Portuguese guns turned out to be more effective (though not more numerous) than Asian guns. European advances in the technology of gunmaking had overcome many of the problems that beset late medieval cannon. In particular, with the development of cast bronze guns, the weight of cannon had been much reduced and, although still prone to this problem, they were much less likely to blow up in the faces of the gunners than the welded pieces that preceded them. Again, Portuguese vessels, built for the inhospitable Atlantic, were more solid than those of their Muslim adversaries.[56] Cipolla puts it in this way: "The gunned ship developed by Atlantic Europe in the course of the fourteenth and fifteenth centuries was the contrivance that made possible the European saga. It was essentially a compact device that allowed a relatively small crew to master unparalleled masses of inanimate energy for movement and destruction."[57]

The cannon, the ship, the master, the gunner, the powder, and the cannonballs—all these formed a relatively stable set of associated entities that achieved relative durability because it was able to dissociate the hostile forces that it encountered without itself being dissociated. It is important to note here that some of these hostile forces were physical (the oceans), while others were social (the Muslims). Technology, as I have suggested, simultaneously associates and dissociates, and the heterogeneous engineering of the Portuguese was designed to handle natural and social forces indifferently and to associate these in an appropriate form. Having said this it is, however, important not to fall into the trap of technological determinism. As was the case for the caravel, the *volta*, and the practice of astronomical navigation, the durability of the armed man-of-war was a function of a collision between the forces of the Portuguese system-builders and those of the seas and, in this case, the Muslims. Thus Boxer argues that the Portuguese "naval and military superiority, where it existed, was relative and limited."[58] It happened that there was no well-armed Muslim shipping in the Indian Ocean. It happened that the Chinese had retired to their coasts. It happened that the Portuguese expeditions were state enterprises, combining the power and organizational ability of the crown

[56]C. R. Boxer, "The Portuguese in the East, 1500–1800," in *Portugal and Brazil: An Introduction* (Oxford, 1953), p. 196.

[57]Cipolla (n. 19 above), p. 137.

[58]Boxer (n. 56 above), pp. 194–97.

with the search for profit. It happened that Muslim merchants traded on their own account and not for their monarchs. It happened that there was little wood available to many of those monarchs in order to build fleets to stop the Portuguese. Under these circumstances, the Portuguese were able to dominate shipping in the Indian Ocean. They were not able (and knowing this never sought) to build up sizable colonies on land. There, with the balance of force weighted against them by cavalry and manpower, they risked crushing defeat.

Conclusion

I started by asking what the sociologist can offer the historian of technology when the latter attempts to understand the nature of successful system-building and suggested that there are two possibilities. One is to add background social factors that are held to be decisive for the nature and direction of the system in question. I suggested that, while there was no knockdown argument for rejecting this option, it was liable to impede investigation of one of the most interesting features of technological change—the fact that this typically involves the creation and combination of *both* technical *and* social elements. I then suggested that there was an alternative: that there is a sociological tradition which focuses precisely on such "heterogeneous engineering" and the way in which heterogeneous engineers associate entities (from people, through skills, to artifacts and natural phenomena) together. This alternative suggests that the system-builder is successful if the consequent heterogeneous network is able to maintain some degree of stability in the face of the attempts of other entities or systems to dissociate it into its component parts. Again, it suggests that the structure of the network (or system) in question reflects not only a concern to build something relatively stable but also the relationship between the forces that it is able to muster and those deployed by its various opponents. We might, if we were to make more use of the metaphor of force, talk of the relative durability or strength of different systems or of different parts of the same systems. Thus, I have attempted to show by empirical example that, in the collisions between different networks, some components are more durable than others, and that the successes achieved by one side or the other are a function of the relative strength of the components in question.

What are the virtues of physical metaphors such as force, strength, and durability? Let me say, first of all, that I am not strongly committed to these terms. Doubtless other metaphors might serve as well or better. I believe, however, that the strength of the vocabulary lies in its capacity to handle, indifferently, the various and heterogeneous elements that have to be assembled by any system-builder. The method, in other words, deals or seeks to deal with the "social," the "economic," the "political," the "technical," the "natural," and the "scientific" in the same terms on the grounds that (in most empirical cases) *all* of these have to be assembled in appropriate ways by those who wish to build a system. Within all of these (usually distinguished) categories, entities, processes, bodies, objects, institutions, or rules may be generated that have the property that they are relatively durable and hence have force with respect to the project of the system-builder. They are "scientific facts," "economic facts," "social facts," or whatever. They form, then, a relatively coercive albeit ultimately revisable scenery which has to be mastered if success is to be achieved. It is for this reason that I have made use of terms such as "element," "entity," and "force" in a way that rides roughshod over conventional category distinctions. My argument is that it is not good enough to use the social as an explanatory last resort. The social, including the "macro-social," has, rather, to be pictured as being in there with everything else if the collisions between "forces" are to be understood.

It may be objected that I have ignored the fact that social forces are associated in different ways than (that is, they have different properties than) physical forces. I have some sympathy with this objection, but my view is that before we can profitably move forward to the analysis of the association of types of forces, it is imperative that we be aware of what it is they have in common and avoid the trap of allocating ultimate explanatory power to our favored category. I would add that there is no reason to suppose that different modes of association, when these are uncovered as they surely will be, will necessarily line up with the standard disciplinary categories that currently structure our explanations.

It may also be objected that I still privilege human beings in my analysis, despite the claim that there is symmetry between social and other forces. Thus I have, for instance, often talked of "system-builders" who seem to lie, in some sense, at the explanatory origins of the systems in question. Again, in doing so, I have distinguished between those forces where there *is* a point of decision making and control (e.g., the Portuguese naval forces) and those where there is not (the Canaries current). Here again, I find myself in some agreement with the objections. However, for at least three reasons I do not think that they undermine the basic symmetry of the approach. First, the system-builder is not necessarily an individual. It can be a corporate body, or, indeed, a nation. Any individual, group, or social entity that is *(a)* a locus of decision making, *(b)* able to speak on behalf of a network of entities, and *(c)* able to speak effectively by mobilizing what it has promised or threatened to mobilize, becomes, in my way of thinking, a system-builder. Think, for instance, of Cabral on the second Portuguese expedition to India. He spoke for Portugal, and he was able to do so because he spoke on behalf of a network of elements—people, vessels, cannon, the monarch—that he could mobilize as necessary. Second, system-builders may be studied themselves, because they are not "primitive" but rather *the end product of sets of system-building strategies* or other forms of mobilization. If King John II or Henry the Navigator built and commanded a system of Portuguese navigation, this was because he was able to speak for and mobilize a wide range of elements. In principle, the methods they used to mobilize these resources may be studied. Thus, though any historical study must have its limits, and it may make sense to take such individuals as unanalyzed system-builders for the purpose of a particular study, in principle there is no question about the feasibility of analyzing the strategies that led to their achieving such a status. Third, it does not, in any case, necessarily follow that a system has a decision-making center: in many systems of forces there may not be a system-builder at all. Thus the natural opponents of the Portuguese—the hostile winds and currents—are a case in point. Though this is, admittedly, a metaphysical commitment, I have worked on the assumption that they were not mobilized by a system-builder from a central point.

The analysis of system-building—or better, system growth—may thus in principle be generalized to all networks of forces and the manner in which these build themselves to form relatively stable wholes. My reason for concentrating on those in which there are identifiable system-builders is thus not one of principle but is entirely practical in character: it is that historians and sociologists typically concern themselves with systems that contain people and social institutions—that is, systems in which there are individuals or groups that make decisions and successfully speak on behalf of others.

In this article I have sought, therefore, to offer a particular, nonreductionist, sociological perspective on three of the major theoretical tools currently used in the history of technology. The first of these is obviously the metaphor of system-building. The emphasis that this approach places on the entrepreneurial activity which juxtaposes heterogeneous elements into a stable system seems to me to be both entirely well-founded and analytically fruitful. At the same time I have sought to emphasize that the shape taken by a system is often a function of conflict. The emphasis on struggle—between Cape Bojador and the galley, between the caravel and the Canaries current, between the size of the Atlantic and navigational practices, between da Gama and the

Muslims—is intended to try to counteract the danger of taking the view that systems are, in some way, self-contained.

Second, I have briefly considered a second set of metaphors, those of adaptation, or variation and selection. At one level it is difficult to find fault with such Darwinian imagery. A working technology or device must, presumably, be one that is adapted. Yet the very fact that it is so unexceptionable suggests that it may do little or no analytical work. It is for this reason that I have sought to reinterpret the metaphor. I have argued that the notion of adaptation can be given real content if the viability of the artifact or practice under scrutiny is seen in network or system terms as an association of forces that is able to maintain a relative degree of stability in the trials of strength and struggles in which it is involved. The notion of adaptability is closely related to this and refers to the capacity of a network or system to retain its shape essentially unaltered in the face of a wide range of outside and potentially dissociating forces. The caravel is a good example of a relatively stable network of elements (timbers, spars, design, sails, master, pilot, crew, stores) capable of coping with and utilizing a wide range of natural forces—currents, deep and shallow water, awkward offshore breezes, winds from a range of quarters, lack of provisioning facilities, and all the rest. The caravel was, indeed, well adapted to its 15th-century task, but the network or system metaphor of associated forces allows us to talk about the way in which its particular features reflect particular aspects of its environment. The lack of "keel" speaks, as it were, for shallow water. The lateen rig likewise "speaks" for the prevalence of adverse winds. The use of a network metaphor emphasizes that the relationships between the "inside" and the "outside" of a technological artifact are close, and it allows them to be specified.

Third, I have sought to link the problem of building successful systems more firmly with the importance of metrication and the role of "technological testing." There is much more to be said here than can possibly be outlined in a short essay. The essential point, however, is that the success of technology is not something that is necessarily obvious. Though this may appear to be a difficult argument to sustain in the face of (say) a vessel foundering on a reef, it is easier to see in those cases where a metric has to be generated in order to make decisions about what should be done next. Constant has described the development of the Prony brake and the consequent metrication that led to an enforceable tradition of testing for water turbines in 19th-century America. I have attempted a sketch of a parallel enterprise—that of the development not only of a method for measuring the latitude but also, and at least as important, a metric for determining the *significance* of any measurement. Accordingly, I share Constant's view that this is a particularly important topic. My intuition is that metrication and the techniques of inscription that this normally requires play a crucial role in determining success in most technological systems.[59]

In sum, then, I suggest that it is too simple to say that a technological artifact whose origins and development are insufficiently understood must be in want of a social or economic explanation. To say this would be no more correct than to argue, as some have done in the past, that incompletely understood social phenomena must be in need of a technological explanation.[60] In this article I have tried to suggest that, although much of sociology privileges the social, there is nonetheless a nonreductionist sociological tradition in the analysis of heterogeneity that parallels the system-building approach that has been developed in the history of technology. The approach makes use of a vocabulary that does not distinguish among the social, the scientific, the technological, the economic, and the political, and makes no a priori assumption that one of these carries greater explanatory weight than all the others. Rather, it displays the strategies of system-building and, in particular, the heterogeneous and conflicting field of forces within which technological problems are posed and solved. I have sought, through empirical argument, to recommend this approach.

[59]On the importance of inscription see Bruno Latour, "Les 'Vues' de l'esprit," *Culture Technique* 14 (1985): 4–29; and Elizabeth Eisenstein, *The Printing Press as an Agent of Change* (Cambridge, 1979).

[60]For a full and critical discussion of the latter position in the context of Marxist historiography, see Donald MacKenzie, "Marx and the Machine," *Technology and Culture* 25 (1984): 473–502.

Muskets and Pendulums: Benjamin Robins, Leonhard Euler, and the Ballistics Revolution

BRETT D. STEELE

Ballistics was revolutionized between 1742 and 1753 by Benjamin Robins (1707–51) and Leonhard Euler (1707–83). As one artillery officer wrote in 1789, "Before Robins, who was in gunnery what the immortal Newton was in philosophy, the founder of a new system deduced from experiment and nature, the service of artillery was a mere matter of chance, founded on no principles, or at best, but erroneous ones."[1] John Pringle, a president of the Royal Society, put it more simply in 1783 by stating that Robins created a "new science."[2] John Nef wrote in 1950 that Robins's work "provides a landmark in the interrelations between knowledge and war."[3] Two engineers more recently described him as being "one of the fathers of aerodynamics," while Thomas P. Hughes referred to him as "a founder of modern gunnery."[4] What did

Dr. Steele is a lecturer in the Department of History at UCLA, and teaches the history and social studies of technology. He completed his dissertation, *The Ballistics Revolution: Military and Scientific Change from Robins to Napoleon* at the University of Minnesota in 1994.

[1]Papacino D'Antoni, *A Treatise on Gunpowder; a Treatise on Fire-Arms; and a Treatise on the Service of Artillery in the Time of War,* trans. Captain Thomson of the Royal Regiment of Artillery (London, 1789), p. xvii.

[2]John Pringle, *Six Discourses . . . on occasion of Six Annual Assignments of Sir Godfrey Copley's Medal* (London, 1783), p. 273.

[3]John Nef, *War and Human Progress: An Essay on the Rise of Industrial Civilization* (Cambridge, Mass., 1950), p. 194.

[4]H. M. Barkla and L. J. Auchterlonie, "The Magnus or Robins Effect on Rotating Spheres," *Journal of Fluid Mechanics* 47 (1971): 437. See Thomas P. Hughes's commentary after David D. Bien's article, "Military Education in 18th Century France: Technical and Non-technical Determinants," in *Science, Technology and Warfare: Proceedings of the 3rd Military History Symposium, USAFA, 8–9 May 1969,* ed. Monte D. Wright and Lawrence J. Paszek (Washington, D.C., 1971), p. 73. No books devoted to the history of ballistics in the 18th century exist in English at present. In other languages, however, there are some useful intellectual histories that include the subject. See M. P. Charbonnier, *Essais sur*

such a seemingly obscure British mathematician and engineer do to generate such acclaim? Charles Hutton, the 18th-century artillery professor at Woolwich, said that Robins's research represents "the first work that can be considered as attempting to establish a practical system of gunnery, and projectiles, on good experiments, on the force of gunpowder, on the resistance of the air, and on the effects of different pieces of artillery."[5] More specifically, Robins invented new instruments which he used to discover and quantify the enormous magnitude of air-resistance force acting on high-speed projectiles and to make the first observations of the sound barrier. He also conducted a theoretical and experimental thermodynamic analysis of interior ballistics, discovered the Magnus (or Robins) effect of fluid mechanics, and established a rational understanding of the rifling phenomenon. Of all his aerodynamic discoveries, the enormous and complex function of air-resistance force encountered by high-speed projectiles caused the greatest sensation in the 18th century. This discovery quantitatively showed the inadequacy of Galileo's projectile theory, which neglected air resistance, and the oversimplification of Sir Isaac Newton's and Christian Huygens's air-resistance theory for projectile motion. Furthermore, Robins's air-resistance experiments made the practical mathematical analysis of high-speed ballistic motion possible.

Robins summarized his initial discoveries in *New Principles of Gunnery,* a short book that was first published in 1742.[6] Euler translated this work into German and added an extensive mathematical analysis in 1745. Using air-resistance values based on Robins's experimental measurements, Euler solved the equations of subsonic ballistic motion in 1753 and summarized some of the results into convenient numerical tables. This was the first published analysis of projectile trajectories to incorporate empirical air-resistance values.

In the short span of eleven years, Robins and Euler dramatically increased the predictive power of ballistics by constructing theoretical and empirical foundations. This effort also marks one of the first

l'histoire de la balistique (Paris, 1928); István Szabó, *Geschichte der mechanischen Prinzipien* (Basel, 1977), esp. the chapter "Die Anfänge der äußeren Ballistik"; and A. P. Mandryka, *Istoriia ballistiki* (Moscow, 1964). The more recent articles by engineers that focus on Robins's ballistics research include H. M. Barkla, "Benjamin Robins and the Resistance of Air," *Annals of Science* 30, no. 1 (March 1973): 107–22; W. Johnson, "Benjamin Robins (Eighteenth Century Founder of Scientific Ballistics): Some European Dimensions and Past and Future Perceptions," *International Journal of Impact Engineering* 12, no. 2 (1992): 293–323.

[5]Nef, p. 195.

[6]Benjamin Robins, *Mathematical Tracts of the Late Benjamin Robins,* ed. James Wilson (London, 1761). *New Principles of Gunnery* is contained in vol. 1.

significant applications of Newtonian mechanics to engineering analysis, as well as the coupling of differential equations with complex experimental measurements, the hallmark of 19th-century physics. Robins's and Euler's work therefore represents a scientific revolution. This revolution has had a strong effect on science, engineering, and warfare ever since. Significant branches of aerodynamics and thermodynamics grew from it. A cannon, after all, is an internal combustion engine, and a cannonball is a flying body. The ballistics revolution generated new theories that offered a rational understanding of gunnery, the technology of controlling gunfire. This made the teaching of calculus and mechanics to artillery and engineering officers increasingly profitable for Western governments during the second half of the 18th century. The increased precision and reduced uncertainty of their artillery fire in battlefields and sieges proved to be a generous return for this educational investment. This change in gunnery was also intimately linked to the simultaneous changes in artillery hardware, organization, and tactics.

Although the ballistics revolution has received little attention from professional historians, it challenges important claims in the history of 18th-century science and technology: that mathematical analysis and scientific experimentation had little interaction and that technology (with the exception of navigation) was not significantly influenced by rational mechanics. The ballistics revolution also highlights important characteristics regarding the historical role of engineering research. In addition to "merely" applying existing scientific knowledge to develop technologically useful theories, engineers have also created fundamental scientific knowledge in direct response to technological needs. By "science," I am specifically referring to that branch of knowledge exemplified by Newtonian celestial mechanics. In Robins's case, as in many others, the creation of scientific knowledge for engineering had profound consequences for science in general.

Ballistics before Robins

Galileo's vacuum or parabolic trajectory theory was the only widely used ballistics theory before 1742. He presented it in *Two New Sciences* (1638), one of the key works of the scientific revolution. While often perceived to have no practical value because it neglects air resistance, Galileo's theory was nevertheless valid for certain gunnery problems. For heavy mortar shells fired at low speeds, the air-resistance force is too small to decelerate the projectile significantly during its relatively short flight.[7] Galileo admitted that his neglect of air resistance made his

[7]Paul-Lawrence Rose, "Galileo's Theory of Ballistics," *British Journal of Science* 4, no. 14 (1968): 156–59.

theory too inaccurate for high-velocity or "violent" shots. Nevertheless, he argued that his theory was valid for low-velocity mortar shells:

> This excessive impetus of violent shots can cause some deformation in the path of a projectile, making the beginning of the parabola less tilted and curved than its end. But this will prejudice our Author little or nothing in practicable operations, his main result being the compilation of a table of what is called the "range" of shots, containing the distances at which balls fired at (extremely) different elevations will fall. Since such shots are made with mortars charged with but little powder, the impetus is not supernatural in these, and the (mortar) shots trace out their paths quite precisely.[8]

While the vocabulary Galileo used sounds archaic, he demonstrated a qualitative comprehension of the effects of air resistance: it deforms the parabolic trajectory at high speeds but is negligible at low velocities. To suggest that one of Galileo's motivations for studying the dynamics of falling bodies was to solve a key problem in military technology may sound cynical. Nevertheless, it is important to remember the extent to which war dominated early modern European life. As Henry Guerlac wrote in his pioneering dissertation, "Science and War in the Old Régime":

> Those who doubt that war and pure science can ever be bedfellows would do well to consider the career and works of Galileo. Living in one of the most war-torn periods in European history, he saw during the course of his long life the civil wars of France, the struggle of the Dutch against their Spanish masters, the defeat of the Armada of Philip, and nearly the whole course of the Thirty Years' War. His work felt the impact of war, and he was no wise averse to capitalizing on the usefulness to the soldier of certain of his discoveries. His teaching, his inventions and his theoretical studies all reveal this tendency.[9]

The validity of the parabolic theory for low-velocity mortar shells helped convince many gunnery textbook authors in the 17th and 18th centuries to incorporate this theory.[10] After all, Western Europeans had recognized the value of uniting mathematical theory with technological practice in the education of military engineers and artillery

[8]Galileo Galilei, *Two New Sciences,* trans. Stillman Drake (Madison, Wisc., 1974), p. 229.

[9]Henry Guerlac, "Science and War in the Old Régime" (Ph.D. diss., Harvard University, 1941), p. 58.

[10]The most famous examples are Nicholas François Blondel, *L'art de jeter les bombes* (Amsterdam, 1669); Surirey de Saint Rémy, *Mémoires d'artillerie* (Amsterdam, 1702).

commanders since the 16th century.[11] The anonymous editor of *The Compleat Gunner* (first published in 1672), a collection of English translations of the most important artillery literature of that age, included a description of Galileo's projectile theory as developed by Evangelista Torricelli, as well as Marin Mersenne's account of his own ballistics experiments.[12] Bernard Forest de Bélidor, the prominent French military engineer, used Galileo's theory to derive extensive range tables for mortar fire in *Le bombardier français* in the 1730s.[13] To use these tables, a gunnery officer first had to determine the shell's range when fired at an angle of 15 degrees with a specific quantity of gunpowder. He then had to find the particular range table where the range for 15 degrees matched his shell's observed range. This range table would then provide the officer with the mortar's range at other elevation angles, provided he maintained the same gunpowder charge. A French military commission, as described in *Le bombardier français,* verified the utility of Galileo's theory for heavy mortar shells fired at low speeds and short ranges (less than 600 yards).[14] Numerous problems existed with maintaining uniform artillery hardware and gunpowder quality, as well as consistent aiming techniques, during the 17th century. Such problems often made it difficult for bombardiers to maintain the consistency assumed in Galileo's theory. Such limitations, however, far from weakening Galileo's influence on early modern gunnery, inspired efforts to improve the consistency of artillery fire in order to take full advantage of his theory's power.[15] Such work was reflected in the 18th-century military interest in mechanical uniformity and interchangeable parts, which culminated in the French artillery reforms of Jean Baptiste Vaquette de Gribeauval.

Although artillery officers did use Galileo's ballistics theory for particular problems, general ballistics theories valid for all artillery

[11]See William Bourne, *The Arte of Shooting in Great Ordnance* (Amsterdam, 1969), originally published in 1587. Also see Martha D. Pollak, *Military Architecture, Cartography and the Representation of the Early Modern European City* (Chicago, 1991).

[12]*The Compleat Gunner in Three Parts* (Yorkshire, 1971), pt. 3, pp. 3–75.

[13]B. Forest de Bélidor, *Le bombardier français, ou nouvelle méthode de jeter les bombes avec précision* (Paris, 1731).

[14]Ibid., pp. 15–18.

[15]The practical difficulty of shooting mortar shells consistently was addressed in de Resson, "Méthode pour tirer les bombes avec succès," *Mémoires de l'Académie Royale des Sciences* (Paris, 1716), pp. 79–86. Only by following the rigorous loading and firing techniques, outlined in this paper, could the necessary consistency be maintained to utilize the parabolic theory, according to de Resson. Perhaps some soldiers listened to de Resson's advice on loading mortars, because in 1717 at the Siege of Belgrade a confident young Polish bombardier literally bet his head to Prince Eugene that he could knock out the Turkish magazine with only three mortar shells. The first two missed, but the third succeeded, with horrendous consequences for the Turks. See C. Duffy, *Fire and Stone: The Science of Fortress Warfare, 1660–1860* (Newton Abbot, U.K., 1975), pp. 122–23.

pieces eluded the great natural philosophers and mathematicians of the scientific revolution. Huygens, Newton, and Johann Bernoulli attempted to analyze projectile motion in a resisting medium.[16] They failed to improve on Galileo's theory for solving gunnery problems for the following reasons. The basic differential equations of projectile motion in the atmosphere are nonlinear and do not have an exact solution.[17] Had these mathematicians succeeded in numerically integrating these equations or deriving linear approximations, their results would have remained worthless for gunnery. These equations also contain two numerical parameters that no one had accurately measured before Robins. These are the projectile's initial or muzzle velocity and the air resistance (the aerodynamic drag acting on a projectile in flight).[18] Without valid numbers for these parameters, the differential equations can yield analytical solutions that may be mathematically interesting but remain worthless for making quantitative scientific predictions. Galileo's idealized ballistics theory therefore remained the only useful ballistics theory for gunnery until the ballistics revolution. It had a similar appeal that basic neoclassical microeconomics theories have today: even though their direct applicability is limited, the rational thinking these theories stimulate is very powerful.

In the early 1950s, A. Rupert Hall made an influential argument that ballistics was useless for gunnery before the 19th century. One reason, he claimed, was that smoothbore artillery was too inaccurate to make a mathematical prediction of its performance possible. He wrote that "the gun itself was so inconsistent in its behavior that great accuracy in preliminary work, even in the lay of the gun itself, was labour in vain."[19] Hall concluded that "it was the engineering ingenuity of the nineteenth century, not the progressive elaboration of dynamical theories originating with Galileo and developed by Newton, that was responsible for the revolution which then, at last occurred, with the introduction of scientific ballistics to gunnery."[20] This idea is flawed regarding both the

[16]See Charbonnier (n. 4 above), chap. 3; A. Rupert Hall, *Ballistics in the Seventeenth Century* (Cambridge, 1952), chaps. 5 and 6; and Szabó (n. 4 above), pp. 199–211, for an analysis of these natural philosophers' ballistics analysis.

[17]A nonlinear differential equation refers to the lack of the additive property, $f(x + z) \neq f(x) + f(z)$, with respect to the dependent variable.

[18]Huygens and Newton demonstrated that air resistance was proportional to the projectile's velocity squared, but they did not establish a numerical proportionality factor appropriate for military projectiles.

[19]Hall, p. 55.

[20]Ibid., p. 71. Hall slightly modified his opinion that ballistics was not useful for gunnery before the 19th century in "Gunnery, Science and the Royal Society," in *The Uses of Science in the Age of Newton,* ed. John Burke (Berkeley, 1983). For example, he stated that "existing military art was incapable of adopting a mathematical theory of projectile flight and

timing and the cause of the ballistics revolution. The revolution was caused, in part, by the engineering ingenuity of the *18th* century *and* by the scientific revolution of Galileo and Newton.

During the 18th century, numerous technological innovations succeeded in increasing the efficiency and precision of artillery fire.[21] The artillery reforms of Austria's Prince Joseph Wenzel von Lichtenstein and France's General Gribeauval that began in the 1740s and 1760s, respectively, depended on numerous innovations to improve the precision of artillery fire. These innovations include Jean de Maritz's boring machine,[22] various aiming instruments including tangent sights, and accurate screws to control the cannon's elevation angle.[23] One indication of such reformed artillery's precision was the French artillerists' boast, at the Battle of Yorktown at the end of the American Revolutionary War, that they could shoot six consecutive shots through a small opening in the British fortifications.[24] B. P. Hughes has measured the accuracy of this artillery, determining that a 12-pound cannon could hit a 6-foot screen 100 percent of the time at 600 yards, 26 percent of the time at 950 yards, and only 15 percent of the time at 1,300 yards.[25] While it remains difficult to determine precisely the accuracy of late-18th-century smoothbore artillery, it was certainly consistent enough to make scientific gunnery calculations useful, not to mention being accurate enough to cause almost half the allied combat casualties suffered during the Napoleonic Wars.[26]

Benjamin Robins

Benjamin Robins, the central figure of the ballistics revolution, has received little notice from professional historians.[27] This contrasts with

---

applying it to practice, nor [so far as one can tell] did it ever attempt to do so, at least before the death of Newton. Therefore, if the learned men developed the theory out of a desire to solve useful problems, they were mistaken" (p. 116).

[21]William H. McNeill, *The Pursuit of Power* (Chicago, 1982), pp. 167–74.

[22]These machines were designed to bore solid-cast cannon accurately enough to achieve consistently straight bores and to minimize windage, the difference between the diameter of the cannon bore and ball. Similar machines were adopted by John Wilkinson to machine James Watt's steam engine pistons and by Count Rumford to disprove the caloric theory of heat.

[23]C. Duffy, *The Army of Maria Theresa: The Armed Forces of Imperial Austria, 1740–1780* (Vancouver and London, 1977), pp. 106, 112.

[24]North Callahan, *George Washington: Soldier and Man* (New York, 1972), p. 243.

[25]B. P. Hughes, *Firepower* (London, 1974), pp. 36–38.

[26]Robert O'Connell, *Of Arms and Men* (New York, 1989), pp. 178–79.

[27]William Johnson, a prominent research engineer in impact mechanics, has done much to reverse this neglect with his recent biographical research on Robins. John Nef is one of the few historians to consider seriously Robins's military significance, in *War and Human Progress* (n. 3 above), while Seymour Mauskopf is the first to consider Robins's significance

his high reputation in the 18th century.[28] He initially attracted attention in mathematics.[29] In 1727, at the age of twenty, he published a demonstration of the last proposition of Newton's *Treatise of Quadratures*[30] in the *Philosophical Transactions of the Royal Society.* He published a refutation of Johann Bernoulli's impact theory the following year. Elected as a fellow to the Royal Society, Robins became a private mathematics tutor to prospective university students. Although he remained active in pure mathematics, a serious interest in military engineering quickly developed. In the 1730s, Robins began a study of fortifications, hydraulics, and ultimately ballistics.[31] He received the Royal Society's Copley Medal in 1747 for his work in ballistics. The depth of Robins's engineering ambition is also reflected in his trips to study Dutch fortifications and his invitation by the prince of Orange to assist in the Dutch defense of Bergen op Zoom in 1747. The French siege of this large fortress was a key conflict in the War of the Austrian Succession. Unfortunately for Robins's engineering career, the fortress fell shortly after his arrival.

During the mid-1730s, Robins demonstrated his mathematical abilities in a debate with James Jurin concerning the nature of limits and infinitesimals in the theory of fluxions. Their argument was ignited by Bishop Berkeley's famous critique of the logical foundations of Newtonian mathematics.[32] Robins also criticized Euler's use of Gottfried Wilhelm Leibniz's differential calculus, which increased his reputation as a staunch defender of Newton. Nevertheless, Robins did not hesitate to describe Newton's errors regarding air resistance. Active in politics as well, Robins wrote a number of political pamphlets in support of the Tories, as well as serving on various political committees. Sir Robert

---

for the history of chemistry, in "Gunpowder and the Chemical Revolution," *Osiris,* 2d ser. (1988), pp. 96–97.

[28]See Leonhard Euler's praise of Robins in H. Brown's English translation of Euler's critique and German translation (*Neue Grundsätze der Artillerie* [Berlin, 1745]) of Robins's *New Principles of Gunnery* titled *The True Principles of Gunnery Investigated and Explored* (London, 1777), p. 49; Charles Hutton's praise in *Tracts on Mathematical and Philosophical Subjects* (London, 1812), tract 34, 2:307; and Pringle's praise, delivered on the occasion of awarding the Copley Medal to Charles Hutton (n. 2 above).

[29]W. Johnson, "Benjamin Robins: New Details of His Life," *Notes and Records of the Royal Society of London* 46, no. 2 (London, 1992): 235–52.

[30]Isaac Newton, *Two Treatises of the Quadrature of Curves, and Analysis of Equations of an Infinite Number of Terms,* explained by John Stewart (London, 1745).

[31]Wilson wrote a short biography on Robins in his introduction to *Mathematical Tracts of the Late Benjamin Robins* (n. 6 above), the main source of the biographical information provided here.

[32]For a description of this controversy, see F. Cajori, *Conceptions of Limits and Fluxions in Great Britain, from Newton to Woodhouse* (Chicago, 1919), pp. 96–148; and Niccolò Guicciardini, *The Development of Newtonian Calculus in Britain* (Cambridge, 1989), pp. 45–46.

Walpole and his Whig supporters were sufficiently antagonized by Robins's critiques that they blocked his application for the professorship of mathematics at the recently established Royal Military Academy at Woolwich in 1741. Robins published *New Principles of Gunnery* in 1742, in part as a response to this rejection.[33] His literary career flourished with the publication of *A Voyage Around the World in the Years 1740–1744 by George Anson, Esq.*, a book Lord Anson commissioned him to ghostwrite.[34] A naval gunnery reformer, Anson aided Robins in his ballistics research.[35] Robins died of a fever in 1751 in India, where he had gone to serve as engineer general and captain of the Madras artillery for the East India Company. In addition to commanding the artillery batteries, he redesigned Fort St. David as a part of the East India Company's military buildup against the French after the War of the Austrian Succession.[36]

Robins's single most influential accomplishment was his invention of the ballistics pendulum, the first reliable instrument that measured projectile velocity.[37] (See fig. 1.) The ballistics pendulum, coupled with a sophisticated knowledge of mathematics and mechanics and a talent for experimentation, gave Robins the means to address fundamental experimental and theoretical ballistics questions. Like any pioneering effort, his work was imperfect, yet it provided a rigorous scientific foundation for ballistics. Robins's interest also extended to artillery design issues. In his article "Practical Maxims relating to the Effects and Management of Artillery, and the Flight of Shells and Shot," he argued for the tactical benefits of decreasing artillery weight and lowering gunpowder charges.[38] These arguments were remarkably similar to those adopted by General Gribeauval in his efforts to reform the French artillery following France's humiliating defeat in the Seven Years' War (1756–63).[39] Robins also published "A Proposal for Increasing the

[33]Robins, *New Principles of Gunnery* (n. 6 above).

[34]For a discussion of the controversy surrounding Robins's work on this book, see W. Johnson, "Benjamin Robins: Two Essays: Sir John Cope's Arraignment and Lord Anson's *A Voyage Round the World*," *International Journal of Impact Engineering* 11, no. 1 (1991): 121–34.

[35]Peter Padfield, *Guns at Sea* (New York, 1974), p. 102.

[36]For a discussion of Robins's life in India, see W. Johnson, "Benjamin Robins (1707–1751): Opting Not to Be a Commissary for Acadia but a Fortifications Engineer in East India," *International Journal of Impact Engineering* 9, no. 4 (1990): 503–25, "In Search of the End of the Life, in India, of Benjamin Robins, F.R.S," *International Journal of Impact Engineering* 11, no. 4 (1991): 547–71, and "Benjamin Robins: New Details of His Life" (n. 29 above).

[37]For a study of the attempts before Robins to use a pendulum to study ballistic motion, see W. Johnson, "The Origin of the Ballistic Pendulum: The Claims of Jacques Casini and Benjamin Robins," *International Journal of Mechanical Science* 32, no. 4 (1990): 345–74.

[38]Benjamin Robins, *New Principles of Gunnery* (London, 1805), pp. 245–78.

[39]For a historical account of the Gribeauval artillery reforms, see Howard Rosen, "The Système Gribeauval: A Study of Technological Development and Institutional Change in Eighteenth Century France" (Ph.D. diss., University of Chicago, 1981).

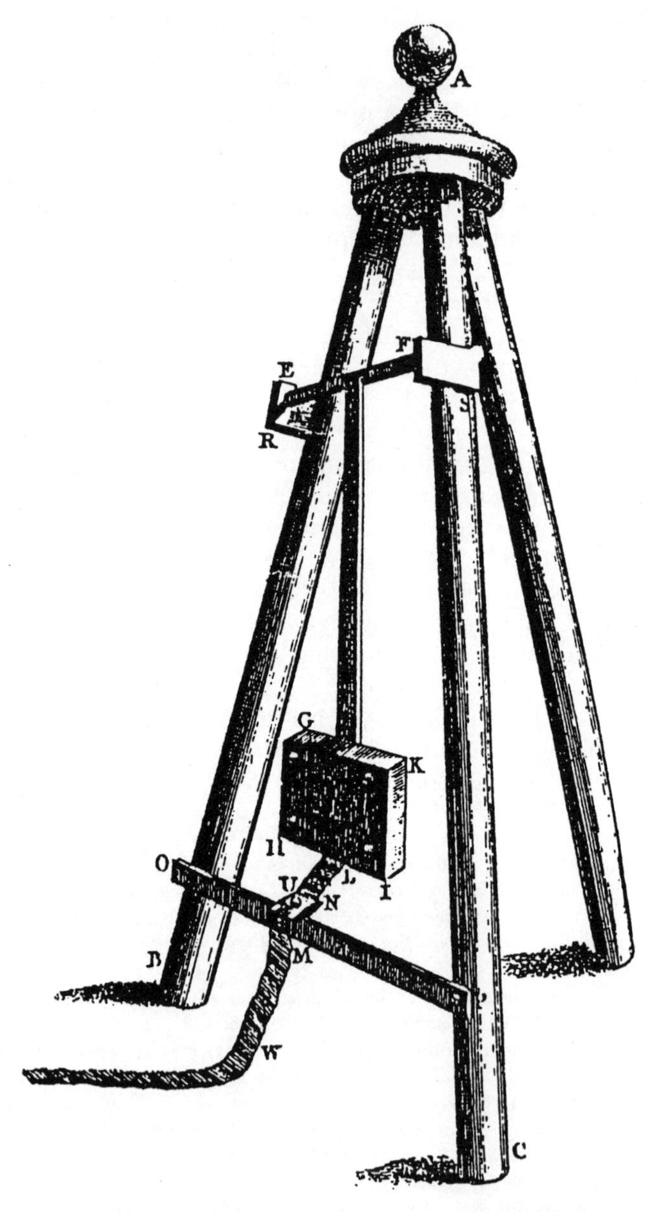

Fig. 1.—Benjamin Robins's 1742 ballistics pendulum. (Benjamin Robins, *Mathematical Tracts of the Late Benjamin Robins,* ed. James Wilson [London, 1761], 1:89.)

Strength of the British Navy," in which he argued for the effectiveness of a high-caliber, lightweight cannon in naval warfare.[40] This proposal helped inspire the invention of the carronade in the 1770s, a weapon the British Navy effectively used during the Napoleonic Wars.[41]

Robins's Interior Ballistics

Robins devoted the first half of *New Principles of Gunnery* to the basic question of interior ballistics: what is a projectile's muzzle velocity as a function of its mass, gunpowder quantity, and barrel geometry? Building on the work of Daniel Bernoulli, who analyzed theoretically the muzzle velocity of a gun fired with compressed air,[42] and on the experimental work of Francis Hauksbee and Stephen Hales, Robins pursued this question by measuring the necessary empirical parameters, setting up the equations of motion, performing the mathematical analysis, and comparing his results with experimental observation. Robins first assumed that gunpowder is instantly transformed into an elastic fluid or compressible gas when ignited. Using high-quality military powder, he established empirically a relationship between gunpowder mass and the quantity of gas generated from the explosion. He also measured this gas's pressure at his estimation of the explosion temperature.[43] With these empirical relationships, Boyle's law (the isothermal law relating gas pressure to its volume), and the thirty-ninth proposition of book 1 of Newton's *Principia,* Robins obtained a solution.[44] He showed that the area beneath the pressure-volume curve representing the gas pressure that pushes the projectile through the barrel (the "work" performed by the gas on the projectile) is equal to the projectile's "kinetic energy" at the muzzle, in the modern definition of those terms.[45] (See fig. 2.) He then easily solved for the bullet's muzzle velocity. The work performed by the gas on the projectile, as Robins showed, depended on the initial gas pressure, which in turn depended on gas temperature—an early recognition of the connection between

[40]Benjamin Robins, "A Proposal for Increasing the Strength of the British Navy," in *New Principles of Gunnery* (n. 38 above), pp. 283–94.

[41]F. L. Robertson, *The Evolution of Naval Armament* (London, 1921), p. 126. Also see John E. Talbott, "The Rise and the Fall of the Carronade," *History Today* 39 (August 1989): 24–30, for a description of this military innovation.

[42]Daniel Bernoulli, *Hydrodynamics* (1738), trans. T. Carmody and H. Kobus (New York, 1968), pp. 264–74. Bernoulli showed that the "work" done by expanding air is equal to the "kinetic energy" of the projectile at the muzzle.

[43]Robins (n. 6 above), p. 70.

[44]The thirty-ninth proposition of book 1 of the *Principia* is an analysis of a body moving under the influence of centripetal forces. See Isaac Newton, *Mathematical Principles* (Berkeley, 1934), 1:125–27.

[45]Robins (n. 6 above), p. 76.

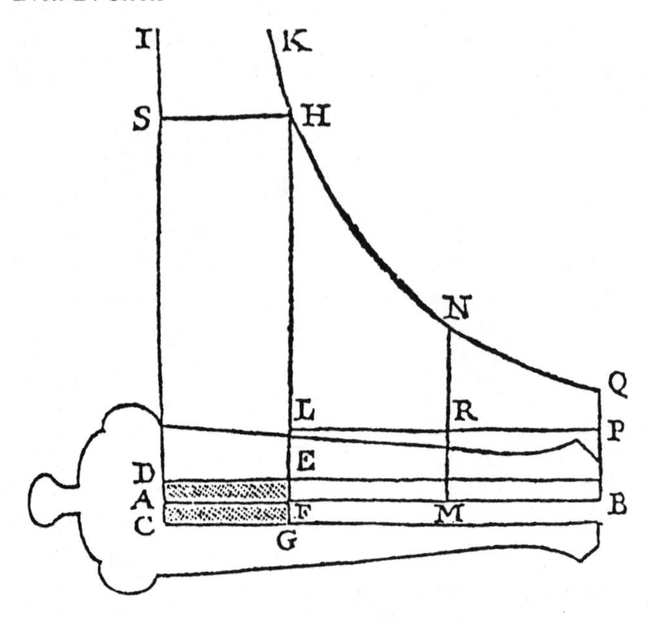

Fig. 2.—Robins's force-versus-displacement diagram, where lines *FH, MN,* and *BQ* represent the force acting on the projectile at the beginning, middle, and end of the barrel, respectively. Robins showed that the area *FHQB* is equal to the projectile's "kinetic energy" at point *B.* The area *DCGE* represents the volume of gunpowder. (Benjamin Robins, *Mathematical Tracts of the Late Benjamin Robins,* ed. James Wilson [London, 1761], 1:75.)

heat and mechanical work.[46] In spite of his numerous assumptions, Robins succeeded in demonstrating his theory's utility for muskets. He compared a musket's theoretical muzzle velocity with experimental observation by using the ballistics pendulum and achieved astonishingly close results.[47]

The ballistics pendulum is a simple instrument to build, but complex to use. It proved to be a revolutionary scientific instrument in the 18th century, because it allowed Robins and his successors to quantitatively measure both muzzle velocity and (by moving the pendulum at progressively greater distances from the gun) the air resistance of a projectile, the two fundamental parameters in the differential equations of ballistic motion. It was used by such prominent 18th-century ballistics researchers as Patrick D'Arcy, General Alessandro Vittorio Papacino D'Antoni, Hutton, and Count Rumford to investigate increasingly

[46]For a closer look at the mathematical analysis in this book, see W. Johnson, "Benjamin Robins' *New Principles of Gunnery,*" *International Journal of Impact Engineering* 4, no. 4 (1986): 205–19; and Barkla (n. 4 above).

[47]Robins (n. 6 above), p. 91.

heavier projectiles with greater accuracy, and it remained the most popular ballistics research instrument until the 1850s.[48] Robins's ballistics pendulum, designed only to measure musket ball velocities, was a simple pendulum consisting of a flat plate connected to a rigid bar that swung from a tripod. Robins measured a musket ball's velocity by shooting the ball into the plate and observing the amplitude of the pendulum's swing. He derived an equation of the bullet's velocity just before impact as a function of the ballistics pendulum's swing angle. This calculation might seem elementary today, but it was sophisticated in the 1740s. It required insight into the conservation of linear momentum, angular momentum, and the dynamics of pendulums and falling bodies.[49]

Robins's theoretical analysis of a projectile's muzzle velocity represents a fundamental thermodynamic analysis of an internal combustion engine. This influenced John Robison's and possibly Sadi Carnot's subsequent thermodynamic analysis of the steam engine. Donald Cardwell credits Davies Gilbert with discovering in the 1790s that the area beneath a pressure-volume curve for a steam engine is equal to the work done by the steam against the cylinder.[50] While this claim may be valid for the steam engine, Robins's and Daniel Bernoulli's analysis shows that this knowledge was well established in the middle of the 18th century for another heat engine, the gun. Whether Davies Gilbert read *New Principles of Gunnery* is not known, but John Robison certainly did.[51] Robison was the author of the 1797 *Encyclopaedia Britannica* article on the steam engine where he essentially repeated Robins's analysis of the work done by expanding elastic fluids in a gun and then applied it to the steam engine.[52] He may have first read about it during his service in the Royal Navy during the Seven Years' War. It is also quite possible that Sadi Carnot, the founder of modern thermodynamics and a graduate of the École Polytechnique and the École d'Application de l'Artillerie et du Génie at Metz, was familiar with Robins's analysis. *New*

[48]Patrick D'Arcy, "Mémoire sur la théorie de l'artillerie, ou sur les effets de la poudre, et sur les conséquences qui en résultent par rapport aux armes à feu," in *Mémoires de l'Académie Royale des Sciences* (Paris, 1751), pp. 45–63; Benjamin Thompson, "New experiments upon gun-powder with occasional observations and practical inferences; to which are added, an account of a new method of determining the velocities of all kinds of military projectiles," *Philosophical Transactions of the Royal Society* (London, 1781), pp. 229–328; Charles Hutton, "The force of fired gun-powder and the initial velocities of cannon balls, determined by experiments," *Philosophical Transactions of the Royal Society* (London, 1778), pp. 50–85. Also see Hutton (n. 28 above), vols. 2 and 3.

[49]See C. Truesdell, *Essays in the History of Mechanics* (New York, 1968), pp. 239–71, for an essay on the history of rigid body dynamics.

[50]Donald Cardwell, *From Watt to Clausius* (Ithaca, N.Y., 1971), p. 79.

[51]John Robison, *A System of Mechanical Philosophy* (Edinburgh, 1822), 1:193–203.

[52]Cardwell, p. 81.

Principles of Gunnery was widely read by the technical officers of that age.[53] The full significance of Robins's thermodynamic analysis remains uncertain, however. Guillaume Amontons used the expansion of heated air to design a theoretical engine well before Robins.[54] Nevertheless, Robins appears to be the first to analyze the work performed by an internal combustion engine and to confirm his calculations by using rigorous experimental methods.

Robins's Exterior Ballistics

Benjamin Robins's principal fame in the 18th century rested on his investigations of exterior ballistics or the mechanics of projectiles in free flight. By using the ballistics pendulum, he discovered the enormous and complex air-resistance forces acting on high-speed projectiles. These forces could initially be as high as 120 times the musket bullet's weight, a surprise to the scientific and engineering communities of the day. Conventional wisdom generally held that air resistance only slightly modified the parabolic trajectory. To validate his claims, Robins gave a demonstration of these air-resistance measurements before the Royal Society, showing how grossly distorted high-speed trajectories were from a parabola. According to the parabolic theory, a 24-pound solid shot fired from a cannon could reach 16 miles, should its actual initial velocity be used.[55] In practice, its maximum range was less than 3 miles because of air resistance. This disparity could not have been determined by simply observing projectile ranges. In Galileo's theory, the projectile's muzzle velocity is deduced from its observed range and elevation angle. The range, however, is affected by air resistance. For high-speed projectiles, this resistance is sufficient to cause a wide discrepancy between such deduced and actual muzzle-velocity values.

Newton and Huygens argued that air resistance was proportional to the square of the projectile's velocity. Robins showed that this was valid only at lower speeds. At velocities greater than 1,100 feet per second, approximately the speed of sound, he concluded that the resistance increased by a factor of three. Robins then commented that "as I have forbore to mix any hypothesis with the plain matters of fact deduced from experiment, I did not therefore animadvert on this remarkable circumstance, that the velocity, at which the moving body shifts its resistance, is nearly the same, with which sound is propagated through the air.... But the exact manner, in which the greater and lesser

[53]Even in the United States, much of *New Principles of Gunnery* was published in vol. 8 of the *Encyclopaedia; or, a Dictionary of Arts, Sciences and Miscellaneous Literature,* ed. T. Dobson (Philadelphia, 1798). See pp. 200 and 201 for Robins's thermodynamic analysis.

[54]Cardwell, p. 19.

[55]Robins (n. 6 above), p. 142.

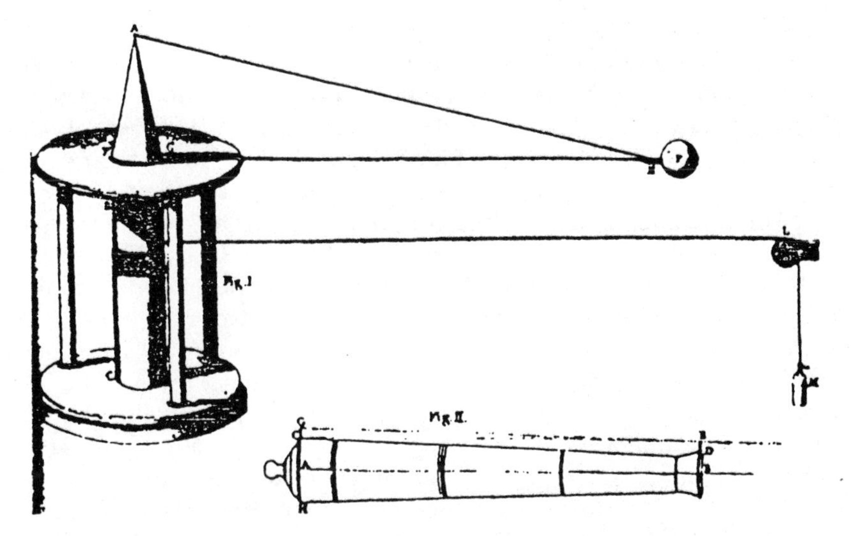

FIG. 3.—Robins's whirling arm. (Benjamin Robins, *Mathematical Tracts of the Late Benjamin Robins,* ed. James Wilson [London, 1761], vol. 1, between pp. 340 and 341.)

resistances shift into each other, must be the subject of farther experimental enquiries."[56] While it would be an exaggeration to suggest that Robins discovered the "sound barrier" in the modern sense of the word, he certainly made the first observations of the phenomenon while conducting the first supersonic aerodynamics experiments.[57]

In order to measure projectile air resistance at velocities too low for the ballistics pendulum, Robins invented the whirling arm (fig. 3). This consisted of a pivoted arm that rotated in the horizontal plane and was powered by a falling weight. The test object was placed at the outer end of the arm. While the results were less complicated to interpret than the ballistics pendulum, effective interpretation of its data also required a knowledge of mechanics.[58] Robins equated the product of the falling weight and its steady-state velocity (the "power" generated by the falling weight) with the "power" dissipated by the arm and projectile in the air. He then solved for the air-resistance force acting on the projectile alone. The whirling arm became the most widely used aeronautical research instrument until the development of the wind tunnel in the late 19th century.[59] John Smeaton adopted it for his windmill power measure-

[56]Ibid., p. 182.

[57]Hugh Dryden, "Supersonic Travel within the Last Two Hundred Years," *Scientific Monthly* (May 1954): 289–90.

[58]Robins (n. 6 above), p. 203.

[59]J. L. Pritchard, "The Dawn of Aeronautics," *Journal of the Royal Aeronautical Society* 61 (1957): 438.

ments.[60] Charles de Borda redesigned it to measure water-resistance forces.[61] George Cayley, the conceptual inventor of the airplane, used the whirling arm to measure both lift and drag forces acting on airfoils. F. W. Wenham, Otto Lilienthal, Horatio Phillips, W. H. Dines, Hiram Maxim, and S. P. Langley continued to develop the whirling arm for experimental airfoil analysis at the end of the 19th century.

With the whirling arm, Robins confirmed that the air-resistance force acting on subsonic spheres is proportional to its velocity squared. He also investigated the behavior of different geometries. Robins demonstrated that two different objects with the same surface area (a pyramid and an inclined plane) moving at the same velocities generated different air-resistance forces. Different air-resistance values also resulted when the short and the long side of an inclined plane faced the direction of motion at the same speeds. These results anticipated later investigations of aspect ratio in airfoils. Robins grasped some of the significance of his unexpected discovery that air resistance is not just a function of speed and surface area. He commented that "surely a matter, on the right knowledge of which all true speculations on ship-building and sailing must necessarily depend, cannot but be deemed, in this country at least, of the highest importance, both to the publick interest of the nation, and to the general benefit of mankind."[62]

Another aerodynamic property that Robins investigated was the lateral deflection of flying projectiles.[63] By firing musket balls into a series of evenly spaced tissue-paper curtains and by observing the resulting bullet holes, he demonstrated the enormous deflections of the bullet's trajectory from the initial direction of motion. At a range of 760 yards, for example, one musket ball deflected over 100 yards to the left of the gun barrel. Robins identified the spin imparted on the musket ball as it struck the musket barrel's side during firing as the cause of its deflection. He theorized that the rotation of the ball in flight disturbed the uniform flow of air past the ball. This rotation created a faster air flow on the side of the ball moving in the direction of the air flow than on the opposite side moving against it.[64] These conditions created a friction or "pressure" force that pushed the ball to the side where the air

[60]J. Smeaton, "Experimental enquiry concerning the natural powers of water and wind, to turn mills and other machines depending on circular motion," *Philosophical Transactions of the Royal Society* (London, 1759), pp. 138–74.

[61]C. de Borda, "Expériences sur la résistance des fluides," in *Memoires de l'Académie Royale des Sciences* (Paris, 1767), p. 495.

[62]Robins (n. 6 above), p. 217.

[63]For a detailed view of this aspect of Robins's work, see Barkla and Auchterlonie (n. 4 above). This phenomenon is readily observed while playing tennis or baseball.

[64]Robins (n. 6 above), pp. 207–8.

flow velocity was the greatest. To demonstrate this hypothesis, he hung a 4 ½-inch wooden ball from an 8-foot double string.[65] By winding up the string, Robins showed that the lateral deflection of the ball from its initial direction of motion coincided with the direction of spinning. The ball deflected to the side where the ball's side moved in the direction of the flowing air.

Robins offered further experimental proof of his theory with a musket.[66] In one of his experiments, he bent a fixed musket barrel a few degrees to the left. The bullet he shot had a trajectory that completely coincided with his theory. The bullet holes in the tissue-paper curtains showed that the bullet first began moving toward the left, in the direction of the musket's deflection. Eventually the bullet reversed its lateral direction of motion and crossed to the right side of the musket. Robins explained this apparent paradox by noting that the deflected musket forced the bullet to rotate from the left to the right. This would cause the air flow to be greater on the right-hand side; hence, the friction or "pressure" force pushed the projectile to the right, which its trajectory demonstrated. This phenomenon is now called the Magnus effect, named after a 19th-century German physicist who investigated it with full knowledge of Robins's work.[67] Barkla and Auchterlonie have argued that it should be renamed the "Robins effect" in recognition of Robins's prior discovery.[68]

Robins used his understanding of the "Robins effect" to explain theoretically why rifles have greater accuracy than muskets and why rifled bullets should not be spherical. He presented these ideas in a paper titled "Of the Nature and Advantage of Rifled Barrel Pieces," which he read before the Royal Society in 1747.[69] He initially critiqued the existing theories that explained the superior accuracy of rifles over smoothbore muskets, including the theory that the rotating motion of the rifles creates less air resistance because it bores through the air, just as a screw bores through wood, and that the rifled bullet receives a greater initial velocity because of its tight fit in the barrel, which causes it to fly straighter. This was utter nonsense, according to Robins. He argued that the rifled bullet's direction of forward motion coincides with its axis of rotation. Therefore, the flow of air passing around the bullet is uniform with respect to this axis. Such a uniform velocity could

[65]Ibid., p. 207.

[66]Ibid., p. 213.

[67]G. Magnus, *Über die Abweichung der Geschosse* (Berlin, 1860), p. 2.

[68]Barkla and Auchterlonie (n. 4 above), p. 438. Also see W. Johnson, "The Magnus Effect: Early Investigation and a Question of Priority," *International Journal of Mechanical Science* 28, no. 12 (1986): 859–72.

[69]Robins (n. 38 above), p. 328.

not create any lateral friction or "pressure" forces, according to Robins's theory. Robins also recognized a problem with using spherical bullets in a rifle. When the trajectory becomes curved, he feared the axis of the sphere's spin would not necessarily coincide with the direction of motion. He suggested using egg-shaped bullets to provide for the same stability seen in arrows. He then concluded this paper with an ominous prediction:

> I shall therefore close this paper with predicting, that whatever state shall thoroughly comprehend the nature and advantages of rifled barrel pieces, and, having facilitated and completed their construction, shall introduce into their armies their general use with a dexterity in the management of them; they will by this means acquire a superiority, which will almost equal any thing, that has been done at any time by the particular excellence of any one kind of arms; and will perhaps fall but little short of the wonderful effects, which histories relate to have been formerly produced by the first inventors of fire-arms.[70]

The use of smoothbore muskets as the standard infantry weapon until the second half of the 19th century was not due to an ignorance of their bullets' random trajectories but rather to the inability to manufacture desirable rifles and to the social lag that existed in military bureaucracies.[71]

Robins's and Euler's Ballistic Trajectory Analysis

Leonhard Euler was the most accomplished mathematician of the 18th century. If he was not the greatest mathematician of all time, he was certainly the most productive. It was Euler, as Clifford Truesdell convincingly argued, who first derived what we now call "Newton's Second Law of Motion."[72] Euler was born in Basel in 1707 and attended the University of Basel with the initial ambition to become a minister. Johann Bernoulli held the chair in mathematics at Basel and gave Euler private lessons. In 1727 Euler accepted a position in physiology at the newly established Saint Petersburg Academy of Sciences but soon became the leading member of the mathematics section. He was extremely productive during his first stay in Russia, which lasted for fourteen years. He worked on a wide range of engineering research and mathematical problems, including problems related to navigation,

[70]Ibid., p. 341.

[71]See Merritt Roe Smith, *Harpers Ferry Armory and the New Technology* (Ithaca, N.Y., 1977), esp. chap. 7.

[72]C. Truesdell, *An Idiot's Fugitive Essays on Science* (New York, 1984), pp. 98–101.

shipbuilding, and pumps, and made fundamental contributions to number theory, mechanics, and differential and integral calculus. It would be difficult to find a topic in the mathematical sciences of the 18th century that Euler did not work on.

Euler's productive career at the Saint Petersburg Academy of Sciences came to an end when Empress Anna died in 1740 and the position of Germans in the Russian government became hazardous. Fortunately for Euler, Frederick the Great had just become king of Prussia and was in the process of reorganizing the Berlin Academy of Sciences. Frederick offered Euler a position, which he accepted in 1741. In addition to performing numerous administrative functions, including serving as the academy's acting president after the death of Pierre-Louis Moreàu de Maupertuis in 1759 and managing numerous technical projects, Euler generated an enormous quantity of research results, including fundamental work in elastic and rigid body mechanics, fluid dynamics, lunar theory, and optics. He also worked on such engineering subjects as water-turbine and gear design in addition to numerous mathematics topics including complex variables, differential equations, and the calculus of variations. An increasingly sour relationship with Frederick the Great and a specific disagreement regarding the financial management of the academy led to Euler's return to Saint Petersburg in 1766. Even total blindness failed to restrict his productivity during his second stay in Russia, which lasted until his death in 1783.

In 1745, in response to a request from Frederick the Great, Euler translated Robins's *New Principles of Gunnery* into German. He added an extensive mathematical commentary which criticized some of Robins's assumptions, pointed out his analytical errors, and addressed many additional areas of artillery and ballistics not covered by Robins. For example, Euler made an early analysis of a pressure vessel, while investigating the theoretical strength of a gun barrel. He derived the standard equation for a pressurized cylinder with unconstrained ends where the maximum stress is the product of the internal pressure and the ratio of the cylinder's radius to wall thickness.[73] Unfortunately, he made a mathematical error and obtained an additional factor of $\pi/2$ as a consequence. Euler also erred when he denied the existence of the Robins effect, arguing instead that the deflections occurred because of imperfections in the bullet's curvature. Nevertheless, in this work he derived the first proof of d'Alembert's paradox in fluid mechanics and also conducted a pioneering mathematical analysis of "supersonic" air resistance.[74]

[73]Euler (n. 28 above), pp. 85–91.

[74]Clifford Truesdell, *Rational Fluid Mechanics, 1687–1765* (Zurich, 1954), pp. 38–41.

Another significant part of Euler's commentaries was his analysis of a ballistic trajectory, which included air-resistance effects.[75] Euler used approximate methods to simplify his analysis of the complex nonlinear differential equations of ballistic motion. Unfortunately, his solution was valid only for military projectiles flying at low velocities. Yet he did obtain a partial solution relating launch angle to range when the air resistance is small. With this solution he proved that the angle required to launch a projectile at its maximum distance was not 45 degrees, as Galileo's theory held, but depended on the air-resistance magnitude instead.

Euler did not organize this analysis into useful gunnery tables. Rather, it was Robins who provided the first tabulated solutions of the differential equations of projectile motion in the atmosphere, thus permitting easy gunnery calculations.[76] Use of this table required only a knowledge of Galileo's ballistics theory. Robins presented these results to the Royal Society in 1746, just a year after Euler published his analysis of these equations of motion in his translation of *New Principles of Gunnery.* Unfortunately, Robins's table was not published until 1761 when James Wilson published *Mathematical Tracts of the Late Benjamin Robins,* a two-volume collection of Robins's research papers and books. Robins did not describe how he derived these ballistics tables; he seemed more concerned with demonstrating their utility, which he accomplished in an article titled "A Comparison of the Experimental Ranges of Cannon and Mortars, with the Theory contained in the preceding Papers," sent to the Royal Society from India in 1750.[77] Robins showed that his ballistics table, while admittedly approximate, was sufficiently accurate for the mortar and cannon pieces of his age.

Robins demonstrated his table's accuracy by comparing its predictions to experimental observations. A 13-inch sea mortar threw a 231-pound shell to a distance of 3,350 yards when loaded with 30 pounds of gunpowder, according to the Woolwich Arsenal reports. Robins's calculation of this range, after accounting for the change in atmospheric pressure, was 3,230 yards—a difference of 120 yards. For a 10-inch mortar firing a 96-pound shell to a distance of 3,350 yards, Robins's calculation was off by 160 yards. Robins pursued his argument by comparing his results with measurements of cannon ranges, as published by authors such as Surirey de Saint Rémy. He showed that, for a cannon firing a 24-pound shot at angles from 4 to 40 degrees and producing ranges from 820 to 2,050 toises (1 toise = 2 yards), respectively, the

[75]Euler (n. 28 above), pp. 294–320.
[76]Robins (n. 6 above), p. 179.
[77]Robins (n. 38 above), p. 231.

error of his theory varied from 46 to 122 toises. After comparing his calculations to a series of French experiments made at Metz in 1740, Robins concluded, that "Our theory differs less from the experiments, than the experiments do from each other."[78] For example, a 24-pound cannon effectively elevated to 5 degrees, loaded with 10 pounds of gunpowder, and fired repeatedly by the French had ranges of 834, 872, 851, 845, 871, and 838 toises. Robins's table for the same situation gives 850 toises. In spite of the approximate nature of this ballistics analysis, especially when Robins took into consideration the changes in air resistance due to atmospheric pressure changes, it represents the first adequate calculation of high-speed projectile range.

In 1753 Euler published the first complete analysis of the equations of ballistic motion in the atmosphere.[79] His basic assumption was that the projectile's air resistance was proportional to its velocity squared. Using Robins's quantitative measurements of muzzle velocity and air resistance, the key parameters essential for achieving any numerical results, Euler provided a fundamental method for handling this complex mechanics problem, a problem perhaps as difficult as the notorious three-body problem in celestial mechanics.[80] Both phenomena are nonlinear without an exact mathematical solution. For the ballistics problem, Euler numerically integrated the equations representing the trajectory's range, altitude, time, and velocity by using the trapezoidal rule.[81] He decided to divide the trajectories into sets of families or species to perform this integration. Each species represents those trajectories whose asymptotes of the ascending section are the same when the trajectory becomes infinitely extended. In other words, each species represents particular combinations of muzzle velocity and elevation angle. Euler calculated the ballistics table for a particular species as an example. With it he could determine velocity, range, maximum altitude, and flight time for a projectile fired at certain muzzle velocities and elevation angles. In 1764, Henning Friedrich, Graf von Graevenitz, a German infantry officer, provided a complete set of ballistics tables by calculating eighteen species of trajectories using

[78]Ibid., p. 239.

[79]L. Euler, "Recherches sur la véritable courbe que décrivent les corps jetés dans l'air, ou dans un autre fluide quelconque," *Mémoires de l'Académie de Berlin* (Berlin, 1753), pp. 321–52. H. Brown translated this paper and included it in *True Principles of Gunnery Investigated and Explored* (n. 28 above), pp. 322–66.

[80]For additional perspectives on Euler's ballistics work, see Szabó (n. 4 above), pp. 211–20; as well as Charbonnier (n. 4 above), pp. 118–41.

[81]The trapezoidal rule is a technique to integrate a function numerically or to find the area beneath a curve by approximating the area with a series of trapezoids. Integration is then a matter of calculating the area of each trapezoid and adding them up.

Euler's method.[82] (See fig. 4.) These ballistics tables proved to be especially valid for mortar fire in the 18th century.[83] (They remained in use at least until World War II for calculating ballistics tables for low-velocity and high-angle mortar fire.)[84] Numerous mathematicians and engineers expanded on Euler's work to analyze more complex ballistic trajectories and to derive simpler approximate theories. In the 18th century, these researchers included Johann Heinrich Lambert, de Borda, G. F. Tempelhoff, and Adrien-Marie Legendre. Vannevar Bush's differential analyzer and the ENIAC computer were developed in part to compute solutions of ballistics equations for high-speed and long-range artillery during the first half of the 20th century.

The Military Response to Robins's and Euler's Ballistics

The new scientific ground opened by Robins's and Euler's revolution in ballistics was rapidly utilized by the European military establishments. Euler's initial motivation to translate *New Principles of Gunnery* was provided by Frederick the Great, who wanted Euler to translate into German the best artillery theory book available in order to increase his artillery officers' competence in gunnery.[85] The French soon took notice of these Prussian and English ballistics developments.[86] *New Principles of Gunnery* was originally translated into French in 1751 by Jean-Baptiste Le Roy. According to Truesdell, Anne Robert Jacques Turgot wrote to Louis XVI in 1774 that "the famous Leonard Euler, one of the greatest mathematicians of Europe, has written two works which could be very useful to the schools of the navy and the artillery. One is a Treatise on the construction and Maneuver of Vessels; the other is a commentary on the principles of artillery of Robins. . . . I propose that Your Majesty order these to be printed."[87] It was 1783 before Jean-Louis Lombard translated Robins and Euler into French. The ballistics revolution offers

[82]H. F. Graevenitz, *Akademischen Abhandlung von der Bahn der Geschützkugeln* (Rostock, 1764).

[83]Jean-Louis Lombard, *Traité du mouvement des projectiles, appliqué au tir des bouches à feu* (Dijon, 1797), pp. 201–22.

[84]E. McShane, J. Kelley, and F. Reno, *Exterior Ballistics* (Denver, 1953), p. 258.

[85]See the introduction to Leonhard Euler, *Neue Grundsätze der Artillerie* (see n. 28), written by F. R. Scherrer for Euler's *Opera Omnia,* 2d ser., vol. 14 (Leipzig and Berlin, 1922). Also see Euler's letter to Frederick the Great regarding Robins's work and his translation plans, in *Correspondance de Leonhard Euler* (Basel, 1986), 4:309.

[86]For an overview of the relationship between science and warfare in 18th-century France, see Charles Gillispie, *Science and Polity in France at the End of the Old Regime* (Princeton, N.J., 1980); and Roger Hahn, "L'enseignement scientifique aux écoles militaires et d'artillerie," in *Écoles techniques et militaires au XVIII siècle,* ed. Roger Hahn and René Taton (Paris, 1986), pp. 514–45. The most extensive analysis of this relationship remains Guerlac's dissertation (n. 9 above).

[87]Truesdell (n. 72 above), p. 337.

33

Die VIII. Art, $\gamma = 35°$, aufsteigender Bogen.

Erhöh. Winkel β	Bogen AG =2, 302585c mult. mit	Weite AF =2, 302585c mult. mit	Höhe FG =2, 302585c mult. mit	Geschwind. in G $=\sqrt{2agc}$ mult. mit
0°	0, 0000000	0, 0000000	0, 0000000	1, 1517744
5°	0, 0536504	0, 0535931	0, 0023402	1, 2298471
10°	0, 1164065	0, 1158121	0, 0105315	1, 3372699
15°	0, 1935934	0, 1911693	0, 0272378	1, 4901162
20°	0, 2952450	0, 2881162	0, 0578050	1, 7219160
25°	0, 4441013	0, 4256416	0, 1147698	2, 1190543
30°	0, 7134214	0, 6645303	0, 2391278	3, 0237457

Niedersteigender Bogen.

Winkel in H	Bogen AH =2, 302585c mult. mit	Weite AE =2, 302585c mult. mit	Höhe EH =2, 302585c mult. mit	Geschwind. in H $=\sqrt{2agc}$ mult. mit
5°	0, 0477456	0, 0477002	0, 0020826	1, 0943354
10°	0, 0916889	0, 0912676	0, 0078183	1, 0523529
15°	0, 1334123	0, 1320020	0, 0168489	1, 0225533
20°	0, 1741257	0, 1708310	0, 0290916	1, 0030180
25°	0, 2149125	0, 2085131	0, 0447000	0, 9923740
30°	0, 2568288	0, 2456933	0, 0640548	0, 9894922
35°	0, 3010101	0, 2829554	0, 0877934	0, 9942320
40°	0, 3487869	0, 3224633	0, 1181088	1, 0062590
45°	0, 4008361	0, 3608380	0, 1532727	1, 0267250
50°	0, 4624001	0, 4024300	0, 1986624	1, 0521799
55°	0, 5336408	0, 4457985	0, 2551789	1, 0862895
60°	0, 6202174	0, 4923160	0, 3281968	1, 1279210
65°	0, 7295507	0, 5418005	0, 4251766	1, 1766132

E Die

FIG. 4.—One of Graevenitz's ballistics tables based on Euler's analysis of projectile motion where air resistance is considered. (H. F. Graevenitz, *Akademischen Abhandlung von der Bahn der Geschützkugeln* [Rostock, 1764], p. 33.)

one explanation of why the École Polytechnique provided such advanced scientific and mathematical education—to prepare cadets with sufficient knowledge to comprehend advanced ballistics textbooks such as the fourth volume of Etienne Bézout's *Cours de mathématiques, à l'usage du corps de l'artillerie* (first published in 1772) or Lombard's *Traité du mouvement des projectiles, appliqué au tir des bouches à feu.*[88] During the Napoleonic Wars, more École Polytechnique graduates served in the artillery than in any other branch of the French military.[89] One of the consequences of providing such an education was the mathematization of physics during the early 19th century, as the work of Augustin Jean Fresnel, Étienne Louis Malus, Joseph Louis Gay-Lussac, Sadi Carnot, and other École Polytechnique graduates so dramatically demonstrated.

Napoléon Bonaparte himself provides an example of the extent to which Robins and Euler influenced military thinking in the late 18th century. As a young artillery lieutenant at the regimental artillery school at Auxonne, he studied Lombard's translation of Euler's *Neue Grundsätze der Artillerie.* Lombard, incidentally, was Napoléon's artillery theory professor. Napoléon's thorough comprehension of such concepts as Robins's interior ballistics theories, the Robins effect, the limitation of Galileo's parabolic theory, the change in the air-resistance function at the speed of sound, and other fundamental ballistics concepts discovered by Robins is evident in his twelve-page summary of *New Principles of Gunnery,* written in 1788.[90] Napoléon's competence in ballistics was also demonstrated when Baron du Teil, the commander of the Auxonne artillery school, chose him to supervise a series of experiments designed to investigate the possibility of shooting mortar shells out of cannon.[91] Napoléon ultimately wrote two memoirs on this subject for du Teil.[92]

Even though an influence clearly exists, it is difficult to measure the precise impact that Robins and Euler had on Napoléon. It is possible to conclude, however, that their ideas contributed to Napoléon's success as a military commander, especially in his utilization of artillery. During the Siege of Toulon, the conflict that thrust Napoléon into national prominence during the French Revolutionary War, his scientific understanding of cannon and mortar fire was an important element in the development of his victorious strategy, which required precise

[88]Etienne Bézout, *Cours de mathématiques, à l'usage du corps de l'artillerie,* vol. 4 (Paris, 1797); Lombard (n. 83 above).

[89]Joachim Fischer, *Napoleon und die Naturwissenschaften* (Stuttgart, 1988), p. 199. According to Fischer, in the years 1804, 1807, 1808, 1809, 1810, 1812, and 1813, more than half the new École Polytechnique graduates served in the artillery. In 1814 and 1815, more than 70 percent served.

[90]N. Bonaparte, *Napoléon inconnu, papiers inédits (1786–1793) publiés par Fréderic Masson et Guido Biagi accompagnés de notes sur la jeunesse de Napoléon (1769–1793) par F. Masson* (Paris, 1895), 1:249–61.

[91]S. Wilkinson, *The Rise of General Bonaparte* (Oxford, 1930), p. 5.

[92]N. Bonaparte, "Mémoire sur la manière de disposer les canons pour le jet des bombes," in *Napoléon inconnu,* pp. 272–78.

information regarding his artillery's effectiveness at different ranges. At the beginning of the siege, Napoléon wrote to the Committee of Public Safety to request that it send an artillery general who can, "if only by his rank, demand respect and deal with a crowd of fools on the staff with whom one has constantly to argue and lay down the law in order to overcome their prejudices and make them take action which theory and practice alike have shown to be axiomatic to any trained officer of this corps."[93] Such a display of confidence from a young artillery officer with little previous combat experience certainly came in part from his theoretical understanding of ballistics.

A military historian likewise argued that one of Napoléon's advantages as a military commander lay in his education in ballistics:

> Trained in the artillery sciences, [Napoléon] had a keen grasp of the principles of physics and the concepts of energy and force. No one understood better than he the relationships of mass, time and the distance that went into the creation of energy. This much emerges from his methods of conducting a campaign or battle. . . . Once his plan developed, Napoleon's understanding of the concepts of energy and force would erupt again. Massive batteries, massive infantry columns and on occasion, massive cavalry columns could break a wall by the concentration of great energy and create rapid concentrations of force at this chosen point of attack, achieving local superiority even if he did not enjoy the overall advantages in numbers.[94]

Perhaps Napoléon referred to such reasoning when he stated, "I judge also that all officers ought to serve in the artillery, which is the arm which can produce the best generals."[95]

Great Britain was also inspired in part by the ballistics revolution to increase the scientific and mathematical education of artillery officers. The British established a military academy for the education of artillery and engineering officers in 1741. Its specific purpose, according to its warrant, was to instruct "the raw and inexperienced People belonging to the Military Branch of this (Ordnance) Office, in the several parts of Mathematicks necessary to qualify them for the Service of the Artillery, and the business of Engineers."[96] From the very beginning, the combina-

[93]John Eldred Howard, ed. and trans., *Letters and Documents of Napoleon,* vol. 1, *The Rise to Power* (New York, 1961), p. 40.

[94]R. Riehn, *1812: Napoleon's Russian Campaign* (New York, 1991), pp. 126–27.

[95]Quoted from F. Downey, *Cannonade: Great Artillery Actions of History* (New York, 1966), p. 333.

[96]O. F. Hogg, *The Royal Arsenal: Its Background, Origin, and Subsequent History* (London, 1963), p. 347.

tion of theory and practice was to be the means of instruction. The original rules for the new academy stated "that an Academy or School shall forthwith be established and opened at the Warren at Woolwich in Kent, for instructing the people of the Military branch of the Ordnance, wherein shall be taught, both in theory and practice, whatever may be necessary or useful to form good Officers of Artillery and perfect Engineers."[97] The quality of the scientific and mathematical education at Woolwich never reached the level of the French military schools. Nevertheless, the mathematics taught by Hutton, as demonstrated by his textbook *A Course of Mathematics, in Two Volumes: Composed and more Especially Designed, for the use of the Gentlemen Cadets in the Royal Military Academy at Woolwich,* shows the cadets had the opportunity to learn enough mathematics to comprehend Robins's and Euler's ballistics work.[98] Hutton seems to have used *New Principles of Gunnery* as a textbook when teaching ballistics during his thirty-four-year tenure at Woolwich beginning in 1773, and he edited a new edition in 1805.[99]

Henry Knox, the commander of the Continental army's artillery during the American War of Independence, was sufficiently impressed with Woolwich's graduates to send an urgent letter to the Continental Congress after the American defeat at Long Island in 1776: "As officers can never act with confidence until they are masters of their profession, an academy established upon a liberal plan would be of the utmost service to the continent, where the whole theory and practice of fortification and gunnery should be taught, to be nearly on the same plan as that at Woolwich—making allowances for differences of circumstances—a place to which our enemies are indebted for the superiority of their artillery to all who have opposed them."[100] Knox tried to maintain the quality of his artillery officers during the Revolution by requesting that "no officer should be appointed to the artillery who does not possess a proper knowledge of the mathematics and other necessary abilities for the nature of the service."[101] He specifically recommended the works of such military engineering and artillery

[97]Ibid., p. 348.

[98]Charles Hutton, *A Course of Mathematics, in Two Volumes: Composed and more Especially Designed, for the use of the Gentlemen Cadets in the Royal Military Academy at Woolwich* (London, 1798).

[99]Robins (n. 38 above). An edition of *New Principles of Gunnery* was published in Great Britain for almost every war in which that country fought from the mid-18th century to the beginning of the 19th century: 1742 (War of the Austrian Succession), 1761 (Seven Years' War/French and Indian War), 1777 (American Revolutionary War), and 1805 (Napoleonic Wars).

[100]William E. Birkhimer, *Historical Sketch of the Organization, Administration, Material and Tactics of the Artillery, United States Army* (Washington, D.C., 1884), p. 4.

[101]Ibid., p. 110.

authors as Sebastien le Préstre de Vauban, Baron Menno van Coehoorn, Nicholas François Blondel, Bélidor, John Muller, and Francis Holliday to John Adams, who was then serving in the Continental Congress.[102] Muller, the professor of artillery and fortifications at Woolwich, wrote *A Treatise of Artillery,* which contains a discussion of Robins's interior ballistics theory. Holliday, an English mathematics instructor, wrote *Introduction to Practical Gunnery, or, The Art of Engineering;* here, he discussed the effect of air resistance on projectiles and the analytical complexity such a phenomenon creates.[103] Even Alexander Hamilton had to demonstrate his mathematical competence to get his commission as an artillery captain from the colony of New York at the beginning of the Revolutionary War. He had learned how to apply his knowledge of mathematics to solve basic gunnery problems from a British army bombardier.[104] Hamilton therefore wrote with considerable authority in 1799 that an American military academy for military engineering and artillery officers should include such topics as calculus, conic sections, chemistry, hydraulics, hydrostatics, pneumatics, and the theory and practice of gunnery and fortification design.[105]

The German states in general and Prussia in particular took advantage of what the ballistics revolution had to offer. This was demonstrated by the ballistics research effort undertaken by Graevenitz, Tempelhoff, Legendre, and Lambert. Tempelhoff is a particularly good example of Robins's and Euler's influence on Prussian military thinking. In addition to his ballistics research and writings, Tempelhoff commanded the Prussian artillery during the French Revolutionary War and was the director of the Prussian Militärakademie der Artillerie from 1791 to 1807. Not only did he write a number of mathematics textbooks for the Prussian artillery corps, with subjects ranging from geometry to differential and integral calculus, but he also wrote *Le bombardier prussien,* a theoretical ballistics work that built on Robins's and Euler's analysis.[106] Frederick the Great praised the work, but he offered his sincerest compliment by restricting its distribution for national security reasons.[107]

The Austrians were also enlightened regarding the scientific education of artillery officers during the latter half of the 18th century. Prince Joseph Wenzel von Lichtenstein (1696–1772), whose reform of the

[102]North Callahan, *Henry Knox: General Washington's General* (New York, 1958), pp. 35–36.

[103]John Muller, *A Treatise of Artillery* (London, 1757), pp. 1–8; Francis Holliday, *Introduction to Practical Gunnery, or, The Art of Engineering* (London, 1756), p. 165.

[104]Robert Hendrickson, *Hamilton I (1757–1789)* (New York, 1976), pp. 92–93.

[105]John C. Hamilton, ed., *The Works of Alexander Hamilton* (New York, 1851), 5:380.

[106]G. F. Tempelhoff, *Le bombardier prussien* (Berlin, 1781).

[107]Oskar Albrecht, "Georg Friedrich von Tempelhoff," *Soldat und Technik* (September 1966), pp. 493–94.

Austrian artillery proved to be so shocking to Frederick the Great during the Seven Years' War and so motivating to General Gribeauval of France, set up an artillery research center at Moldauthein and an artillery school at Bergstadl in the 1740s during the War of the Austrian Succession.[108] The subjects taught included the standard mathematical topics such as arithmetic, geometry, and trigonometry, and such scientific engineering topics as mechanics, hydraulics, optics, fortifications, and ballistics.[109] Lichtenstein initially published the artillery books of Bélidor and others for his artillery school. In 1786, the Austrians set up a corps of bombardiers, who were especially trained to handle mortars, howitzers, and siege cannon.[110] One sees a direct influence of the ballistics revolution on this artillery corps as demonstrated by their textbook *Praktische Anweisung zum Bombenwerfen mittelst dazu eingerichteter Hilfstafeln,* which came from the third volume of *Mathematischen Vorlesungen des Artilleriehauptmanns und Professors der Mathematik bey dem kaiserl. köngl. Bombardierkorps Georg Vega.*[111] The first section of *Praktische Anweisung* addresses the effect of air resistance on military projectiles.

Finally, the Italians. General Alessandro Vittorio Papacino D'Antoni, a major general in the Sardinian army and chief director of the Royal Military Academies of Artillery and Fortification at Turin during the 1760s and 1770s, not to mention a highly regarded research engineer in interior ballistics, wrote extensively on theoretical, experimental, and practical aspects of ballistics and artillery. In his *A Treatise on Gunpowder; a Treatise on Fire-Arms; and a Treatise on the Service of Artillery in the Time of War,* D'Antoni devoted significant attention to Robins's interior ballistics analysis and measurements of air resistance, as well as the advances made in theoretical and experimental ballistics after the ballistics revolution.[112]

Although plenty of evidence exists that shows the influence of the ballistics revolution in 18th-century military education, this only indirectly indicates its influence on the actual battles and sieges of that age. Did artillery officers actually use Euler's and Robins's ballistics in combat, or did they do so only in classrooms to discipline their minds? This is a difficult question and will require considerable archival research to document fully. Nevertheless, some immediate evidence is already at hand. Graevenitz's book, *Akademischen Abhandlung von der*

[108]Duffy, *Army of Maria Theresa* (n. 23 above), p. 105.

[109]Ibid., p. 108.

[110]Anton Dolleczek, *Geschichte der Österreichischen Artillerie* (Graz, 1973). See the section titled "Das Bombardier Corps."

[111]*Mathematischen Vorlesungen des Artilleriehauptmanns und Professors der Mathematik bey dem kaiserl. köngl. Bombardierkorps Georg Vega* (Vienna, 1787).

[112]D'Antoni (n. 1 above), pp. 74–98.

Bahn der Geschützkugeln, where Euler's ballistics theory is presented with a complete set of numerical tables, was published in 1764. Perhaps the fact that Graevenitz took the trouble to numerically integrate the hundreds of equations required to create the tables is the best evidence for its practical utility in combat. What other rational motivation could exist to undertake such a tedious exercise? Graevenitz was not the first to calculate these tables, however. Paul Jacobi, a lieutenant in the Prussian artillery, calculated them and presented them to the Berlin Academy of Sciences soon after Euler published his solution in 1753. Unfortunately, Jacobi was killed in the Seven Years' War, and his manuscript was lost.[113]

Regardless of whether the ballistics revolution was directly felt on the battlefields and sieges of the Seven Years' War, it certainly made an impact on the combat of the French Revolutionary War in the 1790s. The most vivid evidence of this is Lombard's ballistics tables, published in 1787 as *Tables du tir des canons et des obusiers.*[114] French artillery officers used these tables in combat during both the French Revolutionary War and the Napoleonic Wars.[115] In his textbook *Traité du mouvement des projectiles, appliqué au tir des bouches à feu,* Lombard described the approximate interior and exterior ballistics theories he used to calculate the tables.[116] The difference between these tables and those of Bélidor from the 1730s is vivid proof of the ballistics revolution's influence on 18th-century warfare. Instead of using Galileo's parabolic trajectory theory to derive a simple relationship between elevation angle and range, Lombard's tables are all written in terms of muzzle velocity. Each type of cannon has four different tables. The cannon include siege (24-, 16-, 12-, 8-, and 4-pounders), field (12-, 8-, and 4-pounders), and howitzers (8- and 6-pounders). The weight refers to the cannonball. The first table shows the relationship between quantity of gunpowder and muzzle velocity for different strengths of gunpowder. The second table shows the relationship between muzzle velocity and impact velocity for different ranges, whereas the third and fourth tables show the relationship between muzzle velocity, range, and elevation angle below and above the natural point-blank range for the artillery piece in question.

The use of such tables in 18th-century warfare raises another question: if ballistics and range tables were available, why then did European governments take the trouble to provide artillery officers with a

[113]Graevenitz (n. 82 above); see the introduction.

[114]Jean-Louis Lombard, *Tables du tir des canons et des obusiers* (Auxonne, 1787).

[115]C. N. Amanton, *Recherches biographiques, sur le professeur d'artillerie, Lombard* (Dijon, 1802), p. 17. Also see Charbonnier (n. 4 above), p. 187.

[116]Lombard, *Traite' du mouvement des projectiles* (n. 83 above).

mathematical and scientific education? Is it really necessary to know ballistics theory in order to use numerical tables? Napoléon Bonaparte seemed to think so. He wrote in 1801 that some of Lombard's and Robins's ballistics theories are important for the artillery and should therefore be included in the textbooks for the proposed artillery and engineering school at Metz.[117]

Eighteenth-century gunnery tables were designed to speed up calculations, not to replace a solid grasp of the theory. Euler/Graevenitz's table, for example (fig. 4), is quite incomprehensible to someone who does not understand its derivation. Furthermore, there was much more to 18th-century gunnery than simply establishing the correct angle to launch a projectile to a desired distance. When supervising the firing of howitzer or mortar shells, an officer had to estimate the fuse's length to cause an explosion at an optimal point in the shell's trajectory. For example, a howitzer shell needed to explode at just the right point in the air in order to cause the greatest number of casualties.[118] Mortar shells, on the other hand, when used to destroy buildings, needed to explode after they crashed through a roof. Such calculations are relatively easy when using Euler's theory as presented in Graevenitz's tables. Lombard's tables seem much more straightforward than Graevenitz's, and gunners who had no knowledge of their derivation could potentially use them. These tables are based on certain simplifying assumptions, however, and could give misleading results if used under conditions that might violate these assumptions. Hence, only officers with a solid grasp of the theory could use these tables with full confidence, a quality not to be taken for granted during the tremendous confusion and uncertainty of combat. As Matti Lauerma wrote in *L'artillerie de campagne française pendant les guerres de la révolution,* "The lack of fundamental theoretical knowledge apparently diminished the accuracy of artillery fire to a considerable extent."[119]

General D'Antoni strongly believed that a sound knowledge of ballistics in particular, as well as mechanics in general, was essential for an artillery officer. For example, he wrote, "There are other methods of

[117]Napoléon Bonaparte, "Notes sur un project de réglement pour l'école d'artillerie et du génie," June 27, 1801 (no. 5621), *Correspondance de Napoléon I publiée par ordre de l'Empereur Napoléon III* (Paris, 1861), 7:232–33. Napoléon wrote, "Les ouvrages que l'on enseignait à l'école du génie avant la révolution existent et ne laissent rien à désirer. L'aide-mémoire, classé d'une manière convenable, et quelques principes de théorie qui se trouvent dans Lombard et dans Robins, fourniraient un bon ouvrage pour l'artillerie. On a aussi d'excellents traités sur les mines et sur l'art de lever les plans" (pp. 232–33).

[118]Hughes (n. 25 above), p. 34.

[119]Matti Lauerma, *L'artillerie de campagne française pendant les guerres de la révolution* (Helsinki, 1956), p. 57, or in the original French, "Le manque de connaissance théoriques de base diminuait apparemment dans une forte mesure la précision du tir de l'artillerie."

ascertaining the path described by projectiles, and the retarding force of the air; but it is to be presumed that the principles laid down in the course of this treatise will, from their practical utility and easy application, induce the students to exercise themselves in the theory of gunnery, whence they may derive from the use of fire-arms, particularly of mortars, advantages which can by no other means be obtained."[120] D'Antoni gave a specific example why theory was so useful.[121] He stated that there are two different types of mortar bombardments. The first is when the gunner's objective is to launch a shell at a particular target when the force of impact is not consequential, such as when the targets are enemy artillery pieces and troops or when the goal is to set fire to buildings. For such activity, bombardiers without any knowledge of ballistics can be taught to accomplish these goals. For bombardments when the force of the shell's impact is significant, the situation is different. "The second case, to break through casemated buildings, requires much theoretical knowledge in the officer charged with the execution of this piece of service, in order to determine the situation of the mortar, its proper charge and elevation; that the shell may impinge on the object with the greatest possible force."[122] D'Antoni then described the ballistics calculation required to determine the shell's impact velocity and how to calculate the force of impact. Clearly, there were some gunnery problems that could be solved by experience and practical training alone. Nevertheless, certain crucial problems demanded advance mathematical and scientific knowledge to solve effectively. This is just one similarity between 18th-century gunnery and 20th-century engineering.

The most direct evidence for the influence of the ballistics revolution on late-18th-century battlefields comes from the dramatically improved accuracy of ballistics theories that incorporated experimental air-resistance values. Bézout provided a series of tables in the fourth volume of his *Cours de mathématiques* that compared experimental measurements of artillery projectile ranges with calculations by using theories that were derived before and after the ballistics revolution, that is, Galileo's theory and air-resistance theories (fig. 5). The table for mortar-shell range and time of flight consists of columns (from left to right) of the launch angle, theoretical range without air resistance, theoretical range with air resistance, the experimental ranges, and the same division for time of flight. Although the difference between the range predicted by the air-resistance theory and average experimental range was as high as 37

[120]D'Antoni (n. 1 above), p. 98.
[121]Ibid., p. 225.
[122]Ibid., p. 226.

TABLE des Portées de Bombes, calculées 1°. en supposant que l'air ne résiste pas; 2°. ayant égard à la résistance de l'air; & comparées aux Portées observées dans les épreuves faites à La Fère, au mois d'Octobre 1771, par les ordres de M.r le Marquis de Monteynard, Secrétaire d'État ayant le département de la Guerre, & sous la direction de M. de Beauvoir, Brigadier des armées du Roi, Commandant en chef l'école d'Artillerie.

ANGLES de PROJECTION.	PORTÉES CALCULÉES		PORTÉES OBSERVÉES.	DURÉE DES PORTÉES			ANGLES de CHUTE.
	Sans égard à la résistance.	Eu égard à la résistance.		Sans la résistance.	Eu égard à la résistance.	Selon l'Expérience.	
degrés.	toises.	toises.	toises.	secondes.	secondes.	secondes.	degrés.
10	253	227	257. 249. 221. 228.	$4\frac{1}{5}$	$4\frac{1}{20}$	4	14.
20	476	396	440. 424. 394. 398.	$8\frac{3}{10}$	8	$7\frac{1}{3}$	26.
30	640	500	451. 516. 537. 492.	$12\frac{1}{5}$	$11\frac{3}{10}$	$10\frac{3}{4}$	36.
40	728	547	569. 575. 574. 544. 577.	$15\frac{3}{5}$	$14\frac{2}{5}$	$14\frac{4}{5}$	48.
43	738	549	506. 517. 543. 509. 544.	$16\frac{1}{2}$	$15\frac{1}{5}$	14	$50\frac{1}{2}$.
45	739	547	490. 536. 505. 489. 554.	$17\frac{1}{5}$	$15\frac{4}{5}$	$15\frac{1}{5}$	$52\frac{2}{3}$.
50	728	534	481. 512. 488. 507.	$18\frac{1}{5}$	$16\frac{9}{10}$	16	$57\frac{1}{2}$.
60	640	467	457. 424. 457. 448.	21	$19\frac{3}{10}$	$19\frac{1}{3}$	68.
70	476	348	349. 297. 349. 328.	$22\frac{4}{5}$	$20\frac{7}{10}$	22	74.
75	370	277	298. 265. 261. 256.	$23\frac{2}{5}$	$21\frac{7}{10}$	22	78.

Les bombes dont on a fait usage dans ces épreuves, étoient de 11 pouces 10 lignes de diamètre, du poids de 142$^{liv.}$, y compris la terre dont on les avoit remplies; & elles ont été chassées avec 3$^{liv.}$ $\frac{1}{4}$ de poudre.

Fig. 5.—Bézout's range and time-of-flight table comparing Galileo's ballistics theory, ballistics theory with air resistance considered, and experimental observations as a function of firing angle. (Etienne Bézout, *Cours de mathématiques, à l'usage du corps de l'artillerie* [Paris, 1797], p. 456.)

toises or 74 yards (at an angle of 50 degrees and range of about 500 toise or 1,000 yards), the theory failed in only three cases (43, 50, and 60 degrees) to provide a range within the area where the shells landed. Galileo's theory, in contrast, only succeeded at 10 degrees to provide a range within this area. Although Bézout did not provide enough experimental values for a proper statistical error analysis, this table suggests the power available to artillery officers after the ballistics revolution.

Conclusion

The science of mechanics was a significant military technology during the 18th century; ballistics and its effect on gunnery is a case in point. This contradicts the perception that rational mechanics had little effect on early modern mechanical technology.[123] Beginning with Galileo in the 17th century and continuing with Robins and Euler in the 18th century, mechanics was developed, in part, to optimize artillery systems. Galileo's ballistics theory, a first-order approximation and valid only for a restricted set of gunnery problems, was nevertheless the foundation of the science of ballistics. Robins and Euler built on Galileo's work by using advanced theories of mechanics and mathematics coupled with experimental research. The inventor of the first aerodynamic instruments to measure successfully a projectile's velocity and air resistance, Robins discovered such fundamental aerodynamic phenomena as the enormous magnitude of air resistance at high speeds and the Robins effect, as well as making the first observations of the sound barrier. Furthermore, he conducted a pioneering theoretical and experimental study of the work generated by gunpowder gases within a gun barrel. Robins's experimental results made his own, as well as Euler's, analysis of the nonlinear differential equations of ballistic motion possible. They

[123]H. J. M. Bos, "Mathematics and Rational Mechanics," in *The Ferment of Knowledge: Studies in the Historiography of Eighteenth-Century Science,* ed. G. S. Rousseau and Roy Porter (Cambridge, 1980), p. 354. According to Bos, "The actual influence of theoretical results in the case of these (technological) problems is very difficult to assess. As regards the longitude problem the influence was certainly there; in the case of the other problems it is doubtful whether insights gained through mathematical theory effectively influenced practice before the nineteenth century." A corresponding view is found in Thomas L. Hankins, *Science and the Enlightenment* (New York, 1985), p. 23: "Only in astronomy did the new analysis show immediate practical results in the increased precision of astronomical tables and in the creation of new theories concerning the shape and motions of the earth and other heavenly bodies." Contrast this with Thomas Hughes and Gunther Rothenberg's comments regarding the practical utilization of mathematics and mechanics in 18th-century France and Austria (Hughes [n. 4 above], pp. 69–80). Hughes wrote that "science and technology were an integral and practical part of the military culture of eighteenth century France" (p. 74).

used the equation's solutions to construct useful gunnery tables. The ballistics revolution therefore contradicts the popular idea that the experimental and mathematical sciences remained essentially separate until the 19th century. Robins's and Euler's research revolutionized experimental and theoretical ballistics by transforming it into an aerodynamic and thermodynamic science, to use two admittedly anachronistic adjectives. This provided the benefit of increasing the effectiveness of ballistics theory for gunnery practice at the cost of increasing its analytical complexity. As a result, the military powers of 18th-century Europe had to improve their artillery officers' education in mathematics and science to compete effectively in warfare.

The historical development of ballistics demonstrates an important characteristic of engineering: the creation of fundamental scientific knowledge or basic research. As Steven Goldman recently argued, "The crucial point to appreciate is that engineering on its own activity generates knowledge. It does not passively wait for knowledge to be given to it from a different community of practitioners in order for it to attempt complex enterprises."[124] The role engineers have played in the development of the physical sciences before the 20th century has attracted significant, but limited, scholarly attention. Stephen Timoshenko outlined the basic theoretical and experimental research conducted by engineers in elasticity and strength of materials from the Renaissance until the beginning of the 20th century.[125] Terry Reynolds described the basic fluid dynamics theory developed by such 18th-century engineers as de Borda and Lazare Carnot.[126] The structural research conducted by certain U.S. engineers in the 19th century is discussed by Edwin T. Layton in "Mirror-Image Twins."[127] C. Stewart Gillmor's study of Charles-Augustin de Coulomb highlights an 18th-century engineer's contributions to physics and mechanics, while Cardwell shows the similar role engineers such as Sadi Carnot and Emile Clapeyron played in the creation of modern thermodynamics.[128] These authors demonstrate the often unacknowledged fact that scientific discoveries are not made solely by scientists studying natural phenomena and that the history of science is incomplete without taking engineering research into consideration.

[124]Steven L. Goldman, "Philosophy, Engineering, and Western Culture," in *Broad and Narrow Interpretations of Philosophy of Technology*, ed. Paul Durbin (Dordrecht, 1990), p. 141.

[125]Stephen Timoshenko, *History of Strength of Materials* (New York, 1953).

[126]Terry Reynolds, *Stronger than a Hundred Men* (Baltimore, 1983), pp. 231–48.

[127]Edwin T. Layton, "Mirror-Image Twins: The Communities of Science and Technology in 19th-Century America," *Technology and Culture* 12 (1971): 564–80.

[128]Cardwell (n. 50 above), pp. 186–211, 220–29; and C. S. Gillmor, *Coulomb and the Evolution of Physics and Engineering in Eighteenth Century France* (Princeton, N.J., 1971).

Benjamin Robins's ballistics work represents a prime example of engineering research because he deliberately set out to analyze a machine. Yet such research also provides an example of the paradoxical consequences of certain popular assumptions regarding engineering research. While investigating the behavior of bullets fired from muskets, Robins discovered fundamental aerodynamic principles. His study therefore also represents basic research. According to popular assumptions, engineering research should be classified only as applied research, the application of existing scientific ideas to solve technological problems. Since applied research cannot be basic research by definition, Robins's work presents a contradiction. This paradox is created by the rigid assumption that pure scientists hold a monopoly on generating basic research or that engineers can only apply scientific knowledge but that they do not create it. The traditional way to avoid this paradox is to classify any engineer who did especially influential basic research as either a scientist or a mathematician or to dismiss the significance of the discoveries. This is historically misleading, however. The labels of the pure sciences (e.g., "physics," "chemistry," and "biology") and engineering refer to the goals, motivations, values, and ideologies of particular intellectual groups. Such labels are social constructs and do not identify the way these groups generate desirable knowledge. In Robins's case we see an 18th-century mathematician and engineer who applied the science of mechanics, as developed by Galileo, Newton, Huygens, and others, to solve fundamental gunnery problems. Yet he did much more. Since significant parts of the scientific knowledge required to solve these technological problems did not exist, he performed the basic research himself and helped create the modern sciences of thermodynamics and aerodynamics in the process. This is not paradoxical; it reflects the very natures of science and engineering.

The Philosophy of Luddism: The Case of the West of England Woolen Workers, ca. 1790–1809

ADRIAN J. RANDALL

Rees's *Cyclopaedia* in 1819 accounted for the relatively slow rate of technological change in the English woolen industry thus:

> This was owing in a great degree to the circumstances that the manufacture of woollen cloth was rendered very perfect . . . long before the improved system [of machinery] was begun; and there were great numbers of experienced and able workmen trained up for each process who performed their work as well as could be done by machine. The reduction of labour . . . was in this case all that machinery of the most perfect kind could effect; both these were advantages to the public and the manufacturer, but were so directly opposite to the inclination and interest of the able workmen that we find they have made greater and more effectual opposition to the introduction of improvements in the woollen than in any other of our great manufactures.[1]

The vigorous opposition of the English woolen workers "to the introduction of improvements" is well known. The most famous example, of course, was the series of Luddite disturbances in the West Riding of Yorkshire in 1812, when the croppers or shearmen endeavored to stem the rapid rise of the cloth-dressing machinery that threatened their entire livelihood by smashing shearing frames and gig mills, an action that has bequeathed to historians and the wider public a generic term to describe all sorts of labor resistance to technological innovation.[2] But the Yorkshire croppers' violence had been anticipated

ADRIAN RANDALL is now professor of English social history at the University of Birmingham and has published on resistance to new technology, popular protest, and labor conflicts in the Industrial Revolution and on market culture and subsistence riots in the 18th century.

[1]A. Rees, ed., *Cyclopaedia* (London, 1819), vol. 38, s. v. "woollen manufacture."

[2]M. I. Thomis, *The Luddites* (Newton Abbot, 1970); cf. E. P. Thompson, *The Making of the English Working Class* (London, 1963; Harmondsworth, 1969 ed.), ch. 14.

ten years earlier in the West of England in a series of disturbances variously known as the Wiltshire Outrages, and it was the West of England where resistance to mechanization was most general and most effective in delaying the forces of progress. There, unlike in the West Riding of Yorkshire, nearly every new machine sparked an angry reaction that frequently frightened innovators and delayed its adoption for many years.[3]

The reason for the wider antipathy toward machinery in the West of England lay in this region's industrial organization. Here the woolen industry was most highly developed, here the putting-out system, the characteristic organizational form of the preindustrial period, reached its apogee. Here the development of specialist skills enabled the gentlemen clothiers to build up work forces of thousands and capital assets of tens of thousands. But specialization meant that machinery, which in Yorkshire took away just part of the variety of tasks undertaken by a journeyman working for a master clothier, threatened entire trades in the West of England. It is perhaps worth noting the labor-saving effects of some of these innovations. The spinning jenny displaced around nine of ten warp spinners and thirteen of fourteen abb (weft) spinners. The scribbling engine displaced fifteen of sixteen scribblers. With the gig mill one man could do part of the work of a dozen shearmen, while the shearing frame made three of four shearmen redundant. Scribblers constituted around 10 percent of the preindustrial adult male work force, shearmen around 15 percent.[4] Such men found their skills useless, their trade superfluous, when machinery was introduced. It is little wonder that their reaction was hostile. It is with the West of England woolen industry that this article is particularly concerned.

Historians have differed in their interpretation of such resistance to change. Some have echoed the opinion of Josiah Tucker, who be-

[3]For details see J. de L. Mann, *The Cloth Industry in the West of England* (Oxford, 1971), ch. 5; J. L. and B. Hammond, *The Skilled Labourer* (London, 1919), ch. 6; A. Aspinall, *The Early English Trade Unions* (London, 1949); A. J. Randall, "The Shearmen and the Wiltshire Outrages of 1802: Trade Unionism and Industrial Violence," *Social History* 7 (1982): 283–304. "West of England" cloth was produced in the counties of Gloucestershire, Somerset, and Wiltshire. The region had a long tradition of riotous activities by the woolen workers, who responded to any threat to their living standards with vigorous resistance. See Mann book cited and also her "Clothiers and Weavers in Wiltshire during the Eighteenth Century," in L. S. Pressnell, ed., *Studies in the Industrial Revolution* (London, 1960); Randall, "Labour and the Industrial Revolution in the West of England Woollen Industry" (Ph.D. diss., University of Birmingham, 1979).

[4]*British Parliamentary Papers, 1840, Vol. 23,* pt. 8, pp. 439–441; *1840, Vol. 24,* pt. 5, pp. 369–374, Reports from Assistant Handloom Weavers' Commissioners. (*Parliamentary Papers* hereinafter cited as *BPP*.)

moaned "the mistaken Notions of the infatuated Populace, who not being able to see farther than the first Link of the Chain, consider all such Inventions as taking the Bread out of their Mouths; and therefore never fail to break out into Riots and Insurrections whenever such Things are proposed."[5] Even those who, like J. L. and B. Hammond, Duncan Bythell, and Malcolm I. Thomis, have sympathized with the workers' plight have viewed their methods of resisting change as a product of desperation, as "blind vandalism" or as "a throwback to the disorganised activities of a pre-industrial age."[6] I have argued elsewhere that this is an erroneous assessment. The violence with which innovation in the woolen industry was met, far from being mere "vandalism," was carefully controlled and directed and used as part of a wider organized response to detrimental change.[7]

Beyond this, however, as I hope here to demonstrate, the workers' response was in no sense "blind." The woolen workers had a perception of the introduction of machinery very different from the image of progress articulated by the laissez-faire innovators and their advocates. This view was based not merely on the quite understandable special pleading of men threatened with the loss of their livelihoods but also on an essentially 18th-century reading of economic and social relationships, one that placed a premium on stability, regulation, and custom. The woolen workers recognized that the nascent machine economy represented a value system quite alien to this, and they therefore sought not only to combat the machinery itself but also to refute the whole ethic of laissez-faire industrial capitalism that it represented. The struggle over the industrialization of the woolen industry was more than just a physical confrontation. It was an ideological struggle as well.

The case of the woolen industry contradicts the view expressed by Maxine Berg at the start of her stimulating book, *The Machinery Question and the Making of Political Economy*, where she writes, "In the eighteenth century there was no 'Machinery Question.' The introduction of machinery was welcomed "as an indicator of economic expansion" that "would contribute to the general 'improvement' of society." It was not, she says, until after the Napoleonic Wars that "this prospect of a harmonious integration of economic and social improvement was thrown into question." As will be seen, this is not entirely accurate.

[5]J. Tucker, *Instructions for Travellers* (Gloucester, 1757), p. 21. As Dean of Gloucester, Tucker had the West of England woolen workers very much mind.

[6]Hammond (n. 3 above); D. Bythell, *The Handloom Weavers* (Cambridge, 1969); Thomis (n. 2 above). The quotations are taken from Bythell, pp. 180, 181.

[7]Randall (n. 3 above ["The Shearmen . . ."]); see also J. Stevenson, *Popular Disturbances in England, 1700–1870* (London, 1979), chs. 6 and 7.

While Berg is certainly correct to argue that the machinery question became a central facet of the wider "condition of England question" (posed, though never resolved, in the 1820s and 1830s), some of the issues of the industrial, economic, and social consequences of mechanization had already been debated and resolved in connection with the woolen industry long before hostilities with France had come to an end. Yet Berg is correct in asserting that historians have shown little interest "in making connections between [such violent] resistance to the machine and the political and intellectual disputes over technological improvement."[8] It is my purpose here to begin to redress this balance and to indicate the philosophy that informed the machine breakers' actions.

The ideological struggle between innovation and tradition was highlighted in the transformation of the English woolen industry in the years before 1809 because of its peculiar historical development. While all workers threatened with mechanical displacement argued for their own protection, the woolen workers believed that they had a *constitutional* right to be protected from many of the threatened changes. This belief was grounded on a body of legislation enacted by successive governments from the Tudors onward to control the industry that played a key role in the nation's export economy. In 1802 there were found to be some seventy statutes still extant restricting and regulating many aspects of manufacture, organization, and industrial relations.[9] Much of this legislation was archaic; very little of it was still observed to the letter. But this corpus of statute law and the willingness of Parliament as recently as 1756 to interfere directly in their industry encouraged the woolen workers to hope that the old acts might be successfully used to block change and to safeguard their "rights." Thus shearmen and weavers attempted to use the courts to prosecute those who introduced gig mills and weaving shops and who employed "illegal" workmen.[10]

Furthermore, these trades and the scribblers made every effort to persuade the bench and the wider taxpaying public that their opposition to machinery was based in law and that their opponents were

[8]M. Berg, *The Machinery Question and the Making of Political Economy, 1815–1848* (Cambridge, 1980), pp. 1, 15.

[9]*BPP, 1802/3, Vol. 5*, Report from the Select Committee on the Woollen Clothiers' Petition (H.C. 30). The statutes are listed in *BPP, 1806, Vol. 3*, Report of the Select Committee to Consider the State of the Woollen Manufacture (H.C. 268).

[10]Randall (n. 3 above ["The Shearmen . . ."]), pp. 287, 301; *Gloucester City Library*, JF 12.9(i), Notice Concerning Apprenticeship of Weavers, 1802; *BPP, 1806, Vol. 3*, pp. 349, 351; *BPP, 1802/3, Vol. 7* (H.C. 95), pp. 44–45; *Gloucester Journal*, August 9, 1802 (hereinafter cited as *GJ*).

behaving selfishly and illegally. These tactics were not without success. Hence the innovating clothiers in the West of England were faced with a position perhaps unique in the Industrial Revolution. They had not only to overcome vigorous worker opposition to change—they had also to persuade the local authorities that the past policy of control was fallacious and that it was necessary to ignore all such regulations.

In November 1802 leading innovators in the three counties, alarmed by the strength of opposition and by popular support for the workers' often violent actions, converted an association that had been formed to prosecute rioting shearmen into a campaign to get Parliament to repeal *all* the old statutes, thereby removing the workers' justification that their opposition was based in law.[11] This action inevitably drew the battle between "progress" and "tradition" onto a national stage on which both sides sought to demonstrate that only their particular case was in the national interest. For innovators and workers alike, this proved to be a costly and lengthy campaign, involving parliamentary inquiries in 1803 and 1806 and much lobbying, petitioning, and pamphleteering. The propagandist battle of the 1790s and before the select committees in 1803 and 1806 gave an exceptional opportunity for the woolen workers to articulate both their hostility to change and their alternative view of their industry and of the society in which they lived. The introduction of machinery into the West of England woolen industry therefore drew into sharp focus the conflict not just between entrepreneurial innovation and established custom but also that between two very different views of the economy and of society.

This is not to assume that there were *only* two ideologies and that these were distinct and self-contained. Clearly there were many differences and divisions of opinion within the ranks of workers and clothiers alike, as I will indicate. But the advent of a machine economy forced all concerned to determine their attitudes to fundamental questions about the structure, organization, and regulation of their industry. The ideological parameters outlined here are those that the debate itself revealed. Whether or not these ideologies were ultimately proved correct is irrelevant to the purpose of the present article; the fact that the rise of an unfettered machine economy brought eventual national prosperity or involved the total displacement of two major trades, the scribblers and shearmen, is, in this case, of no consequence. Hindsight must not distort our analysis of the perceptions of the protagonists. For in arguing its case each side was doing more than protecting income or capital; each was defining its own view of the nature and legitimacy of

[11]*Gloucester City Library*, JF 13.25, Minutes of the Woollen Manufacturers' Committee, 1802–3.

society. William Reddy has written of the French textile industry of this period that it is necessary to "recognize how deeply any society's organization of its productive activity is informed by its culture, that is, by fundamental conceptions of the nature of the world and of human relations to it."[12] The converse is even more true—culture and conceptions of economic and social relations are informed by experience of the economic process and mediated by one's role within it. It is the contention of this article that such views are worthy of being elucidated in their own right—and that they provide the only way in which we may really understand the nature of the conflicts the Industrial Revolution precipitated.

* * *

The innovators' case for the unfettered introduction of machinery mixed boundless optimism with bottomless pessimism, but its basis was a concept of growth and economic progress. Though modified with time as practice proved some of its larger promises wide of the mark and as the inexorable rise of the West Riding made fears of total eclipse keener, the innovators' premise in essence remained simple. The new woolen machinery was in a direct line from the spade and the plow. It enabled cloth to be produced more cheaply than before. Therefore it must be introduced forthwith without hindrance. Any region that failed to mechanize would inevitably lose its trade to rivals in other regions and overseas, thereby ruining not only the capitalists but also the workers, who would lose their livelihoods and be thrown onto the poor rates.

However, in the years from 1776 to the mid-1790s, machinery was seen as more than just a necessary insurance against imminent ruin. It would also bestow incalculable benefits. Should opposition cease and machinery be generally introduced, claimed the innovators, "there is a rational prospect that the trade will be rendered permanently flourishing." Trade would be "extended and improved, and the wages in many branches of the business increased." Machinery could not harm the poor, claimed Henry Wansey in one of the earliest apologias for machinery written by a clothier: "If there was only a certain quantity of work to do, and that being done, a stop was to be made, the sooner that the work was done, the worse for the labourers. But the principle of manufactures operates differently. The more you can do well, the more you may." The obvious proof of this was the cotton

[12]W. M. Reddy, "The Spinning Jenny in France: Popular Complaints and Elite Misconceptions on the Eve of the Revolution," in *The Consortium on Revolutionary Europe, 1750–1850: Proceedings, 1981* (Athens, Ga., 1981), p. 52.

industry, exports from which had risen ten times. Machinery would enable woolen manufacturers to extend their trade in the same way, and "besides, new countries are opening continually to the trade of Britain." Thus, machinery offered the prospect of continuous growth. Wansey admitted that there was perhaps an ultimate problem that markets might become "overstocked." "There is a point *somewhere* beyond which things cannot grow; but it is not for this Kingdom, under the apprehension of such an event one day or other taking place, to neglect the present means of encreasing its strength."[13]

While earlier promises of unending prosperity gave way to more guarded statements of optimism in the later 1790s and early 1800s, clothiers throughout the period were resolute in their resistance to any argument in favor of the restriction of innovation. Like John Billingsley, they believed that legislative interference that sought to regulate machinery "would be incompatible with wisdom and justice. To allow only its partial establishment would be oppressive: to admit of none would be ruinous because such machinery with its appendant branches of manufacture. . . . is not only susceptible of, but . . . will shortly be in a state of migration." A similar view came from a writer from Marlborough in 1794. Restriction was inherently harmful and could only lead to the rapid migration of the clothiers and the industry.[14] Indeed, the innovators frequently threatened local gentry that, should any attempt be made to "interfere" with their business, they would immediately wind up their operations and move, leaving their work force dependent on the parish. While such threats did little to endear the innovators to the taxpayers, they could not easily be ignored.

By the time of the select committee inquiries of 1803 and 1806, the case for innovation had taken on more decidedly negative characteristics. In their efforts to secure the repeal of the entire regulatory legislative code—efforts that alienated many small clothiers—the innovators now argued that all such restriction was harmful ("impolitic antequated shackles") and irrelevant ("acts which nobody in the trade can or does observe"). At best their enforcement could only occasion great delay and inconvenience and at worst the total loss of the foreign market. The workers' case in favor of controls over machinery, loom shops, and apprenticeship was scarcely worth considering. Machinery was "the principal means by which we have excluded other nations from the foreign trade." To restrict its use was folly. Even if it did lead

[13] *GJ*, October 14, 1776, May 30, 1791, August 22, 1791; J. Billingsley, *A General View of the Agriculture of the County of Somerset* (Bath, 1797), p. 161; *Salisbury and Winchester Journal*, June 7, 1790, August 15, 1791 (hereinafter cited as *SWJ*); H. Wansey, *Wool Encouraged without Exportation* (London, 1791), pp. 67–9.

[14] Billingsley (n. 13 above), p. 162; *SWJ*, June 2, 1794.

to unemployment, the alternative would be disastrous, for without machinery the whole industry would collapse. Restrictions and regulations were unnecessary, for the best protection the consumer and the nation could have was the vested interest of the manufacturer in his product. Parliament, they argued, must free the woolen industry from control and "encourage competition and not monopoly." Only in this way could the industry survive. The prospect of universal prosperity was now a thing of the past.[15]

* * *

The woolen workers repudiated the laissez-faire arguments of economic progress expounded by the innovators. In the introduction of machinery they saw not salvation but that very ruin that their opponents postulated as the consequence of the failure to mechanize. Their belief that regulation and control were essential to safeguard the industry was based on moral, social, and economic grounds and was shared by many small employers and paternalists in the West of England and Yorkshire. Their arguments were echoed by workers in other industries facing similar disruption of custom and culture.

Primarily, of course, the woolen workers emphasized that machinery created unemployment. In 1776 the Somerset workers had asked Parliament to prohibit the spinning jenny, since it threatened the livelihood of many thousands of the industrious poor. When the jenny was finally introduced in the early 1790s, these fears were proved correct. It was not, however, the woolen workers themselves who bore the major impact. Their wives and children could find work at the loom or even working the small jennies in their homes. The real casualties were those wives of agricultural laborers for whom spinning constituted a major secondary source of family income and who received no compensatory employment. Many were forced to apply to the parish for relief, generating a considerable ground swell of antipathy among the taxpayers at large to machinery. When machinery threatened their trades in the 1790s, scribblers and shearmen emphasized that countless hitherto self-supporting families would likewise soon be indigent. The general public sympathy accorded to both these groups, even when they resorted to violence, was in no small way due to the taxpayers' realization that, as in the past, the interests of the clothiers and of the community did not necessarily coincide. The workers deliberately played on such suspicions and enmities, pointing out that while cost-cutting machinery was forcing up poor rates, the price of cloth had not

[15]T. Plomer, *The Speech of Thomas Plomer to the Committee of the House of Commons . . . Relating to the Woollen Trade in 1803* (Gloucester, 1803), pp. 88, 9, 25, 26, 23, 35; 14–15; 70.

fallen, a fact noted by several paternalists. "However much the public interest may be affected by machinery," wrote one, "their private interest is undoubtedly advanced."[16]

In some cases the workers were able to claim that the machinery being introduced was actually prohibited by statute and that the innovators were therefore breaking the law. The shearmen continually claimed that the gig mill was proscribed by the *Act 5 and 6 Edward VI, ch. 22*, "An Act for the putting down of gig mills." The clothiers disputed this, and the act was sufficiently vague to ensure that cases brought by the Wiltshire shearmen against gig mill owners came to nothing. Under *2 and 3 Philip and Mary, ch. 11*, which prohibited clothiers from owning more than one loom or letting looms for hire, weavers were able to bring actions against clothiers for establishing loom shops. They also commenced actions against the nonapprenticed workers who were employed by such men.[17]

The woolen workers were certainly well aware of the old legislation. Nevertheless, they did not intend that the law should necessarily be enforced to the letter; the crucial factor was how custom had interpreted it. For example, while both shearmen and weavers wanted to enforce apprenticeship legislation, this did not mean that they insisted on formal indentures for the seven years. A worker "brought up in the trade" by his father or by a relation was regarded as a legal worker by the community, if not strictly so in law. Indeed, the weavers made it clear that they wanted the law to impinge only on persons coming into the trade in later life without this same kind of schooling in the industry.

Again, while in Wiltshire *any* use of the gig mill was seen as being strictly illegal, in Gloucestershire the gig mill was regarded as being morally legal, if not in fact so, when used on coarse cloth. It had been thus used for a very long time, so that custom had assimilated its presence into the work style of the community. The increasing use of the gig mill in the 1790s on fine-quality cloth was, however, regarded not only as illegal but as an invasion of the customary moral code. Attitudes of moral law in the acceptance of machinery extended

[16] *Journal of the House of Commons* 36:7 (hereinafter cited as *JHC*); F. M. Eden, *The State of the Poor* (London, 1797), 3:796; J. Anstie, *Observations on the Importance and Necessity of Introducing Improved Machinery into the Woollen Manufactory* (London, 1803), pp. 19–20; *JHC* 49:599 and 600, 58:352; *A Letter to the Landholders of the County of Wilts. on the Alarming State of the Poor* (Salisbury, 1793), p. 7. See also D. J. Jeremy, *Henry Wansey and His American Journal, 1794* (Philadelphia, 1970), p. 5, for this hostility to spinning machinery.

[17] *GJ*, July 18, 1796; *Public Record Office*, H.O. 42/66, Read to Pelham, September 6, 1802; 42/83, Read's memoranda, n.d.; *BPP, 1802/3, Vol. 7*, pp. 44–45; *GJ*, August 9, 1802; September 6, 1802; *BPP, 1806, Vol. 3*, p. 342. (Public Record Office hereinafter cited as *PRO*.)

beyond those machines actually banned by statute. Randle Jackson, the cloth workers' counsel, argued that while "the shearing frames are not prohibited in the same distinct form of words" as the gig mills, "they are prohibited in spirit and in fact." The weavers' concern over the legislation relating to apprenticeship and loom ownership reflected this view. They were prepared to concede that aspects of the old code were in need of "revision," provided that the "spirit" of the code was retained.[18]

In a similar vein the workers argued that machinery made inferior cloth. Scribbling engines "tear the wool from its true staple," claimed the Gloucestershire scribblers in 1794, while the shearmen continued from 1794 to 1816 to assert that the gig mill "injures the texture or ground of a fine cloth." Both scribbling engines and gig mills were "driven so violently by the impetuosity of the water" that they could not be adequately regulated. Thus, cloths and yarn were stretched and strained beyond their "true drapery." Such faults could be temporarily disguised but would ultimately show up. Machine-made cloth was thus "deceitfully made," contrary to many of the old statutes, and the machine owners were making large short-term profits at the expense not only of the now-redundant laborer but also of the unwitting customer and the long-term reputation of the trade itself.[19] There was some justification for such claims. Machines could rarely perform the most skilled tasks as expertly as could the skilled man. But in the middle- and lower-quality range of cloths, where market growth was fastest, machines had fewer disadvantages. The clothiers, for their part, claimed that machinery would lead to a better product since it would ensure that standards did not fall when rushed orders were necessary.

The woolen workers foresaw that the introduction of machinery would usher in the factory system and destroy the putting-out system. The weavers and master clothiers candidly admitted to the parliamentary select committees of 1803 and 1806 that their main purpose in pressing for active enforcement of the old legislation was to stem the rise of the factory. The Gloucestershire weavers made this perfectly clear in 1804: "It is our wish that no clothier shall erect or keep any

[18]*JHC* 49:599–600; R. Jackson, *The Speech of Randle Jackson to the Committee of the House of Commons Appointed to Consider the State of the Woollen Manufacture* (London, 1806), p. 30; *BPP, 1806, Vol. 3*, p. 342. It is pertinent to note that the woolen workers' concern was to maintain only that legislation which protected their life-style and work customs. They did not draw attention to legislation against combinations or concerning embezzlement. But the innovators did not intend to repeal such acts.

[19]*JHC* 49:599, 600; 71:431; *BPP, 1802/3, Vol. 7*, pp. 104, 118, 159, 215–6; *BPP, 1806, Vol. 3*, pp. 260, 274, 283, 295, 432; Jackson (n. 18 above), p. 53.

factory for the purpose of carrying on weaving contrary to the statutes, but that we may have the work at our own houses." West Riding master clothiers complained in 1794 that the domestic system "is now in danger of being broken in upon and destroyed by the introduction of modes which have prevailed in other parts of the kingdom which are founded upon a monopoly erected by great capitals and set on foot by cloth merchants becoming cloth makers."[20]

The case against the factory, however, was not merely an assertion of inertia. Much more important, its proponents had strong economic and social reasons for wishing to stem the tide of the factory system. The weavers prized the opportunities that domestic work gave them to employ their entire families, but the family work unit was valued as more than merely an advantageous mode of production. The weavers all believed strongly in the social and moral values of domestic work. The system emphasized the patriarchal roles of work organizer and receiver and dispenser of the collective earnings of the family. The family wage emphasized the coherence of family life and was supposed to strengthen family ties. The father's control over earnings was generally believed to act as a beneficial check on the dangerous temptations of adolescence. Because the children lived and worked under the parents' eyes, the domestic system was felt to give parents a powerful influence in controlling their children's personal development. Parents usually took the major role in providing their children with a rudimentary education, partly from necessity but also because the structure of work made this practicable.[21]

The factory threatened all of that. The development of scribbling and spinning mills was felt to have already shown the trend. William Howard, a master dresser, claimed in 1806 that these developments were responsible for "so many boys running about the streets without any shoes or stockings on and nearly half starved." This was not the result of parental inattention. It was the direct consequence of the parents being forced to leave the home to work in factories. Those children who were employed in factories were often paid their own wages beginning at thirteen or fourteen years of age, thereby giving them an unprecedented economic independence, depriving the father of his key role, and potentially diminishing the family income. The factory not only destroyed parental control and moral authority; it also promoted immorality. Although large factories were few in number in 1806, it was this supposed result that aroused perhaps the most out-

[20]E. Wigley, *The Speech of E. Wigley on Behalf of the Woollen Weavers of Gloucestershire to the Committee of the House of Commons Appointed to Consider the State of the Woollen Manufacture* (Cheltenham, 1806); *BPP, 1806, Vol. 3*, p. 342.

[21]*JHC* 49:275–6; *BPP 1802/3, Vol. 7*, pp. 20, 58; *BPP, 1806, vol. 3*, p. 309.

rage and anger. The domestic system was the embodiment of morality. Therefore the workers assumed that its obverse, the factory, embodied immorality. "The factories are nurseries always of vice and corruption, and often of disease, discontent and disloyalty." They "deprave the morals of our labourers and break up their happy domestic labouring parties." Randle Jackson, the shearmen's counsel in 1806, claimed that the factories were destroying apprenticeship, "a moral and political institution," and breaking down "barriers which have been interposed to check and prevent early licentiousness." This stress on morality was no mere ploy to impress a government traditionally concerned with improving the morality of the lower orders. It was deeply felt. Even some advocates of machinery and factories were worried by such moral implications and sought to play down the supposed evils. John Anstie, whose pamphlet shows him to be a paternalist at heart, agreed that "it is particularly desirable to keep women and children as much as possible employed at home." But, he argued, while there were some particularly bad examples, there were other factories, "where, by strict though temperate discipline and attention to decency, the most perfect order is preserved."[22]

* * *

These arguments against mechanization and structural change were essentially the moral stance that it was wrong for machines to take away men's work and to destroy their customs and culture. Such arguments might be hoped to appeal to the consciences and pockets of the taxpayers and paternalists. The workers' case was more widely based than this, however, for they believed that there were also powerful economic reasons for prohibiting the spread of machinery. The case was put most clearly by and on behalf of the shearmen and croppers, but it was a view widely shared by all woolen workers—and by many traditionalist clothiers and master weavers too, as the parliamentary inquiries were to demonstrate.

The shearmen, claimed their advocate Randle Jackson, did not oppose machinery *in general.* That was a "cruel and malignant falsehood." Like "Looker On," a correspondent of the *Leeds Mercury*, they believed that, in some cases, "Machinery may be called in as an auxiliary to human labour where mechanical contrivance will release [men] from occupations too servile or degrading for rational beings, or too slavish and harrassing to their strength, or abridge, facilitate or expe-

[22]*BPP, 1806, Vol. 3*, p. 308–9; *Observations on Woollen Machinery* (Leeds, 1803), p. 19; *Leeds Mercury*, March 5, 1803; Jackson (n. 18 above), pp. 63, 76; J. Anstie (n.16 above), pp. 55, 57–8.

dite man's labour, still keeping him employed." Nevertheless, "too much machinery may be applied to wool, and there is a point when it ceases to be useful and where it begins to defeat the very end for which it was adopted." The woolen workers believed that "A trade or manufacture is deemed valuable to a country in proportion to the number of hands it employs." Thus, argued Randle Jackson, "The dispensing with manual labour is in itself a great and positive evil: it gives a fatal check to population and, in the language of political economy, deprives the land of a portion of its customers, a nation's best and first consideration." Such an evil could only be justified if "the improvement of the article or the saving in price be such as makes amends to the state for the numbers thrown out of work." But none of these criteria applied: "the foreign market is already in our hands, the price unobjected to, and the customer satisfied . . . in such a case there is no advantage to balance its depopulating effects." Hence, machinery would "lessen our home markets more than increase our foreign ones."[23]

There was an additional reason why mechanization should not be further extended in the woolen industry. Advocates of mechanization cited the growth rate of the unfettered cotton industry, but the traditionalists urged that this was an invalid comparison. As the author of a pamphlet titled *Observations on Woollen Machinery* wrote, "What might be very great improvements in some branches of trade are not so in the woollen; because it cannot be increased beyond the quantity of wool grown." While raw cotton could "be produced ad infinitum," supplies of wool were limited and already stretched to capacity. "A new large mill can be much sooner erected than additional sheep reared to keep it in motion." According to John Anstie, the most persuasive pamphleteer in favor of machinery, many intelligent persons were convinced by this argument, but it was in fact "a mere vulgar error, originating at first either in ignorance or from design." Anstie drew on customs returns to prove that imports of Spanish wool had nearly doubled from 1791 to 1799, while he cited one of Pitt's speeches in asserting that enclosures were increasing the number of domestic flocks. Yet, not only did the traditionalists view the supply of raw wool as being limited, but their whole concept of their market was that it was a static one, already fully supplied. "However powerful the machinery made use of, no more cloth could be produced—no more could be sold than is already sold." This being so, it was clearly folly to displace men

[23]Jackson, p. 73; *Leeds Mercury*, February 5, 12, 1803; *Observations on Woollen Machinery*, pp. 8, 3; *Considerations upon a Bill for Repealing the Whole Code of Laws Respecting the Woollen Manufacture* (1803), pp. 13, 14, 35. While this last pamphlet is anonymous, it is clear from its content, style, and remarks, especially on pages 11–12, that it is Jackson's speech to the 1803 select committee in reply to Plomer.

with machines only to sell the same quantity of cloths cheaper. The buyer country's gain was the producer country's loss. Here was the road to economic dislocation and ruin. This theory of the economic value of hand labor and of selling dear was most ably expounded by Looker On. "It is a well known maxim," he argued,

> that a trade is valuable to a country in proportion to the number of hands it employs. Suppose then, that wool could be converted into cloth solely by machinery . . . it would be little better than exporting the raw material. Everyone knows that the more labour is employed on any article prior to its sale, the more generally profitable. Why then, in contradiction to this plain maxim, seek to diminish the number of people employed . . . ? or why endeavour to sell at 13s what would bring home 18s 6d when in all probability all these sixpences lost to the country would not induce the sale of one yard more?[24]

This concept of a static market and of selling dear abroad was no mere ruse to provide a bogus rationale for resisting machinery. It was a view that enjoyed wide currency throughout the 18th century.[25] But the example of the cotton industry—from which the advocates of machinery drew countless parallels, pointing to the potentialities of ever-expanding markets if the product could be made sufficiently cheap—had showed up the weakness of the old view and was rapidly discrediting it.

On similar lines the workers argued that England's preeminence in the woolen trade lay in the skill of her artisans. Defoe, among others, had frequently stressed this point as justification for England's having higher wage costs than its foreign rivals. But machinery superseded this skill and thereby rendered that national asset worthless. Worse, machinery, unlike a skilled work force, could easily be exported. Machinery "facilitates the transferring of the woollen manufacture to other nations." Thus, foreigners would be able to rival her successfully and seize her trade, leaving English workmen to starve. Proponents of innovation rejected this, claiming that, once unfettered, constant

[24]*Observations on Woollen Machinery*, pp. 6, 10, 9; Anstie (n. 16 above), pp. 37, 34–5, 36–7, 41, 45–7; *Leeds Mercury*, February 5, 12, 1803. *Observations on Woollen Machinery*, p. 5, made the same case.

[25]For example, J. Tucker, *A Brief Essay on the Advantages and Disadvantages which Attend France and Great Britain with Regard to Trade* (London, 1753), and *Instructions for Travellers* (Gloucester, 1757). See also D. C. Coleman, "Labour in the English Economy of the Seventeenth Century," *Economic History Review* 8 (1956): 281; and R. C. Wiles, "The Theory of Wages in Later English Mercantilism," *Economic History Review* 21(1968):113–126.

mechanical improvement would ensure that the woolen industry would remain ahead of all rivals. But it is notable that, in spite of such professions of faith in free-market forces, these same innovators were ardent in seeking to uphold and augment legislation against the exportation of machinery.[26]

Finally, the workers feared that the rise of a machine economy would polarize their industry into opposing groups of "overgrown rich monopolists" and impoverished, subservient workers, with fatal consequences for the "spirit, energy and patriotism" of Old England. The new large mills that began to appear after 1800, housing not just preparatory machinery but frequently finishing machinery and sometimes looms as well, betokened the rise of a new type of employer, the industrial capitalist, who exercised direct control over all aspects of the trade, not being content, as were the "good honest gentlemen" clothiers, to put out work for wages or on commission. These men, "a small but affluent and powerful association of master clothiers who seek the completion of their fortunes in an utter freedom from parliamentary restraint," would, if allowed, destroy all established work customs and the woolen workers' cherished independence.

Many small "inferior" clothiers and employers shared the workers' fears of the economic power and aspirations of the factory masters and sided with the workers in opposing repeal. "The general view of those who employ gig mills is to monopolise the trade," warned one Bradford-on-Avon master clothier in 1803. Such "inferior" clothiers constituted the majority of employers, although they produced much less cloth collectively than did their "respectable" counterparts. Thus, in Chippenham there were said to be some fifteen or sixteen clothiers, but only four or five were in favor of repeal. John Jones, the spokesman for the Wiltshire repealers and himself a factory owner, claimed that seven-eighths of the trade in Wiltshire supported repeal but admitted that the majority of clothiers opposed it. Yet, even among the more prosperous clothiers feelings were divided. There were said to be some thirty "respectable" clothiers in Bradford-on-Avon, but only four were in favor of repeal.[27] The same was true of the West Riding. Master clothiers there had petitioned Parliament to prohibit such "Monopolies and Factories" in 1794, and many master dressers were suspected of giving covert support to the machine-breaking shearmen and croppers in 1802 and 1812.[28]

[26]Jackson (n. 18 above), pp. 12–15; Coleman, p. 281; Wiles, pp. 114–126; Anstie (n. 16 above), pp. 28, 59, 62–3.

[27]*Leeds Mercury*, February 12, 1803; *Observations on Woollen Machinery*, p. 15; *Considerations upon a Bill . . .* , pp. 3, 17; *BPP, 1802/3, Vol. 7*, pp. 194; 133, 339–40, 201, 205.

[28]*JHC* 49:276; *PRO*, H.O. 42/66, Read to Pelham, September 5, 1802; Thomis (n. 2 above), pp. 63–4.

Fears about the monopolistic ambitions of the innovators proved a valuable propagandist weapon at a local level, for they echoed the centuries-old distrust of monopolists in the marketplace and the popular hatred of the forestaller and regrater. In the same way, the workers argued, the innovators' unrestrained greed for increased profits was causing the whole community to suffer. Parliament must therefore reassert the old protective legislative codes and save the woolen workers "from falling sacrifice to the spirit of monopoly, to private cupidity in the guise of public good."[29]

* * *

In the final analysis Josiah Tucker was only partly correct. The woolen workers were indeed prey to some "mistaken Notions." In spite of their belief that machinery must be restrained, the triumph of "progress" was inevitable and ultimately beneficial to all. This was the line that Parliament chose, the Commons enthusiastically in the two select committee inquiries, the Lords more reluctantly. As the larger clothiers had wished, the old regulatory legislation was suspended in 1803 and finally repealed in 1809.[30] Thereafter, resistance to change took a primarily physical form. But the woolen workers were neither ignorant nor stupid. They could indeed see beyond "the first Link in the Chain," and it was toward a future for which they had no liking. Many were correct in believing that machinery would take "the Bread out of their Mouths." Others recognized that it would deprive them of something they held equally dear. The shearmen, for example, knew only too well that no increased work opportunities in other branches of the manufacture could compensate them for the loss of their social status and the redundancy of their one asset, their skill.

Nonetheless, the workers' opposition to change had a wider basis than merely the desire to protect their livelihoods. Machinery and the machine owners represented an ideology quite alien to that to which the woolen workers subscribed. The debate over the mechanization of the woolen industry was at root a conflict between two very different views of industry and society. To the innovators, the vital issues were those of progress and profitability, the need to preserve and enlarge markets regardless of traditional practices and work methods. Theirs was an essentially "rational" economic philosophy that emphasized the liberty of capital to do as it wished with its own. The woolen workers, in opposing this "spirit of monopoly," were in no way challenging the role

[29] E. P. Thompson, "The Moral Economy of the English Crowd in the Eighteenth Century," *Past and Present,* no. 50, February 1971, pp. 76–136; *Observations on Woollen Machinery,* p. 22.

[30] *JHC* 58:641–2, 690; 64:368, 405.

of capital per se. They glorified the putting-out structure of their industry and believed that capital, like labor, had its rights and privileges within it. They were at pains to differentiate the "good, honest gentlemen," those who paid full rates and respected custom, from the rapacious innovators. Theirs was in no sense a class analysis. But to them their industry represented far more than just a means of obtaining an income. Their trades were the source of their status in society and the root of their whole culture. They rejected the right of the entrepreneur to throw this into the melting pot in the search for greater profit margins. Their trades were their property, meriting just the same rights of protection as the clothiers' capital. The economic cake was finite. Capital and labor were entitled to their customary slices. But it was palpably unjust and economically disastrous to allow capital to use its power to seize the laborers' slice and leave them to starve. Ultimately, therefore, the workers held an essentially moral view of economic relationships, a view that, like that of the food rioters, harked back to the old paternalist concept of stability and regulation. Like the food rioters, the woolen workers were to find their case undermined by their rulers' repudiation of such a role and acceptance of the amoral economy of the innovators.

Roads, Railways, and Canals: Technical Choices in 19th-Century Britain

FRANCIS T. EVANS

Between 1760 and 1840 Britain passed from a state of local economies, with poor to middling transport, into a nation with the promise of a national railway system superimposed on a network of good canals and roads. This change has often been examined from an economic point of view, and there are some excellent studies of individual technical elements in the transport system.[1] The excuse for this further essay lies in the lack of an overall technical comparison of the roads, railways, and canals as they competed during the Industrial Revolution. Here the aim will be to concentrate on the technical factors, comparing the advantages and disadvantages as they were understood by contemporaries. The question will usually concern how the methods of transport were expected to behave rather than how individual railways or canals performed. Once the choice among roads, railways, and canals appeared, investors, engineers, and managers needed criteria upon which to base decisions; and these criteria, however crude, were often technical.

When railways were young, there was not a great difference between advanced technical thought and the information digestible by the interested outsider; transport engineering did not require much specialized terminology or mathematics. Consequently, it is reasonable to base this comparison of popular technical expectations upon

Francis Evans is the semiretired professor of the history of technology at Sheffield Hallam University and lectures widely in Britain, Ireland, France, Switzerland, and Russia. His teaching models often appear on BBC TV programs and in exploratories.
The late Paul Nunn and David Tew, also William Moore, are thanked for helpful suggestions.

[1]For general histories of transport see W. T. Jackman (*The Development of Transportation in Modern England* [1916; reprint ed., London, 1962, with a valuable bibliographical introduction by W. H. Chaloner]), H. J. Dyos and D. H. Aldcroft (*British Transport, an Economic Survey from the Seventeenth Century to the Twentieth* [Leicester, 1969]), T. C. Barker and C. I. Savage (*An Economic History of Transport in Britain,* 3d ed. [London, 1974]), and P. S. Bagwell (*The Transport Revolution from 1770* [London, 1974]).

commonly available sources, particularly encyclopaedias. Successive editions of *Encyclopaedia Britannica* and *Chambers's, Rees's, Tomlinson's,* and *Penny* cyclopaedias reflected very faithfully the swings of advantage and disadvantage which characterized the alternative transport systems.[2] This study examines some crucial points pursued further in *Mechanics Magazine,* Priestley's *Navigable Rivers and Canals,* and other sources, but even these are representative of the broad climate rather than of a specialized elite.[3] It is felt that these sources reflect opinion more accurately than generalizations based, for instance, on the best contemporary practice or on the multitudinous patents which throw little more than an amusing sidelight on events. The better encyclopaedias gave clear accounts which were kept up to date in successive editions, and their comparisons illuminate the changing fortunes of roads, railways, and canals, particularly in the interesting period when the canals were in technical trouble, amounting almost to a crisis, and while railways were still working toward a coherent technical structure.

Roads

In the mid-18th century, the English road system looked complete on the map, with adequate connections among villages and towns. But despite the early turnpikes and pioneering work by Tresaguet in France and John Metcalfe in England, much of the system was almost unusable by wheeled vehicles.[4] In many areas, packhorses were the only means of transporting loads. Measurements of the carrying power of a horse varied greatly, but it was never suggested that horsepower was most efficiently employed in carrying instead of pulling loads. The *Britannica* reported that "the most disadvantageous way of employing the power of a horse is to make him carry the load up an inclined plane, for it was observed by de la Hire that *three* men, with 100 pounds each, will go faster up an inclined plane than a horse with 300 pounds. When the horse walks on a good road, and is loaded with about two hundred weight, he may easily travel 25 miles in the

[2]*New and Universal Dictionary of the Arts and Sciences* (London, 1751); Abraham Rees, ed., *Chambers' Cyclopaedia* (London, vol. 1, 1786); *Encyclopaedia Britannica,* 3d ed. (Edinburgh, 1797), 4th ed. (Edinburgh, 1810); Abraham Rees, ed., *Rees's Cyclopaedia* (London, 1819); *Encyclopaedia Britannica,* 4th ed., suppl. (Edinburgh, 1824); *Penny Cyclopaedia* (London, 1833), suppl. (London, 1845); *Tomlinson's Cyclopaedia of the Useful Arts* (London and New York, 1854); *Encyclopaedia Britannica,* 8th ed. (Edinburgh, 1853).

[3]*Mechanics Magazine* (Weekly from August 30, 1823). J. Priestly, *Navigable Rivers and Canals of Great Britain* (London, 1831; reprint ed., Newton Abbot, 1969).

[4]P. Mantoux, *The Industrial Revolution in the Eighteenth Century* (London, 1966), pp. 112–20.

space of seven or eight hours."[5] De la Hire's 2 hundredweight represented a cavalryman and his equipment, but packhorses could carry much more.[6] Often they went in long lines, single file, along the narrow roads. In 1739, two gentlemen traveled from Edinburgh to Grantham, only 110 miles from London, mostly on "a narrow causeway, with an unmade soft road on each side of it; . . . they met from time to time gangs of thirty to forty packhorses."[7] Here, and in plenty of other parts of Britain, there was not room to pass, even with packhorses. A road system like this, with parts virtually untouched since the departure of the Romans, made the more efficient use of horsepower with wagons difficult.

Yet underlying these defects was a more fundamental one. The very complete parliamentary act of 1773,[8] a major public policy statement on roads and their upkeep, shows the problem clearly by its omissions; it was simply not understood how to make a durable road over which a vehicle could travel easily; and the act was more concerned with protecting the roads against vehicles. It is so complete in other respects that the failure must be set down to ignorance. This lack of knowledge was not peculiar to legislators, since contemporary encyclopaedias, otherwise strong on technical matters, betray the same ignorance of road building.[9] They merely defined the term "road," describe how the Romans built them, and, in lengthier works, enumerate the major Roman routes.

The general ignorance of the principles of good road building persisted till the century's end, but it was moderated in a number of ways. Although there was no body of doctrine like that held by the engineers of the *ponts et chaussées* in France, individuals like Metcalfe built soundly; and in other places sheer hard work improved surfaces, flattened slopes, or straightened curves. John Harriott's road harrow

[5] *Encyclopaedia Britannica,* 4th ed., s.v. "Mechanics." However, William Marshall gives an instance where two packhorses and a boy were nearly comparable with a wagon and three horses, "for the dispatch made, by packhorses properly used, is such as no one, who has not seen it, would readily apprehend" (*Rural Economy of the West of England* [London, 1796; reprint ed., 1970], 1: 115).

[6] *Encyclopaedia Britannica,* 3d ed., S. V. "Horse."

[7] Quoted in J. Copeland, *Roads and Their Traffic, 1750–1850* (Newton Abbot, 1968), p. 12.

[8] Parliamentary Act, 1773 (7 Geo. 3, cap. 42).

[9] *New and Universal Dictionary of the Arts and Sciences,* s.v. "Road." *Encyclopaedia Britannica,* 3d ed., s.v. "Road." *Chambers' Cyclopaedia,* vol. 4, 1783, s.v. "Road." The articles "Highway" and "Turnpike" in *Chambers'* give the statutory provisions for regulating the surveying of roads and the legal requirements for vehicles but say nothing about construction methods.

is a good example of the simple improvements taking place in the late 18th century.[10]

The degree of confusion in all this pragmatic concern for roads shows up in the *Rees's Cyclopaedia* article, "Roads in Rural Economy."[11] The writer favored a moderately convex road of one-in-twelve camber, since excessive curvature defeated its own object; on highly convex roads, vehicles bore down too heavily on the outside wheel, overturned, or most usually, drove astride the crown of the road. In either case, deep furrows resulted and retained the water which the convexity was intended to dispel. Yet the article also conceded that Bakewell and Wilkes had made great improvements with their concave and sloping roads, which allowed water to flush surfaces clean and smooth. A further confusing factor was the variability of road-making materials and of the soils to be traversed. Parts of Lancashire, for instance, had limestone close at hand, which knit together for a good surface. But elsewhere in the county, stone and gravel were not hard enough to support the increasing weights and volume of transport, so costly paving stones were imported from Wales. Turnpike road near Manchester, for example, cost £2,000 a mile to pave. Perhaps the important point was not that a confusing picture emerged, but that the many detailed descriptions of roads in each county showed the scope of the problem and made possible a more fundamental approach.

In any case, despite the disagreements apparent in Rees, the late 18th century brought a great increase in road travel. "Flying Machines"—coaches with springs—were in service on many routes, and fast, regular mail coaches were making passenger travel and postal services a commonplace rather than an adventure into the unknown.[12] Much of southwest England, especially Cornwall, continued to use packhorses, and travellers like Arthur Young had plenty of road horror stories to tell. But that did not stop Parson Woodforde confidently making for Bath or London from Oxford in a day, or even from London to Bath in the same time.[13] G. L. Turnbull has

[10]"A man and a boy and two horses will do three miles in length in one day completely, harrowing down the quarters and drawing the stones together, which by means of the mould boards are dropped into the rut" (*Transactions of the Society for the Encouragement of Arts, Manufactures, and Commerce* [London, 1789], 7: 198).

[11]*Rees's Cyclopaedia*, s.v. "Road in Rural Economy." The article relies heavily on essays on road building from communications of the Board of Agriculture. See, e.g., William Marshall (*Review of the Reports to the Board of Agriculture from the Northern Department* [York, 1808]; *Rural Economy of Yorkshire* [London, 1788], 1: 180–93).

[12]For improvements in coach times, see Harold W. Hart, "Some Notes on Coach Travel" (*Journal of Transport History* 4 [1959–60]: 147–49).

[13]James Woodforde, *Diary of a Country Parson*, ed. J. Beresford (Oxford, 1963). See entries for May 23, 1765; Sept. 14, 1766; June 28, 1793.

given a quantitative idea of the frequency and volume of the road carrying trade during the Industrial Revolution in Britain. Around 1780, there were 126 carrier departures weekly from Birmingham, 156 from Manchester, and 126.5 from Bristol. More than half of the services from these and other major towns stopped within 30 miles, but each town had a frequent London service—18 departures weekly in the case of Birmingham. This clearly indicates a national network, but the defects of road transport were still there: perhaps 4 tons per wagon load and a three-week round trip between London and Liverpool. A great expenditure of horsepower and time was needed to move a small volume of goods.[14]

The *Rees's Cyclopaedia* article on roads appeared in 1819, the same year as John Macadam's report on his system of road building and mending to the House of Commons. Five years later *Encyclopaedia Britannica* published a new article on road making by Thomas Young. As might be expected, Young's approach was rational, scientific, and coherent. Beginning from the premise that "the grand object of all modern roads is the accommodation of wheel carriages," Young analyzed the natures of wheels and roads as a connected problem.[15] Previously, as we have seen, the tendency was to separate the problems of wheels and roads on the assumption that roads were inherently weak and that they needed protection against the traffic running along them. Young considered the function of a wheel and deduced the need for smooth, hard roads. There followed a discussion of stone pavements and gravel roads and a brief account of the confused state of knowledge revealed in the Board of Agriculture papers, but Young concluded that Macadam's simple and economical system had superseded all these other real or imaginary improvements.[16]

John Loudon Macadam himself was well aware of the shortcomings of his predecessors—indeed, the road-building profession had "become contemptible in the greatest degree" and "perfectly adapted to . . . the most ignorant day labourer," while the system was worked by "surveyors selected from the lowest and most illiterate class of the community."[17] The recent increases in commerce had merely led to

[14]G. L. Turnbull, "Provincial Road Carrying in England in the Eighteenth Century," *Journal of Transport History,* n.s. 4 (February 1977): 26–30.

[15]*Encyclopaedia Britannica,* 4th ed., suppl., s.v. "Road Making."

[16]"But it may be said of roads as of governments 'that which is best administered is best,' whether a very smooth pavement not too slippery, or a very hard gravel road not worn into great inequalities" (ibid.). Oddly, Young does not mention Telford's system, though the latter's great report to Parliament on the Holyhead Road was published in 1820. See also *Repertory of Arts, Manufactures and Agriculture,* 2d ser. (London, 1820), 38: 16, 82, 149, 213. By the 8th ed. of *Encyclopaedia Britannica* (1859) Telford's methods were more highly regarded.

[17]*Repertory of Arts, Manufactures and Agriculture,* 26: 40. This reprints Macadam's

great quantities of stone being thrown onto the subsoil, leaving the roads still barely passable. The core of his argument was that "Nothing has been written on the subject of the surface of roads, or the means of making them proper for the easy passage of carriages, though volumes have been published to recommend many useless and many vexatious restrictions on the carriages themselves." Macadam's methods for producing a "strong, smooth, solid surface" have often been described, but his reasoning is worth repeating. A surface of compacted 6-ounce stones would not cause jolting and shaking of the carriage and consequent repercussion upon the surface "which is the real cause of the present bad state of the roads of Great Britain." A firm hand was needed to establish the new standards of road engineering, but until that time, he concluded that "whatever carriages the law may compel men to draw through such roads (for at present they do not travel over them) must continue to act as ploughs."[18]

Macadam may have underrated Metcalfe and overrated the durability of his own roads, but he was right in emphasizing the hard, smooth surface. Thomas Young had calculated the resistance offered by a soft road to the passage of a wheel. Normally a horse exerted a pull of about 200 pounds. If a 3-ton wagon with wheels 4 feet in diameter sank just 1 inch into the surface, it required a force equal to at least one-seventeenth, and more likely one-ninth, the weight of the wagon to pull it along. This meant an additional pull of between 400 and 700 pounds—in other words, two or three extra horses. A 2-inch sinkage of the wheel would add half as much again, say three or five more horses.[19] These calculations were borne out by experiment. McNeill found that a wagon which needed a pull of 33 pounds on well-made pavement, or 46 pounds on 6 inches of broken stone laid over large stone pavement, required four times the pull—147 pounds—on thick gravel laid on earth. These and other contemporary calculations offer a rough but useful agreement on how much horsepower was needed on the bad, old roads.[20] On a poor 18th-century road, a horse could scarcely pull 1 ton, while a better road increased this to 2 tons.

---

papers to the House of Commons. Marshall likewise remarked "The surveyors of roads, in general, are as uninformed, or as inattentive, about the REPAIRING of roads as they are about the forming of them" (*Rural Economy of Yorkshire,* 1: 185).

[18]*Repertory of Arts, Manufactures and Agriculture,* 36:40.

[19]*Encyclopaedia Britannica,* 4th ed., suppl., s.v. "Road Making."

[20]*Penny Cyclopaedia,* 1833, s.v. "Road." See also H. Law, *Rudiments of the Art of Constructing and Repairing Common Roads* (1855; reprint ed. Bath, 1970), pp. 56–59.

Canals

The horse which moved 1 ton on land could pull 30 tons or more in a floating barge. This was the great advantage of canals.[21] Britain's long indented coastline made her fortunate in her natural waterways, and from 1750 she began to emulate and then surpass the canal systems of France and Holland. By the 1820s some 2,200 miles of canal had been built and "there was no place in England south of Durham that was distant 15 miles from water communication."[22] Canals offered more than a great savings in horsepower. Fragile loads like pottery suffered far less risk as waterborne freight than if they were subjected to the jolting of the roads, and the dense but yielding medium of water sustained weighty loads which could not be carried on the roads. For these reasons the artificial waterways spread, growing straighter, wider, and deeper as capital and profits furthered their expansion.

When we consider the level of hydraulic skill possessed by the canal builders, we appreciate better the magnitude of the technical switch as railways replaced canals. Even in 1800, most engineers were oriented toward hydraulic rather than steam power, and the common steam engine was a massive, slow pump. Watt's rotating engines were only fifteen years old, and compact high-pressure engines were almost unknown. Meanwhile, engineers were employing a developed hydraulic technology.[23] At Chemnitz, hydraulically created air pressure forced drainage water up 96 feet from the mines, and Montgolfier's ingenious hydraulic ram also raised water with scarcely any moving parts. Smeaton's experiments in 1753 on waterwheels and French achievements, like Poncelet's efficient undershot wheels, are paralleled by Joseph Bramah's beer pump, flushing lavatory, and hydraulic press.[24] Their sureness of touch and self-confidence can be seen in many of Brindley's works, like Wet Earth Colliery, where he supplied an underground waterwheel from an inverted siphon which took

[21]T. S. Willan (*River Navigation in England, 1600–1750* [Oxford, 1936]). See also A. W. Skempton ("The Engineers of the English River Navigations 1620–1760," *Transactions of the Newcomen Society* [1954], pp. 25–53). Skempton gives the typical loads for a single horse as: packhorse, ⅛ ton; stage wagon on "soft roads," ⅝ ton; stage wagon on macadam roads, 2 tons; wagon on iron rails, 8 tons; barge on river, 30 tons; barge on canal, 50 tons. These figures pretty faithfully reflect the order of difference among modes of transport, apart from the difference between edge rails and tramways, which appeared later.

[22]*Tomlinson's Cyclopaedia*, 1854, s.v. "Navigation, Inland."

[23]D. S. L. Cardwell, *Watt to Clausius* (London, 1971), p. 71. See also J. Reynolds (*Windmills and Watermills* [London, 1970]) for a well-illustrated account.

[24]O. Gregory, *A Treatise of Mechanics* (London, 1826), 2: 224, 241, 128.

water from the Irwell River and passed underneath the river to reach the mine.[25]

Britain's canal system presented massive opportunities for applying hydraulic techniques. Great canal terminals like Bugsworth had many wharves and feeder tramways, resembling in function the later railway marshaling yards. Constructional feats accentuated the water engineering. Thomas Telford's cast-iron aqueduct at Pontcysyllte stood 120 feet high on its masonry pillars and carried the canal 1,007 feet in nineteen spans.[26] Of the 42 miles of tunnels, Standedge was the longest, at 5,415 yards, and took sixteen years to build. Canals ran miles underground in the Duke of Bridgewater's collieries, where they had inclined planes to pass barges from one level of canal to another. Thus the nuisance of water to be drained from the workings was ingeniously turned into the benefit of a useful transport system. Few canal engineers had such unwanted water near at hand, and their successful completion of a canal depended on a reliable supply of water. They sought springs and streams, built reservoirs, and, as the last resort, sometimes set up pumping machinery to fill their artificial navigations.

Subordinate Railways

The early iron railways were an ancillary to this impressive maturing of British canals. Although, as we shall show, the canals were to reach a near-crisis state through inherent technical problems which proved intractable, the railways which served mineral traffic and fed the truck network of canals were, at first, not viewed as any challenge. The cost of constructing and maintaining canals limited their successful application to the main arteries of trade, and short runs of railway were a useful supplement. Wooden rails had long been used to facilitate mine transport, and when cast-iron rails were introduced at Coalbrookdale in 1767 their use spread quickly—a horse could pull five or six times as much on these as on a common road. However, they were far inferior to canals in reducing traction.

The early iron railways were short, local, and private, so that there was no call for standardization of gauge or pattern.[27] South Wales and Shropshire generally preferred tramways, sometimes called plate-

[25]A. G. Banks and R. B. Scholfield, *Brindley at Wet Earth Colliery* (Newton Abbot, 1968), pp. 42–72.

[26]N. Cossons, *Industrial Archaeology* (London, 1976), pp. 361–62. See also *Rees's Cyclopaedia* s.v. "Canals."

[27]Brian Morgan states that "at least fifty variants were in use in Shropshire alone" (*Railways: Civil Engineering* [London 1971], p. 15). Morgan gives a good survey of early types of rail (pp. 10–24).

ways, whose raised edge kept the plain wheel on the track. They were popular after 1787, largely on the grounds that suitable road wagons could also use the rails. At first they were cast iron, and their structural weakness led to fishbellied flanges underneath and a variety of other modifications. The other kind—the edge rail—was earlier and at first was simply a cast-iron bar set on edge. This called for trucks with flanged wheels, although these were unsuitable for ordinary roads. The advantages of the edge rail were that it did not hold stones and dirt or offer as much friction as the flat tramplate with its raised flange. It was soon realized, too, that edge rails formed a stronger load-carrying beam, since their material could be disposed to give greater depth, but even so the plateways persisted in South Wales, the Surrey railway, and some other places. Wrought-iron edge rails began to replace cast iron at Sir John Hope's colliery near Edinburgh, and in 1820 John Birkinshaw of Bedlington patented his new wedge-sectioned rails which he rolled from puddled iron[28] (see fig. 1). At first the edge rails only predominated in northern England and most of Scotland, but on a plateway or tramroad, a horse could only pull 5 or 6 tons, whereas on Sir John Hope's edge railway 10.5 tons could be

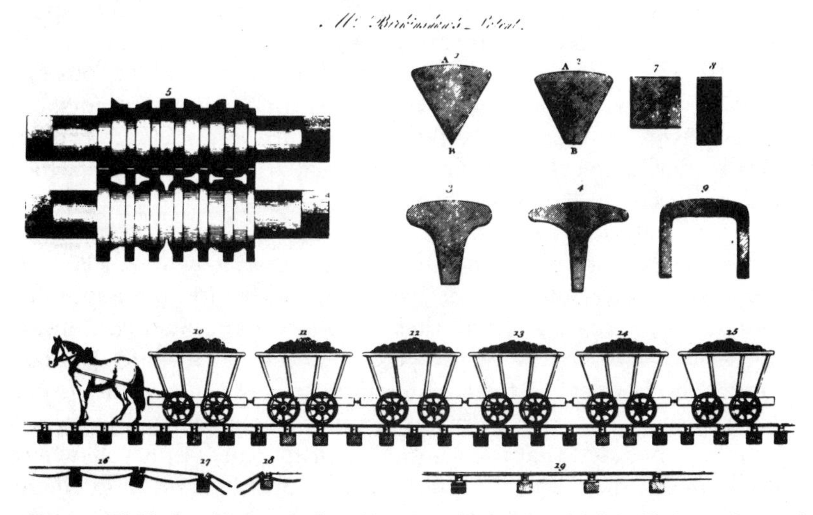

FIG. 1.—Birkinshaw's patent rails, 1821. Note the fractured fish-bellied, cast-iron rail. (*Repertory of Arts, Manufactures and Agriculture,* 2d ser. [London, 1821] 39: 206.)

[28]*Repertory of Arts, Manufactures and Agriculture,* 39: 206–212. These rails were very successful. "We have seen one of these patent rails at Sir John Hope's colliery, and it certainly forms the most perfect iron rail which has hitherto been contrived; combining in its form the qualities of lightness, strength and durability." *Encyclopaedia Britannica,* 4th ed., suppl., s.v. "Railways."

taken at 4 miles an hour. This was a great improvement, though it still left canals with a comfortable margin of superiority, since on a canal a horse pulled three times what it could manage on the best of the railways.

The new railways were usually no more than 10 miles long, in keeping with their roles as private lines or feeders to the canals. The Hay Railway with only 24 miles was the longest. Their independent public life began with the Surrey Iron Railway of 1803 from the Thames at Wandsworth to Croydon. This was followed in 1806 by the Oystermouth Railway in South Wales. Rails suited the mineral traffic in hilly regions, where canal locks would be too frequent, and careful surveying could save animal power by downward inclines on the loaded run.[29] South Wales soon acquired over 300 miles of tramway.

In the early years of railways, before 1825, engineers were remarkably free to devise alternative systems. H. R. Palmer's overhead monorail proposals claimed the advantages of freedom from difficulty during snow and ease of construction over uneven ground.[30] A stone tramway was built into London's commercial road for 2 miles to the West India docks. There were 22 feet of macadamized surface for horses and light carriages; 9 feet of stone paving carried stagecoaches; and a tramway of heavy stone blocks, their outer edges 7 feet apart, was to take the heavy wagon traffic. On this tramway a horse could pull 8–10 tons. The early 19th century thought radically about systems and did not separate ideas sharply into compartments. Sir George Cayley, for example, dubbed his proposals for caterpillar tracks a "new universal railway"—a logical use of the word which would not occur to many modern minds (see fig. 2).

Opinion continued to favor iron rails, and a common argument was that the deeply worn stone road blocks recently uncovered at Pompei showed that even ancient Rome had found difficulty in maintaining such a surface. Apart from all these localized experiments, railways had their broad dreamers too, like Thomas Gray and William James,[31] who both called for a national system. But it was still normal in 1824 to see canals as the greatest economizers of horsepower, and *Britannica* concluded that the railway "is principally applicable where trade is considerable and the length of conveyance short, and chiefly useful therefore in transferring the mineral wealth of the Kingdom from the mines to the nearest land or water communication, whether

[29]See *Rees's Cyclopaedia,* s.v. "Canal." Various authors quoted here were in agreement that descending rail traffic needed less horse traction than canals.

[30]*Mechanics Magazine* 3 (December 25, 1824): 213.

[31]T. Gray, *Observations on a General Iron Railway* (London, 1821). See also, E. M. S. Paine, *The Two James's and the Two Stephensons* (1861; reprint ed., Newton Abbot, 1961).

sea, river or canal."[32] *Rees's Cyclopaedia,* 1819, put the railways among the canals in its thorough gazeteer of British canals. This, after all, was the sensible place for a minor system which common opinion saw as a supplement, not competitor, to canals. Twelve years later, Priestley's *Navigable Rivers and Canals* also put railways in with canals but, as we shall see, had a quite different view of the railways' potential.

Canals' Internal Problems

Although water transport remained the natural choice for a main system until the middle 1820s, serious drawbacks slowly emerged arising from the problems of water supply, the nature of locks, and the behavior of a barge in water. As soon as canals needed to change

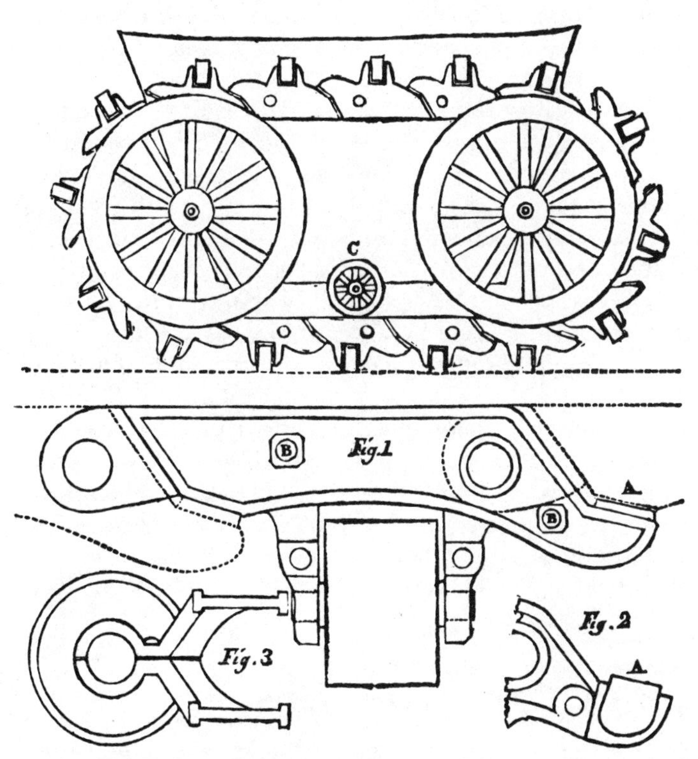

FIG. 2.—Cayley's caterpillar "railway", 1826. The links do not flex inwards, emphasizing the concept of a continuous rail. ("Sir George Cayley's Patent Universal Railway," *Mechanics Magazine* [January 28, 1826], p. 225.)

[32]*Encyclopaedia Britannica,* 4th ed. suppl., s.v. "Railways."

level, some considerable discontinuity in the journey resulted. Brilliant feats of surveying reduced these discontinuities to a minimum: near Coventry there were 73 level miles, Manchester had 70 miles, and there were at least twenty sections of canal with more than 10 miles of level pound. Even the unlikely terrain between Abergavenny and Brecon, where the Welsh mountains begin, had 14 miles free from locks. A lock could easily take 50,000 gallons—over 220 tons—of water. Only one lockful was needed for a pair of boats if they were travelling in opposite directions, but a series of barges following each other up or down needed a lockful for each vessel. The canals found themselves in serious competition with industry which still relied mostly on waterpower, for whose high torque and low speeds many industrial plants were designed.[33] To make matters worse, the depth of locks was limited by the great pressure of water, which at 30 feet exerts a force of more than 2,000 pounds per square foot, "but as the lower gate is strained in proportion to the depth of water which it supports, when the perpendicular height of water exceeds 12 or 13 feet, more locks than one become necessary."[34]

In broken country staircases of as many as thirty locks were built, and the water expenditure—especially in time of drought or when crossing a watershed like the Pennines—became formidable. Anderson, arguing against locks, pointed out that even with good locks only one boat could pass each ten minutes. "In short, in a canal constructed with water locks, not more than six boats on average can be passed in an hour, so that beyond that point all commerce must be stopped."[35] The constrictions imposed by water shortages and the hindrance of locks set a sharp limit on this main transport system of a fast growing economy.[36]

Engineers made a major inventive effort in the thirty years after 1790 to alleviate the water shortage or to do away with canal locks. Their suggestions fall into three broad categories, which take up no fewer than eleven quarto pages of *Rees's Cyclopaedia:* devices for saving water, methods of lifting boats or loads, and the use of inclined planes.[37] Ordinary locks were vulnerable enough to rough usage,[38]

[33]Cardwell, p. 71.

[34]*Encyclopaedia Britannica,* 4th ed. s.v. "Canals."

[35]Ibid. See also A. P. Usher (*History of Mechanical Inventions,* rev. ed. [Cambridge, Mass., 1954], p. 7) for the capacity of early transport systems.

[36]Roads could compete in speed with canals for the transport of less bulky items. There was a profitable "fly van" service between London and Manchester, taking 36 hours (G. L. Turnbull, "Pickfords and the Canal Carrying Trade 1780–1850," *Journal of Transport History* 6 [March 1973]: 5–29).

[37]D. H. Tew, "Canal Lifts and Inclines," *Transactions of the Newcomen Society* 28 (1951–52): 35–38.

[38]R. Harris, *Canals and Their Architecture* (London, 1969), p. 78.

and some of the suggested machines asked too much of a hasty or clumsy canal employee. But, as we shall see, some of the schemes were worthless, even assuming the technical means of achieving them in 1800, and few gave promising results.

A number of methods of saving water were attempted. Where barges varied in size, narrow and wide locks were set side by side to avoid using more water than necessary. Another approach was to build side pounds, which could store the first water drained from a lock; only the bottom portion was passed into the lower level of the canal, and the first part of refilling the lock was undertaken from the side pounds. Playfair's design used no fewer than ten side pounds, each taking 1-foot depth from the lock, and only one-sixth of the water was lost. There were simpler arrangements, but all added substantially to the time spent passing through locks. Lawson Huddleston and Robert Salmon independently came up with methods of raising and lowering the lock water by means of huge plungers forced mechanically into the lock, so that no water needed to pass into the lower pound; these we may safely list as improbable.

Lifting devices offered a promising line of development, for it was well within the power of 18th-century technology to lift 20 tons or so (see fig. 3). It was another matter, however, to lift, move, and lower boats the 12 feet between pounds often, quickly, and safely.[39] Lift designs were put forward by Anderson (1794), Fussell (1798), Rowland and Pickering (1794), and Robert Fulton (1794). Most lifts used caissons of water which counterbalanced each other even if only one held a barge, and lifting was achieved by letting water out of the ascending caisson. This had the advantage that a barge constantly supported by water did not need to be so strong as one made to be suspended by a few points. A crane inside the Duke of Bridgewater's canal tunnel at the Worsley colliery worked by the weight of water tubs which were filled by an underground stream. The crane had a brake wheel "of sufficient size that a man who stands before the lever thereof has his two hands at liberty to pull the lines which connect with the valves, and give signals to those below, while by lunging or stepping forward with his breast against the lever, he can in an instant stop the machinery."[40] The crane moved boxes through a height of 180 feet. Brindley also made the water tubs pump water from the lower to the upper canal when they were not needed for the crane. Apart from Brindley's crane, and an early handworked one on the Churprinz Canal in Saxony, there is no evidence that any of these lifts worked satisfactorily.[41] Later, in the 1870s, a good lift was built at

[39] Tew, pp. 35–38.
[40] *Rees's Cyclopaedia,* s.v. "Canals."
[41] Tew, p. 46.

Anderton, but the canals needed successful solutions in 1830, and lifts were not among them. Robert Weldon made the most remarkable attempt to supersede pound locks with his "hydrostatick lock," which was tried in 1794 in Shropshire and built in 1797 at Combe Hay on the Somerset coal canal. The barge was floated into a great tubular wooden caisson. When this was shut, water was pumped in and the caisson sank to the lower level where it butted against the exit. The doors were opened and the barge floated out. It is noteworthy that when the brickwork bulged, the sinking caisson experiment was dropped. An inclined plane, and, later, pound locks were substituted. The new locks for the 138-foot rise forced the company to add £45,000 to the £80,000 capital they had originally sought for the canal.[42]

Inclined planes turned out better than either the water savers or the lifts. William Reynolds made the first successful attempt at Ketley in Shropshire, where boats twenty feet long entered a lock and settled on wheeled cradles, on which they descended 73 feet on rails. This plane

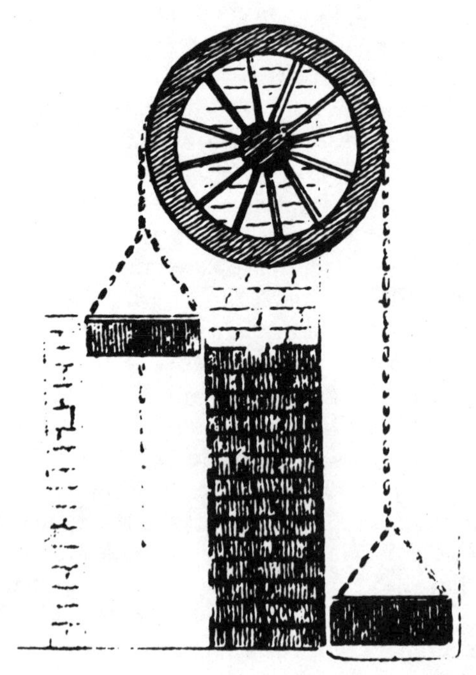

FIG. 3.—Anderson's design for a canal lift, 1801. (*Encyclopaedia Britannica,* 3d ed., suppl. [London, 1801] 1: 152.)

[42]J. Priestly, pp. 580–81 (see n. 3 above).

could pass 400 tons a day, with counterbalanced loads requiring no outside power. Five more planes were built near Coalbrookdale, of which the largest—the Hay incline—still remains. The Hay plane rises 207 feet in a horizontal distance of 1,000 feet. Instead of a lock, the boat and its cradle are wound out of a basin in the top pound before traveling down. An ingenious arrangement of wheels and two rail gauges allowed the boat to remain horizontal on both legs of the journey. The Hay incline saved building twenty-seven locks and passed a pair of boats in three and a half minutes.[43] The group of planes at Hay worked well for as long as they were needed, from 1793 till 1907. Even more remarkable was the underground plane, rising 107 feet, which joined different levels of the Duke of Bridgewater's Worsley mine.[44] A few dozen inclined planes were built, and many worked successfully. It seems fair to comment, however, that this sole successful alternative to locks on canals before 1840 was essentially an application of railway principles rather than a true hydraulic solution.

It is easy in retrospect to see that hydraulic engineering was reaching a crisis which did not touch the coal-powered technologies. Water was a limited resource which gave both power and transport. To the canal engineer it seemed that transport should take precedence over industrial needs, since this would facilitate the transition of industry to steam power by means of easier coal supplies. *Rees's Cyclopaedia* argued in 1819 that canals and their extension "remove one of the principal objections to steam engines, by enabling new mines of coals to be daily opened, and the products thereof, as well as of the old mines, to be regularly and cheaply conveyed to every situation where engines can be wanted."[45] This seems almost a wistful treason against the hydraulic tradition. If steam transport and steam-powered industry had not triumphed, no doubt the hydraulic tradition would have produced an integrated system to optimize the limited water resources. But the *Cyclopaedia* further maintained that:

> we would not, however, be supposed to recommend the annihilation of water mills; on the contrary it hath long appeared to us that their number and their power might . . . be increased, and yet all the purposes of canals be fully answered, and those most capital improvements of irrigation and drainage at the same time be extended, to very large tracts of land; for this purpose it would be necessary, that an entire valley of considerable extent, that has

[43]B. Trinder, *Hay Inclined Plane* (Ironbridge: Ironbridge Gorge Museum Trust, 1973), p. 4.

[44]*Rees's Cyclopaedia* (s.v. "Canals"); see also Cossons, p. 345.

[45]*Rees's Cyclopaedia,* s.v. "Canals."

> a good stream of water through it, as the *Colne* or the *Lea* near London for instance, should be put under a system of improvements.[46]

Improvement meant drainage and the exploitation of springs; the conversion of mills from undershot to overshot, to double efficiency; and better exploitation of waterhead, by transmission in pipes under pressure and siphoning. Had the demon of unlimited fossil fuel not beckoned, perhaps the paradise of living on renewable resources would have come into the future then, instead of now! But all this was said in 1819, and in 1824 George Buchanan could still write that "On some of the railways near Newcastle, the waggons are drawn by means of a steam engine working in a waggon by itself, the wheels of which are driven by the engine. But this application of steam has not yet arrived at such perfection as to have brought it into general use."[47]

The New View of Railways

If, despite their problems, canals retained their position as the most favored form of bulk transport until well into the 1820s, we must ask how and when they lost this position to locomotive-powered railways. If we seek practical demonstration of the locomotive's effectiveness, then it is almost indisputable that *Locomotion* on the Stockton and Darlington, 1825, or the performance of *Rocket* at the Rainhill trials in 1829 were the two decisive events. It is also important to ask, however, at what point the theoretical advantages of railways became manifest as a part of popular technical opinion. When was the ideological victory? It is not necessary in this context to do more than mention the locomotive builders: Trevithick in 1804 at Pen-y-darren; Blenkinsop and Murray in 1812 with toothed wheels and rack rails, but also two cylinders to smooth the power output; Hedley's *Puffing Billy,* 1813; and finally the series of George Stephenson designs beginning with *Blucher* in 1814. However, these men did not create public opinion, and our own familiarity with locomotives should not make us forget how alien the steam locomotive was in an age when traction was synonymous with horses. Public opinion was perhaps first alerted by William James, whom L. T. C. Rolt called "the John the Baptist of railways."[48]

In 1822 William James wrote: "In comparison with navigable canals, generally speaking, articles may be moved by this improved en-

[46] Ibid.
[47] *Encyclopaedia Britannica,* 4th ed. suppl., s.v."Railway."
[48] L. T. C. Rolt, "Introduction," in Paine, p. v.

gine system three times as fast, at one third the expense, and with the advance of only one-seventh the capital in the construction."[49] James is not expressing common knowledge in this report on the possibilities of a railway line in East Anglia. Lord Hardwicke replied: "Though I am very sensible of the advantage of iron railways, yet I should certainly prefer the execution of the canal, for which the Act was obtained with so much difficulty. . . ."[50] Even dealing with a railway company, James could be equally unconvincing. At the first shareholders' meeting of the Stratford and Moreton Railway, July 13, 1821, James (a member of the committee of management) recommended Birkinshaw's malleable iron rails and Stephenson's locomotives; the new railway never used any power but horses and nearly chose cast-iron rails.[51] Simmons has shown that the directors probably chose correctly, but the incident certainly indicates that locomotives on rails were not yet the conventional wisdom.

In 1824 *Mechanics Magazine* reprinted from the *Scotsman* a serialized unsigned article on railways which constitutes an ideological breakthrough, stating in a detailed and sound argument the case for locomotive railways as the rational choice for a transport system.[52] It begins with the familiar argument that a horse on a canal will pull as much as thirty horses with wagons or 120 packhorses can carry, and it admits that, though cheaper to build, the railway hitherto had lacked some of the canal's advantages. It then further argues that "We are quite satisfied, however, that the introduction of the locomotive steam power has given a decided advantage to railways [which will lead to] almost boundless improvement and is destined, perhaps, to work a greater change in the state of civil society, than even the grand discovery of navigation."[53]

Two main reasons were put forward for this swing of the technical balance in favor of railways. In the first place, a horse exerted its best power at low speeds. At a dead pull a horse exerts a traction equal to 150 pounds which is reduced to less than one-half when he travels 8 miles per hour. At 12 miles per hour his whole strenght is expended in carrying forward his own body, and his power of traction ceases. At 2 miles per hour, this still meant that a traction of 100 pounds would pull 90,000 pounds on a canal and only 30,000 on a railway. In the

[49]Paine, p. 32.

[50]Ibid., p. 35.

[51]J. Simmons, "For and against the Locomotive," *Journal of Transport History* 2 (May 1956): 141–51.

[52]*Mechanics Magazine,* 3 (December 25, 1824): 211–13, (January 1, 1825): 237–40, (January 8, 1825): 245–48.

[53]Ibid.

second place, however, the resistance of the canal rises with the square of the barge's speed, so that, at 4 miles per hour the 100-pound traction would only move 22,500 pounds. At the same time, the friction of a railway (excluding air resistance) did not rise with velocity—"the friction is the same for all velocities."[54] The conclusion was clear: at horse speeds, a canal could compete with railways; at steam speeds it could not, and there was little power loss on railways at increasing velocities. This new comparison, unfavorable to canals, became the accepted view.[55]

The *Penny Cyclopaedia* gave similar traction figures, but included roads as well. At 13.5 miles per hour, even a turnpike road was better than a canal.[56] Joseph Priestley's preface to his compendium of canals and railways recognized the same situation.[57] Thus, from about 1825 the viewpoint had changed, and the rising opinion was that steam

[54]Ibid. This conclusion partly arose from Thomas Young's lectures on natural philosophy in 1807: "It is possible that roads paved with iron may hereafter be employed for the purpose of expeditious travelling, since there is scarcely any resistance to be overcome, except that of the air; and such roads will allow the velocity to be increased almost without limit."

[55]For the importance subsequently attached to this see *Mechanics Magazine* (42 [April 12, 1845]: 254–56). This gives comment from the *Scotsman* on the importance of that magazine's predictions about railway speeds in December 1824. "The novelty and startling nature of these doctrines caused a great sensation. The papers were reprinted as a pamphlet in February 1825, and extensively circulated. They were copied into a great many English papers, metropolitan and provincial, and most of those which did not copy the entire gave the more important results in an abstract. They were translated into French, and we believe into German, and they were reprinted in the United States."

[56]The *Penny Cyclopaedia* gave this table (s.v. "Canals"):

WEIGHTS MOVED BY THE APPLICATION OF EQUAL FORCES

Velocity of Motion (mph)	On a Canal (lb.)	On a Level Railway (lb.)	On a Level Turnpike-Road (lb.)
2.5	55,500	14,400	1,800
3	38,542	14,400	1,800
3.5	28,316	14,400	1,800
4	21,680	14,400	1,800
5	13,875	14,400	1,800
6	9,635	14,400	1,800
7	7,080	14,400	1,800
8	5,420	14,400	1,800
9	4,282	14,400	1,800
10	3,468	14,400	1,800
13.5	1,900	14,400	1,800

[57]Priestley, preface, p. xii (see n. 3 above).

railways were potentially superior to canals. This is not to say that railways had indisputably acquired machines that fulfilled expectations. Brandreth's *Cycloped*, an engine propelled by a horse on a treadmill, would yet be entered for the Rainhill trails, and discussion of the merits of pneumatic railways, stationary engines, and steam locomotives still lay in the future.

Technical Reaction by Canals

The reactions of canal companies support the view that an ideological change came in the mid 1820s. Until this time, the canals' technical aspirations were largely directed toward solving internal problems having to do with locks and water supplies. After this, their problems were set externally and they tried to emulate the railways. They did this by concentrating on increasing speeds and applying steam power—before there was serious widespread competition from railways. On the Paisley and Ardrossan canal, and near Birmingham, light barges carried passengers at more than 9 miles an hour. The Paisley passage boats were 70 feet long and weighed only 32 hundredweight, drawn by two half-blood horses which could comfortably work 12 miles a day in stages of 4 miles. Up to seven miles an hour, a high wave built up ahead and the going was hard, but at nine miles an hour the tractive force diminished and the boat rose on a wave behind the bows.[58] Sir John Rennie commented that these swift boats had come too late, but earlier use might have retarded the adoption of railways.[59] This was not so. In 1832, the *Mechanics Magazine* was already convinced that the 10 miles an hour of horses was not sufficient and that steam promised little for canals.[60] Nevertheless steam was tried. The *Victoria* locomotive pulled passenger barges on the Forth and Clyde at 20 miles an hour, and one American enthusiast even wanted to put the rails on the canal bed and float the load.[61]

John Lake designed another rail-canal hybrid consisting of piles

[58]*Mechanics Magazine* 20 (Marsh 29, 1834): 423–25.

[59]*Transactions of the Institution of Civil Engineers* ([London, 1846] 5: 78) quoted in *Encyclopaedia Britannica*, 8th ed. s.v. "Navigation Inland."

[60]"The grand point on which the inferiority of canals turns, is the rate of speed. Although it is true . . . that passengers are conveyed on the Ardrossan Canal at not half the rate charged per mile on the Liverpool Railway, it is to be remembered that it is for a rate of speed one-half less than that realised on the railway" (*Mechanics Magazine* 18 [November 10, 1832]: 92).

[61]*Mechanics Magazine* 31 (August 31, 1839): 413; 32 (November 2, 1839): 76; 46 (May 8, 1847): 453.

supporting rails along each side of the canal. A steam engine in the tug drove wheels which were pressed down onto the rails by levers. Thus the tug was propelled without reaction against the water. The Lake system also did away with locks. Instead, there were inclined planes with rollers, and the rails gave way to a rack up which the tug could wind itself on toothed wheels. The rack continued along the upper pound to allow the tug to pull up its train of barges.[62] A successful trial was made at Leighton Buzzard in 1852, and there seemed promise that 10 horsepower could work 1,000 tons at 3 miles an hour. Yet again, there was hope that "the canal interest of this country may be restored to its former magnitude and importance, and raise its head from beneath the oppressive load of railway ascendancy."[63]

The more conventional response to railway challenge was using paddle or screw tugs. The Duke of Bridgewater had tried this before 1800, but the boat "went slowly and the paddles made sad work with the bottom of the canal, and also threw the water on the bank."[64] Steam towing was certainly preferable to tacking a frigate through the Caledonian Canal, but the difficulty came in applying the principle to the narrower waterways. William Fairbairn wrote a tract about steamboats for canals and constructed stern-wheelers on the American plan for the Forth and Clyde company. He thought that stern wheels would not damage the banks like side paddles and named forty-five canals as suitable for his new boats.[65] The Birmingham and Liverpool had two daily trains each way pulled by towboats with disc engines, and the Union Canal tried mounting screw propellers each side of the bow to prevent wash.[66] Puddled clay banks were certainly vulnerable to damage, but it cannot have been easy to combine a worthwhile payload with a bulky steam engine or, alternatively, to tug the barges through locks. Thus, throughout the 1830s and 1840s there came claims that some new system or other would allow successful steam working on canals, but in practice the canals never competed with railway speeds. Nevertheless, they survived and even increased their

[62]*Mechanics Magazine* 57 (July 3, 1852): 2, (July 10, 1852): 30.

[63]*Mechanics Magazine* 57 (November 13, 1852): 384–87; *Tomlinson's Cyclopaedia,* s.v. "Navigation, Inland" (see n. 2 above).

[64]*Mechanics Magazine,* 40 (April 6, 1844): 239. F. Mullineux, *The Duke of Bridgewater's Canal* (Eccles: Eccles and District History Society, 1959), pp. 26–27. William Symington also tried out a steam canal tug in 1788 and 1789, before his *Charlotte Dundas* of 1802.

[65]William Fairbairn, *Remarks on Canal Navigation* (London, 1831). See also *Mechanics Magazine* (16 [October 15, 1831]: 33–38) for a discussion of this.

[66]*Mechanics Magazine* 41 (October 26, 1844): 288. Bishopp's disc engine was a rotary engine whose "piston" was a swashing disc. It was very compact, hence, presumably, its promise for canal work (see *Tomlinson's Cyclopaedia,* s.v. "Steam"). *Mechanics Magazine* 41 (July 20, 1844): 48.

carrying for a while, helped by the fact that manufacturers and carriers had integrated their operations with the canal network, while the smaller railway companies were slow to develop transfer arrangements. Soon after 1850, however, the railways began to carry a greater volume of goods than the canals.[67]

Road Transport Improves

The evidence of technical comparisons among transport systems cannot be stretched to suggest that complete victory was possible for one or another, for they were not in simple competition. Certainly, canals could never have replaced road transport as effectively as rail could supplant canal, and there would always have to be some residual road system to supplement either of the more inflexible and costly channels. In the late 18th century, road transport improved in parallel with canals, especially providing better passenger services. Times, reliability, and the completeness of services all improved steadily until the mid-1830s,[68] and these organizational changes were accompanied by technical improvements in roads and vehicles. The railway locomotive after 1840 destroyed long-distance stagecoach services more completely than rail replaced canal goods transport, but road and vehicle improvements proved more than enough for survival in local work. It might even be suggested that the improvements in road transport before 1840, though not sufficient to challenge long-distance rail transport for the next half century, nonetheless laid a basis for later victories.

With Telford and Macadam improving the roads, vehicle designers could begin to make real improvements. It was never likely that roads would be completely replaced by rail or canal, though wooden rails were laid on some farms. Previously a vehicle had done well to survive its journeys over the appalling surfaces, but as the latter improved, designers could look for speed, safety, economy of effort, and even comfort for passengers. The stagecoach and many forms of lighter vehicles—phaetons, chariots, landaus, britzskas, chaises, curricles, and so on—all came to be better designed. Weight distribution, friction, and the function of wheels were analyzed in the "Mechanics" articles of early 19th-century *Britannicas,* and the *Penny Cyclopaedia* articles on "Springs," "Carriages," and "Steam Road Carriages" give more evidence of the strong contemporary interest, as well as providing a popular, balanced account of the state of knowledge. The plentiful patents connected with improvements in vehicles are less

[67]Bagwell, pp. 110–14 (see n. 1 above).

[68]A. Bates (*Directory of Stage Coach Services, 1836* [Newton Abbot, 1969]) gives a useful compendium of stagecoach timetables.

instructive, however, and usually throw little more than an amusing sidelight on the innovation taking place.

The wheel was argued about most and changed least. The practice of "dishing" wheels,—that is, of arranging the spokes in a flat cone pointing in toward the vehicle—was condemned by theoreticians as an unnecessary source of weakness.[69] Lewis Gompertz went so far as to design an alternative to the wheel (see fig. 4), but his elaborate arrangement of cranks and cams working rotating legs, like the early attempts at caterpillar tracks or Boydell's plate-laying wheels, had nothing to offer to the new light, fast traffic appearing on Britain's roads. Such designs belonged on the old roads or on cross-country vehicles.[70] Even Jones's and Sir George Cayley's elegant tension wheels, forerunners of the modern wire wheels, were not generally adopted. The wooden-spoked wheel went on until the days of motorcars, sometimes in the form of Hancock's improved artillery wheel, but most often as the traditional, maligned, dished wheel.

Around 1800, legislation and road trustees mistakenly encouraged vehicle owners to use wide-rimmed wheels which, tilting outward from the side of the wagon, needed conical rims to make them lie flat on the road. The grinding and rubbing action of such a wheel did far more harm than good to road surfaces. Improvements in wheels, however, grew from a clearer understanding of their function.

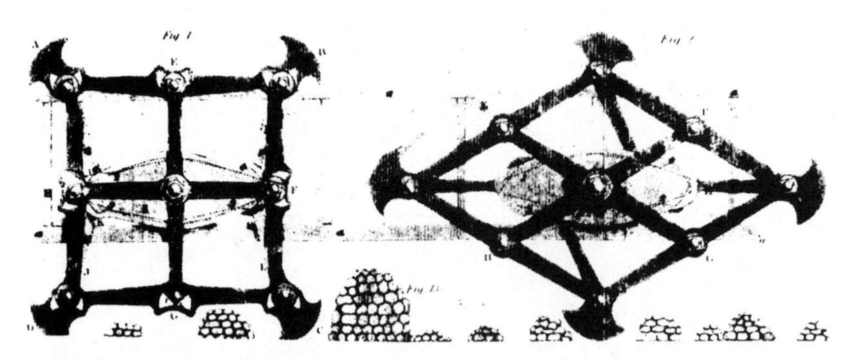

FIG. 4.—Gompertz's substitute for carriage wheels, 1821. A complex mechanism to simulate walking. (*Repertory of Arts, Manufactures and Agriculture,* 2d ser. [London, 1821] 39:10.)

[69] C. Sturt, *The Wheelwright's Shop* (1923; reprint ed. Cambridge, 1963), pp. 91–95. *Encyclopaedia Britannica* (4th ed., s.v. "Mechanics") strongly opposed dished wheels as "more expensive, more injurious to the roads, more liable to be broken by accidents, and less durable in general."

[70] Lewis Gompertz, "Description of Two Improved Substitutes for Carriage Wheels, or Scapers," *Repertory of Arts, Manufactures and Agriculture,* 39: 10–35 (see n. 16 above).

Wheels reduce wasted effort by rolling rather than rubbing over the road; instead, friction is transferred to the small, smooth area of contact between the wheel and its axle. The higher speeds possible on the new roads made this friction more noticeable; but it is worth remembering that after 1800 there was a general need in engineering to reduce friction in faster, lighter machinery and in the new high-pressure engines. Even if Gamett's "friction rollers," a forerunner of the modern roller bearing, were unnecessarily refined for their day, Collinge's patent axle of 1792 was certainly not. In this device, the axles and the wheelbox which turned on them were accurately ground together and their rotation pumped a supply of oil to the rubbing surfaces, so that a well-made Collinge axle could run for 5,000 miles without attention.[71] Such axles went on to mount the wheels of early motorcars a century later.

Wheels also smoothed the effect of bumps in the road by acting as a lever which allowed the carriage to ride less sharply over the obstruction. The larger the wheel, the less a given stone deflected the path of the carriage. If a wheel was 48 inches in diameter and had to be pulled by a horizontal force over a 7-inch bump then the horse had to exert a pull equal to the whole weight of the carriage; a smaller wheel made an even larger pull necessary. Hence, it was concluded, wheels should be as large as possible—but here the trouble began. A four-wheeled vehicle needed steering at the front wheels, and this was done by pivotting the two-wheeled axle about its middle. Swinging the axle like this brings the wheels rubbing up against the sides of the wagon or carriage—unless the wheels are small. The vehicle designer's dilemma was how to have big wheels and still steer the carriage. W. Bridges Adams proposed his curious "equivotol" rotary carriages, arguing that if large wheels could not be mounted on a pivotting beam, then the beam should be abolished. Adams pivoted the vehicle itself, so that the front end with big wheels was hinged to the back end which had equally big wheels; to go round a corner, the carriage bent in the middle.[72] Rudolph Ackermann's improved steering was no more successful when first introduced, and this elegant system for linking the front wheels so that each followed the true curve of the carriage's turn (see fig. 5) did not replace the swinging-beam axle for nearly a century.[73] By some freak of tradition, when early gasoline-powered cars with Ackermann steering were built, they still had front wheels smaller than the rear ones.

[71] A. Bird, *Roads and Vehicles* (London, 1969), p. 77. W. Bridges Adams, *English Pleasure Carriages* (London, 1837; reprint ed., Bath, 1971), pp. 110–14.

[72] Ibid., pp. 264–73.

[73] *Repertory of Arts, Manufacturers and Agriculture,* 34: 70–74.

The simplest way to fit large wheels was on two-wheeled carriages and carts which needed no steering gear at all; they merely followed the horse by the wheel on the ouside of the curve traveling faster than the inside wheel. Such carts and carriages dealt with obstacles well and cornered smoothly, but needed to be loaded carefully since their weight tended to bear down on, or lift, the horse through the leverage

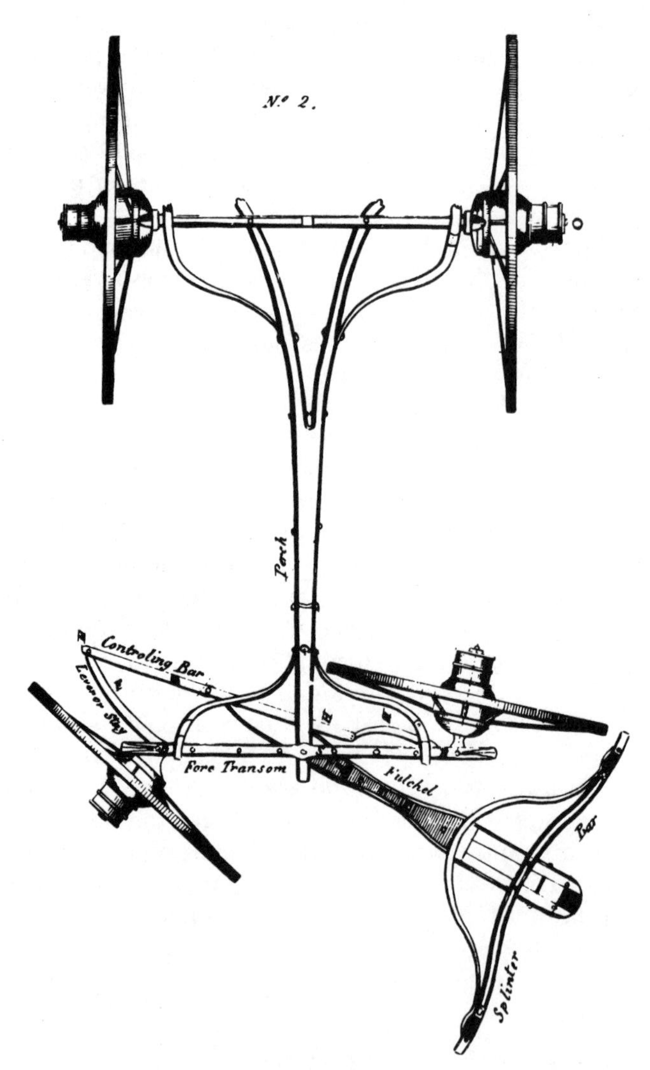

FIG. 5.—Ackerman's steering, 1818. The "controlling bar" places the front wheels at a tangent to the center, O, of the turning circle. (*Repertory of Arts, Manufactures and Agriculture,* 2d ser. [London, 1819] 34:70.)

of the shafts when they were on a hill. The superiority of the cart over the four-wheeled wagon seemed to be unquestionable by 1850 when Pusey reported: "It is proved beyond question, that the Scotch and Northumbrian farmers, by using one-horse carts, save one half of the horses which south-country farmers still string on to their three-horse waggons and three-horse dung carts, or dung pots as they are called . . . last year at Grantham, in a public trial, *five* horses with five carts were matched against five waggons with *ten* horses, and the five horses beat the ten by two loads."[74] Two-wheelers were best for concentrated loads, like dung, but four-wheelers gave the room for lighter, bulky materials. A large farmer would usually have both. The analytical approach was not always welcome, however, and George Sturt and his workmen clung to their heavy designs until 1920. Sturt's classic of technical reaction, *The Wheelwright's Shop,* needs to be read and contrasted with W. Bridges Adams's *English Pleasure Carriages* in order to see how slowly changes make their way against settled tradition.[75]

Improved springs provided the main step forward in carriage building. Earlier vehicles had used leather, wood, and occasionally steel springs to protect passengers against shocks, but around 1800 springs were used to improve performance. The difference was that carriages were designed to mount as much as possible of the total weight on the springs, so that when the vehicle hit a bump, the springs absorbed the vertical movement and limited the shock. Without them, and with small wheels, the horse was virtually lifting the weight over the bump, as we have seen. Lovell Edgeworth investigated springs, concluding that, "Whatever permits the load to rise gradually over an obstacle without obstructing the velocity of the carriage, will tend to facilitate its draught, and the application of springs has this effect to a very considerable degree.[76] A load of 8 hundredweight on springs was easier to pull than 5 hundredweight without. Unsprung, four-wheeled carriages were vulnerable to "discordant motions of fore and hind wheels" and likely to strain their frames by the weight being thrown sharply upon three wheels. Both two- and four-wheeled carriages without springs were subject to shocks from the road causing "violent vibratory motion [and] the wheels to leap from one prominence to another . . . tending to the rapid destruction of the vehicle and extremely unpleasant to the riders."[77]

If British roads adapted to spring carriages were like this, it is hard to imagine the roads in Canada, France, and the United States whose

[74]*Encyclopaedia Britannica,* 8th ed., s.v. "Agriculture."
[75]Cf. Sturt (n. 68 above) and Bird (n. 71 above).
[76]*Penny Cyclopaedia,* s.v. "Spring Carriages."
[77]Ibid.

"imperfect condition precludes the possibility of using steel springs with a due regard for economy."[78] Springs brought a great change for stagecoach passengers, as previously "the danger of sitting on the coach was then never hazarded by outside passengers; they were stuffed in straw in a huge clumsy basket that was fastened precisely over the hind axletree of the coach."[79] Now, with springs, passengers moved to the roof of the coach and at the same time the horses managed to draw a greater load. Obadiah Elliot's elliptical springs were a particularly neat arrangement, cutting down the unsprung weight and lowering the coach's center of gravity. Bagwell considers their importance to coaching equal to the introduction of multitubular boilers for railway engines.[80]

Road Steamers: A Technical Failure

It is not hard to find reasons why a great deal of time and ability were devoted between 1825 and 1840 to establishing steam travel on common roads. Steam was coming to its maturity, a victorious period in which the basic advances such as condensing, high-pressure working, and compounding were giving engineers a flexible technology. Marine engineering produced many different engine configurations; steam hammers, agricultural applications, textile mills, waterworks, and other industrial applications produced more. John Herapath wrote to the Duke of Wellington in 1829 that "Adam Smith, I believe it is, informs us that one horse consumes for food as much land as would maintain eight persons . . . every steam coach constantly in work would save that from horses, which would maintain one thousand four hundred and forty more of the human species."[81] There is a long list of steam-carriage builders including Goldsworthy Gurney, Francis Macerone, Richard Roberts, the Heaton brothers, William James, Dr. Church, F. H. Hills, William Scott Russell, Burstall, and Walter Hancock. Hancock deserves special mention, for he came nearest to succeeding; but their efforts all failed in one way or another, and the steam road carriage was to lie dormant for twenty years after 1840.

Conflicting reasons are given for these failures. Outside influences such as the hostility of magistrates, heavy tolls, and the opposition of

[78]Ibid.

[79]Ibid.

[80]Bagwell, pp. 92–93 (see n. 1 above).

[81]*Mechanics Magazine* 12 (October 31, 1829): 169. See also F. M. L. Thompson, "Nineteenth-Century Horse Sense," *Economic History Review,* 2d ser. 29 (February 1976): 60.

horse-owning interests are usually blamed.[82] Was it really the case that external reactionary forces stopped the steam coaches? John Copeland recounts the problems encountered with tolls and coach owners and notes the difficulties inventors met in raising finance, but he also points out that "mechanical and operational difficulties were still being experienced in the late 1840's."[83] Anthony Bird goes further in discounting the external causes for steam-coach failure and emphasizes the mechanical and metallurgical ills the steamers suffered.[84] W. Worby Beaumont earlier condemned steam carriages on technical grounds. Although "they were creations which were much more than the roughing out of a good working arrangement", they were too heavy; wore out quickly; used too much water; vibrated from the slow, large cylinder engines; and were smelly and hot.[85] Fletcher, in his technical history, *Steam on Common Roads,* was more generous to the steam pioneers, but we may note of Beaumont and Fletcher that the first was an internal-combustion engine man and the other a steam engineer.[86]

These disagreements point to the interesting historical problems presented by the failures of the first generation of road steamers. There is a great deal of contemporary material in the newspapers, issues of *Mechanics Magazine,* and pamphlets by inventors; but much of the information is so prejudiced or incomplete that it needs to be treated very cautiously. Exaggerated claims began with Julius Griffith's pioneering carriage of 1821, which never left Bramah's works and they reached a climax with Gurney and Macerone vaunting themselves and vilifying others. These enthusiasts could be very persuasive. The steam-carriage group convinced a House of Commons committee in 1831 that all problems were solved and the future rosy for steam locomotion: hills could be climbed, boilers were safe and adequate, and no danger or annoyance awaited the public and their animals on the roads. Historians have relied heavily on this 1831 committee as evidence for the viability of steam locomotion, but there was little practical achievement before or after to justify its assertions. A continuous story of disillusioned inventors, breakdowns, and explosions is a more reliable testimony to the real practicability of steam carriages in the 1830s. Yet some inventors stood aside with dignity

[82]Charles Singer et al., eds., *Oxford History of Technology* (Oxford, 1958) 5: 420; Jackman, p. 335 (see n. 1 above).

[83]Copeland, pp. 180–83 (see n. 7 above).

[84]Bird, pp. 154–91 (see n. 71 above).

[85]W. Worby Beaumont, *Motor Vehicles* (London and Philadelphia, 1900), p. 24.

[86]William Fletcher, *Steam on Common Roads* (1891; reprint ed., Newton Abbot, 1972).

from such mendacity. The Heaton brothers did their best, but "have been compelled to doubt the possibility of steam locomotion on common roads at *an average speed of ten miles an hour,* the wear and tear of the machinery, with other expenses, being so great as to exceed any probable receipts."[87] Richard Roberts of Manchester likewise retired with good grace. Walter Hancock—who worked longer, built nine vehicles, and came nearer to success than the others—never descended to abuse; his letters and book are restrained and truthful.[88] There is no doubt that Hancock's machines, after his first three-wheeler experiment, ran long distances with fair reliability. He made trips to Brighton, to Cambridge, and once took the Stratford Cricket Club to a match; he carried more than 12,000 passengers on a regular service which ran for twenty weeks in 1836 (see fig. 6). Yet in July 1839, we still find Hancock calling out for a proper trial and backers: "I have sufficient confidence in my own (carriages) not to shrink from a fair trial with any rival whatsoever—only *I am now ready,* and therefore no delay whatever need take place as to any sufficient trial to satisfy parties really in earnest to begin."[89] It is sad that Hancock, who financed his own experiments and came so near success, did not attract capital as easily as, for example, Gurney. But it is difficult to judge whether Hancock's carriages were reliable and economical enough, unless perhaps by building an exact replica—for which it

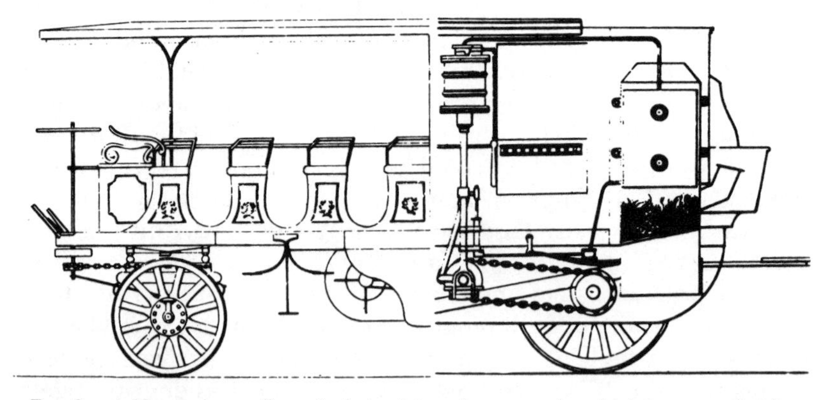

FIG. 6.—A Hancock omnibus. A chain drives the rear axle, which is suspended by a radius arm and springs. (W. Hancock, *Narrative of Twelve Years' Experiments* [London 1838; reprint ed., London, 1975], plate.)

[87]*Mechanics Magazine* 21 (May 3, 1834): 79. Horse-drawn stagecoaches often averaged 10 miles an hour. See H. Hart, p. 149 (see n. 12 above).

[88]See Walter Hancock, *Narrative of Twelve Years Experiments, 1824–1836* (London, 1838; reprint ed. with an introductory essay by F. James, Hampshire, 1975).

[89]*Mechanics Magazine* 31 (July 20, 1839): 283.

would be very hard to find complete plans, identical materials, and working tolerances.

Certainly there was no contemporary agreement that high tolls or opposition from horse owners were decisive obstacles to successful steam road locomotion. The editor of *Mechanics Magazine* remarked acidly, "We have been informed that Sir Charles Dance, the proprietor of that carriage, [Gurney's Cheltenham-Gloucester coaches] does not hesitate to say in private that he was infinitely obliged to the road trustees for furnishing him with so plausible a pretext as they did for abandonning a losing concern."[90] Hancock wrote in 1839 that "the whole evidence of Clerks of Trusts, and others connected therewith, goes to prove . . . that tolls are and will be diminished in consequence of the rivalry of railways."[91] The *Penny Cyclopaedia* bears this out further.[92] A similar doubt exists about the hostility which, it was claimed, the coaching interest showed to steam. A leading coach proprietor, Benjamin Horne, who was interviewed by a parliamentary committee on tolls in 1836 argued that "You think you should be able to compete with steam carriages: Do you think you should beat them?—I fancy so. I only hope that steam carriages will be on the high road instead of being on railroads; there is every probability of our coaches doing very well if they draw about half the weight. With a tramroad we could maintain about twelve miles an hour very easily."[93] If there is reason to doubt that turnpike or coach owners, whose livelihoods were threatened by the railways, were principally responsible for suppressing steam carriages, it is worth turning to the technical aspects of their history.

It was naturally tempting to repeat the railways' success on common roads, but a steam carriage could not just be a steam engine on wheels. The ease with which railway locomotives developed contrasts strongly with the difficulties of steam carriages. A good example of this is the simplicity and effectiveness of Stephenson's *Rocket.* Railways were, it turned out, ideally suited to steam locomotion, and the *Penny Cyclopaedia* explained the difference.[94] Railways, unlike roads, bore an almost unlimited weight, and the locomotive ran on hard smooth surfaces (needing little springing), with few changes of level to surmount. Flanged wheels spared railway engines the need for steering,

[90] *Mechanics Magazine* 18 (November 24, 1832): 122.

[91] *Mechanics Magazine* 31 (July 20, 1839): 280.

[92] *Penny Cyclopaedia* (s.v. "Steam Carriages") offers a valuable and balanced account of steam road vehicles before 1840.

[93] *Minutes of Evidence before a Select Committee of the House of Lords on the Tolls on Steam Carriages Bill* (London, 1836).

[94] *Penny Cyclopaedia,* s.v. "Steam Carriages."

and the bends were too slight to need differential gears between the inside and outside wheels. A road carriage faced not only these problems but more. It had to be light to spare the road surface and carry a good proportion of payload; yet all the propulsive machinery, fuel, water, passengers or freight had to travel on the one carriage with perhaps a single trailer. By contrast, a railway locomotive with massive machinery could have its tender and then a great string of wagons. The design of horse-drawn carriages was equally little help to road locomotive designers. Unlike carriage wheels, the road locomotive's wheels had to transmit the thrust of an engine to the road. Its body had to carry a hot, heavy, vibrating steam engine and boiler as well as the driver, passengers, and useful load. All this had to be sprung against the jolting of the road, and it was not easy to carry the engine's power to the wheels while bouncing on top of springs. Thus, the designer had to achieve simultaneous major improvements in steering, suspension, transmission, boiler, and engine, and one can see to what extent Hancock, James, Hills, and the others were working at the limit of contemporary technology.

The *Penny Cyclopaedia* suggested that the time had come to combine the best elements of the steam carriages so far designed.[95] Admittedly, the achievements were impressive. Hancock's multiple-plate boilers were noted for their lightness and steaming capacity; they were fed air by a fan and consumed exhaust steam, making for a very silent vehicle. Hancock, Russell and others overcame the problems of springing by a radius arm and chain drive from a separate crankshaft. Many other carriages suffered because their road axle was also the crankshaft, making it impossible to have a gear change or effective springing, and the machines were overstrained by road shocks, resistance from loose gravel, or steep inclines. Redmund used the Ackermann steering system, though it was perhaps anachronistic even in 1832 to operate it by means of reins. The *Penny Cyclopaedia*'s list of individual technical advances can be extended. The Heatons, (see fig. 7) and W. H. James gave their vehicles sets of speed-changing gears driven by chains. Roberts in Manchester fitted his driving axle with a "jack-in-the-box," or differential gear, which he had used for cotton-spinning machinery. Hills proposed a crankshaft running in oil and a condenser. The condensing idea was especially important. Heavy water consumption, usually 100 pounds per mile, was a perpetual headache on cross-country trips and even in London; and it is very doubtful whether muddy village duck ponds were a lesser evil than obstructive London water companies. When Hancock hoped to go 60

[95]Ibid.

Fig. 7.—Heaten brothers' steam drag, 1833. The transmission and gearing are bulky. ("Heaton Brothers' Steam Drag," *Mechanics Magazine* 19 [September 14, 1833], p. 417.)

miles to Reading without a stop he must have had 2 tons of water in his trailer, and on trips like this a condenser would have saved many times its own weight. Hancock's strong wheels were another successful innovation. If the best of these ideas had been combined in one design, it might have been financially successful. But the *Penny Cyclopaedia*'s advice was not taken before the steam carriages began their sleep of twenty years about 1840. If a technology does not include a method of making correct choices, there is a defect in the entire system, whatever its piecemeal merits.

We are also left with the impression, however, that the steam carriages were lacking, despite all efforts, in a number of technical respects. Constructors spoke more readily of propelling 4 tons at 20 miles an hour than of the difficulty of retarding such inertial motion. Crankshafts, hand welded and forged from bundles of wrought-iron bars, were sometimes flawed and may partly explain the axle breakages which Gurney and Russell blamed on loose stones laid by malicious opponents. Boilers exploded (even Hancock's), hills overstrained machinery, and carriages were ignominiously towed home by horses. Hancock got to Brighton in 1833, but there "unluckily an inferior part of the mechanism, technically called a clutch, gave way and led to the fracture of a cogged wheel, which gave motion to the centrifugal fire funnel, and the carriage was brought to a dead stand."[96] Similar foreign technical failures reinforce the conclusion that it was not just British turnpike tolls or local opposition which stultified progress. The difficulties of Dietz in France, inventors in Prussia and Denmark, and J. R. Fisher of New York all support the conclusion that it was technology rather than social or economic institutions which stymied steam coaches.[97] It is also not fair to blame those who put their money into two-horse omnibuses and thereby starved steam coaches of capital.[98] When steam succeeded on roads after 1860 the situation was quite different, though even then it was preeminently the large slow traction engines, built for massive strength rather than speed, which succeeded. And when Serpollet built his successful steam cars at the end of the century, even greater changes had taken place in materials and manufacturing methods. The successful White or Stanley steam cars came from an advanced technology using liquid fuels, machine tools, steels, and boilers unknown in Hancock's day. They had more in common with their gasoline-engined contemporaries than with the slow-revving juggernauts of the 1830s.

[96]*Mechanics Magazine* 19 (September 11, 1833): 448.
[97]Fletcher, p. 147.
[98]Charles E. Lee, *The Horse Bus as a Vehicle* (London, 1968).

Conclusion

Three general conclusions follow from this study of transport in the industrial revolution. There were fairly clear technical criteria for comparing the various forms of transport, and comparisons were available in popular accounts like encyclopaedia articles. Furthermore, despite many local exceptions, the general outcome of development was broadly congruent with the technical merits of the systems as they appeared to be emerging at the time. Railways began as feeders to canals and only came to the forefront with new ideas as well as new locomotives after 1825. Canals predominated, despite their difficulties, until their general inferiority was shown; and steam road vehicles, which could reasonably have been expected to succeed, turned out to have unexpected drawbacks.

Thus, there seems to be a strong argument for laying greater emphasis than is normally done on technical factors, even though many cases exist of personal or local rivalries dictating the choice of system. Contemporaries were plainly interested in technical comparisons, and further work is needed to evaluate their influence. The great floods of investment in railways after 1840 were doubtless due to a "bandwagon" effect and the expectation of high profits based on past experience, without any regard to careful technical knowledge. But before the bandwagon investment, presumably, technical expectations could have been decisive.

Second, it is tempting to argue that something like a paradigm shift occurred around 1825 in the changing ideas about horses and canals compared with railways and steam. The canals' change in conduct, before and after 1825, and the way the new ideas preceded the great mania for building railways seem to support this. However, there must be reservations about using the paradigm idea, since many ambiguities existed. Horse-powered railways remained; steam-powered canal boats were built, and for decades there was an extraordinary flexibility in assembling combinations of power sources and transport channels—railed, flat, or tubular, water, iron, or stone. A strong theoretical case for steam railways was created, but, equally, engineers continued to choose the combination of technical means which suited the particular situation. Concentration on locomotives, like the standard gauge, only came later with a rational national system. On balance, I would argue that engineering pragmatism, fundamental to the success of the profession, renders concepts like paradigm shifts of dubious usefulness in the history of technology.

Finally, there is much to respect in the way the pre-Victorians held public debate on the best methods of transport. It is interesting to reflect that a similar debate is being held again today, partly because the fossil fuels and materials which carried the early 19th century out of the hydraulic tradition are now scarcer. Thus we find ourselves reconsidering those decisions of the 1820s which excluded renewable resources.

The Expansion and Diffusion of Western Industrial Technology (ca. 1850 to present)

Technology and the Industrial West

TERRY S. REYNOLDS AND
STEPHEN H. CUTCLIFFE

Between 1750 and 1850 Britain pioneered in the development of a host of interrelated technological and organizational changes that led to the emergence of modern industrial society. These included the emergence of coal as the primary thermal fuel, steam power, centralized factory production, mechanized manufacturing, the railroad, centralized banking, and limited liability stock companies. By 1820 Britain dominated world trade, and its technological and industrial preeminence was widely recognized.

One of the major themes in the history of Western technology from the mid-19th century onward is diffusion: diffusion of industrial technology from Britain to other Western nations and to the rest of the world, as well as diffusion of technology developed in one industry to other industries. The transfer of a technology from the environment in which it was created to other environments was not an easy task, in part because technologies are tied up in systems. In the essay which opened this volume ("Technology and History: 'Kranzberg's Laws'"), Melvin Kranzberg expressed his "third law" as "Technology comes in packages, big and small." All technologies, he explained, involve multiple processes and components. If one component, whether technical or social, of the package (the system) is modified, the other components of the system must be transformed as well for the system to continue to function. Technology transfer usually involved entire systems (or packages), rather than simply components, and the different geographical, cultural, or economic contexts to which a technology was transferred almost inevitably altered some element in its system. Hence, the diffusion of industrial technology from Britain to other countries and from one industry to another was a complex and difficult process, with success depending on the receiving culture's or industry's values, especially its ability to accommodate change.

Nathan Rosenberg's essay, "Economic Development and the Transfer of Technology: Some Historical Perspectives," demonstrates this point. He fleshes out the argument that success or failure in transferring a technology to a new culture or even to a new industry depends on context, that is, on the complex of institutions, shared values, and traditions in the culture or industry to which a technology was being transferred. Rosenberg also points out that one of the key problems in diffusion, whether between countries, between civilizations, or between industries both in the 19th century and today is that "know-how"—the capacity to use technical knowledge—is largely acquired by direct exposure to and participation in the work process and hence is largely dependent on the movement of people.

One of the first countries to profit from the diffusion of British technology was the United States. The close cultural ties between the two nations, no doubt, contributed heavily to American success since British craftsmen with appropriate "know-how" were much more likely to migrate to the United States, which shared cultural background and language, than elsewhere. As early as the 1790s America began to acquire and modify British industrial technologies, and by the 1850s the United States had developed an industrial tradition of its own.[1]

Rosenberg's essay details how elements of the American social and institutional context influenced the unique way in which American industrial technology took British technological systems and modified them. For example, British consumers, whether individual or industrial, usually imposed their preferences on producers, significantly influencing the way the product was made. In America, in contrast, where aristocratic tastes had much less influence and where the skill levels of craftsmen were lower, process, not product or consumer, ruled. American manufacturers designed products for ease of manufacture and low cost (which meant simple, standardized products), not to meet the tastes of individual consumers.

British industrial technology diffused to other nations besides America in the 19th century. On the European continent, France, Germany, Russia, Italy, and a host of smaller nations sought to acquire elements of British technology, with, as one might expect from Rosenberg's essay, varying success. Prevalent social values, the complex of existing institutions, and natural resource endowments all influenced their success in taking and modifying British technological systems. Moreover, the pattern of industrialization differed. In Britain, the textile and coal industries had played a major role in initiating industrialization and spurring technical change and self-sustained economic growth. In America, as Rosenberg points out, the growth of a dynamic capital goods industry was vital. On the European conti-

[1] For a good account of the transfer of textile technology from Britain to America and American adaptation of the British technological "package" to its own environment, see David J. Jeremy, *Transatlantic Industrial Revolution: The Diffusion of Textile Technologies between Britain and America, 1790–1830s* (Cambridge, Mass.: Merrimack Valley Textile Museum and MIT Press, 1981).

nent, in contrast, the steam railroad and the state played the key important roles in initiating industrialization.[2]

Before the 19th century ended, moreover, non-Western civilizations had begun to seek Western industrial technology. In countries like China and Japan, military technology often served as the "leading edge of modernization" according to Barton Hacker's "Military Technology and Modernization in 19th-Century China and Japan." There was good reason for this. By the mid-19th century, growing technological prowess and national rivalries had encouraged Western nations to farther expand their commercial and political domination over non-Western regions. As early as the 16th century, the West's cannon-equipped, oceangoing sailing vessels had secured dominance of the world's waterborne trade, but that dominance had not penetrated far inland. Europeans were too few; the superiority of their land-based military technology was too narrow; local governments were too strong. In the late 19th century, these factors changed in favor of the West.

The politically fragmented nature of Western civilization had encouraged the continued development of military technology. By the late 19th century, Western powers had replaced muzzle-loading muskets with improved breech-loading, repeating rifles, had begun to experiment with machine guns, had replaced loose gunpowder with sealed cartridges, and had evolved lighter weight and more mobile artillery. At the same time, Western industrial technology had replaced the slow, weather-dependent, wooden sailing vessel with iron, steam-powered ships and with armored, steam-powered gunboats capable of penetrating rivers. In China and Japan, however, strong ties to traditional arms (e.g., the samurai's sword in Japan),[3] isolationist attitudes, and centralized political systems had brought peace and economic growth but had left them with naval and land-based military technologies little different from those they had possessed when the West made initial contact in the 16th century. By the mid-19th century both Far Eastern civilizations were hopelessly outclassed militarily, and, beginning in the 1840s, Western nations began to take advantage of the situation to secure increasingly more humiliating

[2]Tom Kemp, *Industrialization in Nineteenth-Century Europe,* 2d ed. (London and New York: Longman, 1985), provides a good overview of the different patterns of industrialization in various European counties.

[3]Japanese culture so valued the sword and what it symbolized that, after rapidly assimilating the gun in the 16th century, it turned back to the sword in the 17th. See Noel Perrin, *Giving Up the Gun: Japan's Reversion to the Sword, 1543–1879* (Boston: David R. Godine, 1979).

territorial and trade concessions. By 1900 Western technology had brought most of Africa and Asia into Western colonial empires.[4]

Hacker's essay reviews the sharply divergent responses of China and Japan to Western technology and, in the process, further reemphasizes the importance of social and political context to the history of technology and the pitfalls of taking a simplistic attitude toward technology transfer. China attempted to adopt only Western military hardware, without the other elements of that technological "package" (to use Kranzberg's term)—for example, the supporting capital goods and metals industries and the system of technological education. Chinese officials, on the one hand, feared development of a fully Western military and industrial complex would undermine traditional Chinese society and its values. Japanese officials, on the other hand, accepted the price of adopting Western military technology, embracing the military and social systems in which Western military technology was enmeshed. As a result, by 1900 Japan was the most militarily and industrially powerful nation in Asia and was able to maintain its political and economic independence from Western imperial powers and expand its interests at China's expense. China slowly descended into chaos and foreign domination.

In the late 19th and early 20th centuries, the West's industrialization not only threatened the cultural values of civilizations thousands of miles from the mills of Manchester and the blast furnaces of Belgium, but it also undermined traditional ways of doing things at many levels in the West. Randall's essay earlier in this volume pointed out how the emerging machine economy destroyed the traditional culture of English woolen workers. The impact of industrialization went much further. For instance, the rise of the steam-powered factory permitted the centralization of industry in locations favorable to trade, economically benefitting some regions, hurting many others. The development of railroad systems, by making possible the import of food and raw materials to centralized locations and the dispersion of finished products to a wide market, further encouraged centralization of production. The result was the very rapid growth of industrial cities. Traditional forms of city government faltered under the resulting pressures placed on local water supplies and traditional waste disposal methods.[5]

[4] See Daniel R. Headrick, *The Tools of Empire: Technology and European Imperialism in the Nineteenth Century* (New York and Oxford: Oxford University Press, 1981).

[5] For a general overview of the impact of the "Industrial Revolution," see John W. Osborne, *The Silent Revolution: The Industrial Revolution as a Source of Cultural Change* (New York: Charles Scribner's Sons, 1970); or Peter N. Stearns, *The Industrial Revolution*

Ruth Schwartz Cowan's "The 'Industrial Revolution' in the Home: Household Technology and Social Change in the 20th Century" provides another example of the spreading impact of industrial technology in the West: its impact on housewives, perhaps the largest category of workers. Cowan details the series of new technologies that penetrated middle-class American households in the early 20th century. Vacuum cleaners replaced hand beating of carpets, automatic washers and electric irons made "blue Mondays" easier, prepared foods lessened the cooking burden, the automobile made shopping easier and faster. Cowan convincingly argues that these technological changes had just as extensive an effect on the household as the introduction of automatic spinning machines, power looms, and the factory system had on textile production in the late 18th century.

Focusing on middle-class urban housewives, who probably "first felt the impact of the new household technology," Cowan found that the coming of a machine economy changed the structure of the household workforce, its skills, and its ideology just as extensively as it had changed those elements for British textile workers earlier, but not in the manner postulated by the standard sociological model for the impact of modern technology on family life. For example, Cowan found that mechanization and other household changes did *not* decrease the amount of time spent on housework, as sociologists had long theorized, and may in fact have increased it.

The social context in which the technological revolution in the home occurred explains this apparent paradox. The decline of immigration from Europe with the outbreak of World War I in 1914 and the passage of more restrictive immigration laws in the 1920s compelled middle-class American housewives to take on functions once delegated to servants, such as cooking and child care. The coming of the automobile gave housewives greater freedom of movement, but it coincided with the decline of delivery services. This development required housewives to spend up to a full day a week shopping themselves. At the same time, the new technologies raised expectations. Since a washing machine could wash clothes faster, housewives were expected to provide their families with cleaner clothes by washing more often, and they were made, through advertising, to feel guilty if they did not. The time gained on the one hand was taken

---

in World History (Boulder, Colo.: Westview, 1993). One good study of the impact of technology and urban growth on the underlying technological infrastructure of a Western city is Donald Reid, *Paris Sewers and Sewermen: Realities and Representations* (Cambridge, Mass.: Harvard University Press, 1991).

away with the other; the diffusion of industrial technology into the home had the same mixed effects it did in the factory.[6]

Clearly, by 1900, if not earlier, industrial technology was steadily diffusing into all aspects of Western life from industrial iron furnaces to domestic clothes irons. In the same era, the pace of technological change significantly increased. Two interrelated factors strongly contributed: the increased application of scientific concepts to technology and the institutionalization of invention by industry in the industrial research laboratory.

Although there were scattered instances of the successful application of scientific methods to technological problems before 1850 (e.g., see Steele's essay earlier in this volume on the ballistics revolution) and institutionalized industrial research,[7] both appeared on a large and permanent scale only in the late-19th century. Between 1880 and 1890, for example, the German dye industry successfully institutionalized scientific research, sharply accelerating the rate at which new dyes were introduced. In the fourteen years before the Bayer Company firmly established its industrial laboratories in 1884, it had produced twelve dyestuff innovations; in the fourteen years that followed, Bayer produced 151.[8]

Lynwood Bryant's essay on "The Development of the Diesel Engine" in this volume provides another example of the increasing use of theoretical science as an aid to technological development. Diesel, a German engineer, around 1890 proposed a heat engine that would use the Carnot cycle, a theoretical model for the optimum operation of a heat engine. With logic based on scientific abstractions and theoretical calculations, he convinced himself and several financial backers that he could build a universal engine, one that could use any fuel and burn it so efficiently that no ignition or cooling system would be needed.

[6] Cowan provides a more detailed study in *More Work for Mother: The Ironies of Household Technology from the Open Hearth to the Microwave* (New York: Basic, 1983).

[7] For a rare pre-1850 precedent for industrial research, see Edwin T. Layton, "Scientific Technology, 1845–1900: The Hydraulic Turbine and the Origins of American Industrial Research," *Technology and Culture* 20 (1979): 64–89.

[8] John J. Beer, "Coal Tar Dye Manufacture and the Origins of the Modern Industrial Research Laboratory," *Isis* 49 (1958): 123–31, and *The Emergence of the German Dye Industry* (Urbana: University of Illinois Press, 1959). See also George Meyer-Thurow, "The Industrialization of Invention: A Case Study from the German Chemical Industry," *Isis* 73 (1982): 363–81. The figures for Bayer dyestuffs innovations are on p. 381 of this article. For an American example of a successful early industrial research laboratory, see George Wise, "A New Role for Professional Scientists in Industry: Industrial Research at General Electric, 1900–1916," *Technology and Culture* 21 (1980): 409–29.

Bryant describes Diesel's efforts to bring the engine to reality in each of the three traditional stages of technological change: invention, development, and innovation (or commercialization). He points out, however, that there are no sharp breaks between the three stages, that elements of all three activities exist in each stage. Bryant's narrative also suggests that while theoretical science provided Diesel with the idea that initiated his engine development, Diesel quickly found that reducing his idea to practice (i.e., the development stage) required far more effort than the original, scientifically derived idea. The conflict between the scientific abstractions with which Diesel had originally worked and the realities of the world (the tolerances attainable by machine tools, the heat carrying capacities of available materials, and so on) eventually forced him to abandon several of the essential elements that were to have given the engine competitive advantages over all other prime movers. In a sense, in the development stage Diesel had to reinvent large segments of the engine to accommodate limitations imposed by the real world.

Lillian Hoddeson's essay on "The Emergence of Basic Research in the Bell Telephone System, 1875–1915" provides yet another example of science's growing role in Western technological development, for Alexander Graham Bell and Thomas Watson, she points out, drew heavily on science as a guide in their early telephone experiments, even if their success was defined in technological rather than scientific terms. Her essay also illustrates Western industry's institutionalization of technological change with the emergence of the industrial research laboratory. Industrial research laboratories emerged in the United States at about the same time as in Germany. Thomas Edison, who established a laboratory at Menlo Park, New Jersey, in 1878 was an independent inventor, not affiliated with any industrial corporation, but he intended his facility to turn out new inventions at a regular pace.[9] Research and development facilities directly tied to industrial companies followed very shortly thereafter.

One company that early incorporated R&D facilities into its operations was the Bell Telephone Company. Hoddeson describes how practical problems and the growing scale of its technological systems slowly led the Bell Company toward increased support of, first, practical research and, then, basic research. As early as 1884 the firm had a "Mechanical Department," expected to study and report on various elements of telephone systems and on possible inventions for com-

[9]Thomas P. Hughes, "Edison's Method," in *Technology at the Turning Point* ed. W. B. Pickett (San Francisco: San Francisco Press, 1977), sees Edison's Menlo Park establishment as a forerunner of later American industrial research facilities.

mercial use. As the Bell Company's telephone system expanded and problems grew more complex, the company increasingly recognized the desirability of having a permanent staff of not only engineers but also academically trained scientists who could be directed into "basic" research that promised returns to the company.

Between 1890 and 1915 scientists and engineers employed by Bell undertook several specific scientific studies requiring knowledge of theoretical physics and advanced mathematics that solved or helped solve key commercial problems, especially in the long-distance service area. These successes strengthened support for basic research in Bell. By 1907 basic research had acquired a permanent role in the company. In 1911 Bell Telephone created a formal basic research arm. By 1925, when Bell spun off a new corporation, Bell Telephone Laboratories, to replace the engineering department of its Western Electric Company, the new R&D division had over 3,000 persons on staff.[10]

German success in synthetic dyes and heat engine technology (the diesel engine), American development of the telephone and new manufacturing processes, and both German and American pioneering accomplishments in establishing successful, industry-based R&D laboratories were but a few of the indicators that the technical and industrial superiority enjoyed by Britain in the first half of the nineteenth century had begun to dissipate by 1900.[11] America's replacement of Britain as the West's leading industrial power very early in the 20th century was clearly signaled by the rapid rise of the American automotive industry and its mass production techniques. As Ro-

[10] See also Leonard S. Reich, "Industrial Research and the Pursuit of Corporate Security: The Early Years of Bell Labs," *Business History Review* 54 (1980): 504–9. There is a large literature on the emergence of industrial research, especially in the United States. For example, Nathan Rosenberg, *Technology and American Economic Growth* (White Plains, N.Y.: M. E. Sharpe, 1971), discusses the economic impact of industrial R&D in chap. 7; David F. Noble, *America by Design: Science, Technology, and the Rise of Corporate Capitalism* (New York: A. A. Knopf, 1977), provides an overview of the emergence of corporate control of invention in select industries in the early 1900s in chap. 7; and Alfred D. Chandler, *The Visible Hand: The Managerial Revolution in American Business* (Cambridge, Mass.: Harvard University Press, 1977), analyzes the introduction of research into American industry from a managerial prospective in chap. 13.

[11] For discussion of Britain's comparative technological and economic decline in the late 19th century, see E. J. Hobsbawm, *Industry and Empire*, Pelican Economic History of Britain, vol. 3 (Harmondsworth: Penguin, 1969), pp. 172–94; D. S. L. Cardwell, *Turning Points in Western Technology* (New York: Science History Publications, 1972), pp. 190–95; and David S. Landes, *The Unbound Prometheus: Technological Change and Industrial Development in Western Europe from 1750 to the Present* (Cambridge: Cambridge University Press, 1969), pp. 326–58.

senberg's essay, discussed earlier, suggests, American manufacturers, by focusing on adapting the product to the machine, pioneered in the development of low-cost, standardized, mass-produced products. The pinnacle of American achievement in production technology was the introduction of the automobile assembly line in 1914. By freezing a standardized design and using interchangeable parts and a host of highly specialized machine tools, American companies were able to produce very complex products like the automobile in numbers far above and at a cost far below those produced using more traditional batch methods with skilled craftsmen operating general purpose machine tools.[12] By the 1920s the American automotive industry dwarfed the British automotive industry, and American automobile companies were exporting their products and their productive technologies worldwide. By the 1920s the diffusion of mass production from the automotive industry into other industries had made America the unchallenged industrial superpower of the Western world.

World War II (1939–45), as Jonathan Zeitlin's "Flexibility and Mass Production at War: Aircraft Manufacture in Britain, the United States, and Germany, 1939–1945" indicates, began to bring out some of the limitations of American mass production techniques. The full benefits of American-style mass production required extensive planning and capital investment in specialized machine tools and then freezing designs for long production runs of a standardized product. Military effectiveness, however, required more than the ability to manufacture a standardized product in large numbers (quantity); it also required modifying and improving designs quickly to take advantage of what was learned under battle conditions or rapidly introducing improved new products (quality).

Zeitlin compares British, American, and German aircraft manufacturing during World War II. After war broke out all three quickly discovered the central dilemma of wartime aircraft manufacture: "the need to balance the qualitative gains obtainable through design modifications against the quantitative losses resulting from interruption to continuous production runs." All three countries responded differently to the dilemma, again illustrating the theme that social context plays a major role in determining what technological choices a society makes. In quantitative terms, American aircraft production dwarfed both Britain and Germany; in qualitative terms, Zeitlin maintains,

[12] For American development of mass production, see David Hounshell, *From the American System to Mass Production: The Development of Manufacturing Technology in the United States* (Baltimore, Md.: Johns Hopkins University Press, 1983).

the American pattern of temporarily freezing designs and making retrospective modifications was less impressive than the British policy of continuous improvement. In Germany, where radically different social and political conditions existed, the trade-off between quality and quantity was managed more poorly than in either the United States or Britain, oscillating sharply from emphasis on one to emphasis on the other and negatively affecting both.

World War II sharply changed the political context of the Western world. By destroying the heartland of Western civilization, it prepared the groundwork for the emergence to political prominence of the West's peripheral powers: the decentralized, capitalist United States and the heavily centralized, communist Soviet Union. In population, resources, and industrial potential, these states dwarfed the nation-states that had initiated the West's technological and industrial development. Soviet and American struggles for influence over ruined Europe had, by 1948, set off a "cold war" that very quickly spread worldwide as the colonial empires of the older European powers—Britain, France, Spain, Portugal, Holland, and Italy—collapsed and as the two superpowers vied for influence.

Technologically, World War II, directly or indirectly, had three major effects. First, it solidly reaffirmed America's industrial dominance and spread American mass production techniques throughout the Western world. Second, the growing technological nature of warfare significantly increased the scale of state involvement in promoting, directing, and managing technological change. Third, the destructiveness of the weaponry developed in World War II, notably, nuclear bombs, began to dampen the West's previously almost unbridled enthusiasm for technological change.

America's role as the "arsenal of democracy" during World War II and the destruction of much of the productive potential of its industrial rivals left the United States as the unchallenged industrial and technological power not only of the Western world but of the entire world after 1945. The desire to imitate American industrial success and the role America played in reconstructing western Europe after World War II rapidly diffused American-style mass production technology. As Zeitlin observes, the enormous productivity gains the American system offered obscured the lessons about its inflexibility in handling rapid change offered by aircraft manufacturing during World War II. For several decades this would not matter. With no effective foreign competition, American manufacturers continued to pursue the pattern that had brought them to the pinnacle. For example, during the 1950s the American "Big Three" auto manufacturers standardized and specialized more than ever before, pursuing productivity and superficial style changes and ignoring real innovation and flexibility.

In the late 20th century, however, overreliance on traditional mass production techniques left American industries, and the British who had imitated them, vulnerable to challenges from foreign manufacturers, especially the Japanese, who had adopted productive systems that were more adaptable to product change and innovation. Such systems enjoyed the advantage in the more competitive and turbulent global markets that emerged in the 1970s and 1980s as reconstructed European economies and a host of Asian nations that had established industrial economies suddenly began to compete for international trade.

A second spin-off of World War II was the growth of command technology: the "force-feeding" of technological change in certain directions through state mobilization of human, financial, and natural resources. While having precedents in World War I and in the interwar Soviet infatuation with large-scale technology, command technology reached new levels in World War II and carried over into the postwar world.[13] Examples of command technology during the war included aircraft development and manufacturing in all of the major belligerent states, German development of synthetic fuels and ballistic missiles, British development of radar, and the American synthetic rubber program. But the most notable example of World War II command technology was America's Manhattan project, the project which led to the development of the atomic bomb.[14]

Command technology played a major role in the cold war between the Soviet and American superpowers that dominated international relations through most of the late-20th century. After World War II, the Soviet Union mobilized its scientific and technical resources to duplicate American nuclear weapons and, as the struggle for prestige accelerated, to initiate a space-exploration program. The Soviet space program in 1956 launched the world's first artificial satellite. The United States countered by mobilizing its R&D resources to initiate competitive space programs in both the military and civilian spheres, culminating in a manned lunar landing in 1969.[15]

[13]For examples of interwar Soviet command technology, see Paul R. Josephson, "'Projects of the Century' in Soviet History: Large-Scale Technologies from Lenin to Gorbachev," *Technology and Culture* 36 (1995): 519–59.

[14]See Richard Rhodes, *The Making of the Atomic Bomb* (New York: Simon & Schuster, 1986), for an excellent account of America's atomic bomb project.

[15]A good account of Soviet-American rivalry in space is Walter A. McDougall, . . . *the Heavens and the Earth: A Political History of the Space Age* (New York: Basic, 1985).

Walter A. McDougall's "Space-Age Europe: Gaullism, Euro-Gaullism, and the American Dilemma" demonstrates that the use of command technology in the late 20th century was not a monopoly of the two superpowers. Other states, fearing permanent political, military, and economic marginalization, pursued command technologies of their own. France was the most notable of these. In the 19th century, France, once a central figure in the West's industrial economy, fell behind first Britain and then Germany and the United States, at least in part because French society, which valued conservatism, protection, and the small firm, had not adjusted well to the industrial age's stress on expansion, competition, and concentration. Fearing that France would be relegated to permanent second-class status and convinced that military and economic independence required an institutionalized, permanent technological revolution, France's president Charles de Gaulle set out in the 1950s and 1960s to transform France into an "R&D state" by "force-feeding . . . technological change, with the state itself as managerial czar."

Focusing on one of the primary arenas in which late-20th-century Western command technologies operated—space programs—McDougall describes not only France's efforts, but the efforts of other European states, to cooperate in space in order to match the resources and talent that the United States and Soviet Union were able to bring to bear.[16] In an age in which nationalism was a dominant ideology, cooperation proved difficult to achieve, and, as McDougall points out, only through a partial nationalization of the international effort were Europeans able to begin, in the 1970s, to make significant progress in cooperative efforts in space.

In the closing sections of his essay, McDougall compares the American space shuttle and the European Ariane systems for commercial use of space. He notes that European countries, by taking less advanced, "outmoded" technologies and making them commercially viable and by focusing on niche markets, have been able to seriously threaten the commercial viability of the much more advanced and technologically sophisticated (but also much more expensive) American space shuttle. As a result, McDougall concludes, the United States in the late 20th century found itself in a position comparable to late-19th-century Britain. Its overall leadership remained intact, but the United States was being outmaneuvered in specific markets by others. To make matters worse, McDougall notes, American cultural values

[16] France's efforts at command technology were also significant in the area of nuclear power. See Gabrielle Hecht, "Political Designs: Nuclear Reactors and National Policy in Postwar France," *Technology and Culture* 35 (1994): 657–85.

are less conducive to state-driven technological change than its rivals', which may make it more difficult for America to adapt in a world where technological and industrial development seems to be becoming more technocratic.

A third major effect of World War II on technology was the exacerbation of doubts about the value of technological change, some of which had begun to emerge in the great economic depression of the 1930s. In part due to the long-standing belief that humanity had the right, if not the duty, to dominate and manipulate nature, a belief that can be traced far back into Judeo-Christian history, the Western world had traditionally embraced technological change uncritically. The development and use of nuclear weapons in World War II, and the ensuing cold war buildup of enough nuclear weapons to destroy all of humanity, began to raise the level of ambivalence toward technology in the West.

Developments in the 1960s raised levels of ambivalence further. Industrial societies faced growing evidence that a host of technologies from insecticides to automobiles had begun to have clear and measurable effects on environmental quality. By the 1970s, in many Western nations, ecological and related social movements had grown strong enough to block the spread of certain technologies. In Sweden, for example, public opinion forced passage of legislation aimed at dismantling Sweden's nuclear power capacity. Even in the United States, the West's leading technological power, attitudes toward technology shifted significantly. In the early 1970s, for example, the American congress killed the SST (supersonic transport) project, then viewed as the logical next step in the advance of jet transportation.[17] J. Samuel Walker's essay, "Nuclear Power and the Environment: The Atomic Energy Commission and Thermal Pollution, 1965–71," provides a good case study of the growing Western ambivalence toward technology.

[17]Thomas Parke Hughes, ed., *Changing Attitudes toward American Technology* (New York: Harper & Row, 1975), provides a collection of sources that illustrate the growing ambivalence of Americans toward technology. See also Samuel P. Hays, *Beauty, Health, and Permanence: Environmental Politics in the United States* (Cambridge and New York: Cambridge University Press, 1987). For the growing strength of environmental movements in other portions of the West, see, e.g., Russell J. Dalton, "The Environmental Movement in Western Europe," in *Environmental Politics in the International Arena,* ed. Sheldon Kamieniecki (Albany: State University of New York Press, 1993), pp. 41–68; or Ferdinand Müeller-Rommel, *New Politics in Western Europe: The Rise and Success of Green Parties and Alternative Lists* (Boulder, Colo.: Westview, 1989). For nuclear power specifically, a good overview is Dorothy Nelkin and Michael Pollack, *The Atom Besieged* (Cambridge, Mass.: MIT Press, 1981).

In the 1950s nuclear power had not yet been completely demonized in the West. Nuclear weapons threatened the very survival of humanity, but civilian uses of nuclear power offered the dream of inexpensive electric power, power "too cheap to meter," that could be used to elevate the living standards of everyone on Earth. By the 1960s it was clear that early views on the cost of nuclear power had been overly optimistic, to say the least, but civilian nuclear power still seemed to be a major boon. In America, for instance, it offered a way out of the dilemma posed by simultaneously growing demands for electricity and for protection of the environment from the noxious emissions of coal-fired electric power plants. Nuclear power plants, because they did not emit noxious gases, promised to solve both the energy and the environmental "crises."

That was not to be. In the mid-1960s, the "discovery" that the water used in cooling nuclear plants and discharged back into streams raised the temperatures of streams by several degrees, possibly endangering aquatic life, raised concerns about the environmental friendliness of nuclear plants, sparked public protests at nuclear plant sites and led to federal legislation—all before scientific studies had actually determined whether "thermal pollution" had serious impact on the environment or not. Studies would later suggest that its impact was minor. Thus, Walker's essay on thermal pollution in many respects illustrates Kranzberg's "fourth law" (from the opening essay in this volume), which maintains that even when technology is a prime element in a public issue—as it was in this case—nontechnical factors take precedence in technology policy decisions.

The thermal pollution issue had disappeared by the early 1970s, but, Walker concludes, it laid the foundations for widespread skepticism about the environmental benefits of nuclear power. The doubts it stirred laid the foundations for future protests over the dangers of nuclear radiation wastes and reactor safety and ultimately halted the expansion of nuclear power plants not only in the United States but in many other Western nations as well. Increased public concerns over the environmental effects of technology also led to a host of legislative remedies in Western countries, creating a number of new bureaucracies charged with regulating technologies that would once have been left to develop relatively unfettered in the belief that technological development could only, in the long run, yield good.

* * *

In the early 19th century, technological change in the West had been largely a product of individuals working outside of formal institutional frameworks with minimal guidance from science and minimal interference from a culture that generally regarded new technology favorably. The state played an important role in the development of new technology only in a few areas, mostly military related. However, by the end of the 20th century, the context in which Western technology operated had changed sharply, in large part due to the expansion of economies from local and regional to global, but also because of the increasingly visible connections between economic strength and military strength and the increasingly technological nature of modern warfare. Invention now commonly occurred not in isolated garrets but in industrial research laboratories where teams of corporate employees used expensive instrumentation and applied sophisticated theoretical and mathematical tools. The state's role in technology had multiplied enormously, with selected technologies—not always military—being force-fed or heavily subsidized because they were deemed in the national interest.

Paradoxically, just as the state became a primary instrument of technological change on a broader and larger scale than ever before, growing ambivalence toward new technologies and growing concern about technology's impact on the environment also led to sharp increases in the role of the state in controlling and regulating technology. Which role—the state as a promoter of technology or the state as a regulator and controller of technology—will dominate the future of Western technology has yet to be determined. In view of long-standing Western infatuation with technology, however, it is likely that in Westernized societies the state's role as promoter and supporter of technology will, in the long run, take precedence over its role as regulator.

Economic Development and the Transfer of Technology: Some Historical Perspectives

NATHAN ROSENBERG

I

Back in an earlier, more naïve day, we managed to allow ourselves to believe that there was a purely technological solution—a cheap "technological fix"—to the problems of poverty and economic backwardness which beset most of the human race. In the Point Four of his 1949 inaugural address President Truman spoke optimistically of the incalculable benefits which technical assistance could bring to improving the desperate plight of the poor throughout the world "We must embark," he said, "on a bold new program for making the benefits of our scientific advances and industrial progress available for the improvement and growth of underdeveloped areas."

Unfortunately we exaggerated from the outset what could be accomplished solely by making Western techniques available. In some measure the present mood of disillusion with foreign aid in general is due to the unrealistically high expectations which we once attached to such programs (including their effectiveness as weapons of politics and diplomacy). We now realize that there are a host of difficulties—institutional and otherwise—which hamper the successful adoption of foreign technology. I would like to contribute some observations of my own on the obstacles in the way of a successful transfer of technology. I will also, before I am done, go one pessimistic step further, and suggest that there are serious problems, not just in facilitating the transfer but in the very nature of the technology which we presently have to offer and its possibly limited relevance (especially in agriculture) to the problems of poor countries. But it should be emphasized at the outset that the main thrust of this paper is to pose problems for the international transfer of technology which emerge from an examination of earlier—primarily 19th-century—mechanisms through which such transfers have taken

NATHAN ROSENBERG is Fairleigh S. Dickenson, Jr. Professor of Public Policy, in the Department of Economics at Stanford University.

place. I do not propose to offer authoritative conclusions or recommendations but rather—and much more modestly—to use historical experience to sharpen and to enlarge our appreciation for the difficulties which may still lie ahead.

I propose, first, to discuss the role of the capital goods industries in the 19th century as centers for the creation and diffusion of new techniques. I start from the proposition that technical change is not, and never has been, a random phenomenon. In both the United States and the United Kingdom in the 19th century, technological change became institutionalized in a very special way—that is, in the emergence of a group of specialized firms which were uniquely oriented toward the solution of certain kinds of technical problems. The rapid rate of technological change was completely inseparable from these capital goods firms. In fact, I would regard the emergence of such firms as the fundamental institutional innovation of the 19th century from the point of view of the industrialization process. I am particularly interested in the factors which affected the success and failure of these industries in the context of a developing industrial economy. In general, I shall argue that these industries performed certain critical functions which accounted for the speed of the growth process in the 19th century. But I am anxious to free myself of the charge of suggesting that poor countries today will have to tread exactly the same path. The most important reason why poor countries may *not* have to tread the same path as their industrial predecessors is precisely that industrial countries have already done so. One of the advantages of not taking the lead in economic development is that, once an objective has been reached and clearly demarcated, other and easier routes to attain that objective may become obvious. Or, to put the point a little differently, there is no reason to believe that the optimal path in the development of a new technology is the same as the optimal path for transferring and adapting that technology, once it has been developed.[1] In fact, I want to insist on this point. Economic growth has never been a process of mere replication.

[1] For a more formal treatment of the problems of technological leadership and "followership," see Edward Ames and Nathan Rosenberg, "Changing Technological Leadership and Economic Growth," *Economic Journal*, vol. 73, no. 289 (March 1963). A recent study of the diffusion of new technologies in six West European countries suggested that, for five important new processes, "countries which are pioneers tend to have *slower* speeds of diffusion. This result is consistent with the hypothesis that the pioneer faces all sorts of teething troubles, new problems associated with the new technique; these are likely to be—partly and gradually—solved by the time others adopt it. It is therefore not necessarily desirable to be the first to introduce a new technique" (G. F. Ray, "The Diffusion of New Technology: A Study of Ten Processes in Nine Industries," *National Institute Economic Review* [May 1969], p. 82).

Although one can identify certain broad phenomena and functions common to all rapidly growing economies (changes in the composition of output and labor force, urbanization) the actual paths to development show very wide institutional variations. Certainly the latecomers on the development scene (e.g., Germany, Japan, Russia) followed very different paths and sequences than did pioneering England.[2]

Perhaps it should also be said that the policy implications of what follows are not at all clear—at least not to me. It is not my intention to press the analysis of technology transfer in the last third of the 20th century onto a 19th-century Procrustean bed. Rather, I want to draw upon the experience of the 19th and early 20th centuries as a way of suggesting what are likely to be problems of considerable interest and importance.

II

I have been very much interested in problems of the transmission of technology in the 19th century, not only from one country to another but from one industry to another. The problems involved are not entirely unrelated. Simply because transmitting technologies from advanced to underdeveloped countries presents many unique problems, we should not ignore the fact that transmitting technology from one use to another always presents certain elements of novelty.[3]

In both the United States and United Kingdom, to a very surprising degree, innovations in the area of machine technology were transmitted from existing uses to new uses by a very personal mechanism. Impersonal forms of communication—such as trade journals—were often important in generating interest but hardly ever provided enough of the highly specific information required for the successful transmission of a new productive technique.

There are remarkable similarities, in the United States and the United Kingdom, in the pattern of transmission. A dominant pattern in both countries went as follows: A firm develops a new technique or process in response to a problem in a particular industry—say firearms manufacture. It later becomes apparent that this technique is applicable to typewriters. The technique is "transferred" to the typewriter industry

[2] See Alexander Gerschenkron, "Economic Backwardness in Historical Perspective," in *The Progress of Underdeveloped Areas*, ed. B. F. Hoselitz (Chicago, 1952).

[3] See A. P. Usher, *A History of Mechanical Inventions*, rev. ed. (Cambridge, Mass., 1954), esp. chap. 4; Nathan Rosenberg, "Technological Change in the Machine Tool Industry, 1840–1910," *Journal of Economic History* 23, no. 4 (December 1963): 414–46; Nathan Rosenberg, ed., *The American System of Manufactures* (Edinburgh, 1969).

by the firm which developed the technique actually undertaking the production of typewriters (Remington). In other words, a very common mode of technology transfer *between industries* took place as a result of firms adding to, or switching, their product lines.

An alternative way of stating this is to say that, in the area of machine technology, firms became increasingly specialized by *process* rather than product. In both countries we find firms undertaking a combination or a sequence of products which can be understood only in terms of their common underlying technical processes. We can find many examples of firms in each country undergoing the same product sequences over time—for example, firearms, sewing machines, bicycles—or undertaking the same product combinations. If we examine historically the operation of multiproduct firms in both countries, we find that they frequently arrived at the combination through the development of a specialized skill at a particular process—for example, hydraulics (Armstrong, Tangyes, Fawcetts)—or an unusual expertise in the property of a material—for example, William Fairbairn in Manchester and William Sellers in Philadelphia both undertook activities where the structural properties of iron were critical to success—bridges, iron ships, multistory dwellings, cranes, etc. The center for the transmission of relevant knowledge and techniques from one industry to another, and for the application of known techniques to new uses, was, to a very considerable degree, the individual firm.

Where the transfer of technology involved places geographically distant from one another, the reliance upon the migration of trained personnel (at least temporarily) was very strong. This was true even within the United States. It would be possible to trace the diffusion of machine tool technology in America in the 19th century in terms of a genealogical table showing the movement of a very small number of highly skilled mechanics from the places where they received their training (apprenticeship and early "on-the-job" learning) to the firms where they eventually brought their valuable skills; often these were new firms established by the mechanic himself.[4]

I have been impressed by the extent to which the transfer of technological skills—even between two countries so apparently "close together" as Britain and the United States in the mid-19th century—was dependent upon the transfer of skilled personnel.[5] This was demonstrated

[4] Joseph W. Roe, *English and American Tool Builders* (New Haven, Conn., 1916), in fact contains several such genealogies.

[5] See the illuminating account of a 16th-century transfer in Warren C. Scoville, "Minority Migrations and the Diffusion of Technology," *Journal of Economic His-*

by several striking episodes. For example, when the British government purchased a large quantity of American gun-making machinery in the 1850s and introduced it into the Enfield Arsenal, it was found to be absolutely essential to employ a large number of American machinists and supervisory personnel. At about the same time, Samuel Colt, impressed with the backwardness of British firearms technology as compared with his own plant in Hartford, set up a factory in London for the purpose of producing firearms with American machinery and British workers. Although Colt had been fantastically successful in the United States—where he ran the largest private armory in the world—his London plant had to be closed down, apparently because of the difficulties encountered in employing British workers with unfamiliar American machinery.[6]

There is much evidence that the transmission of industrial technology from England to the Continent in the first half of the 19th century was also heavily dependent upon the same sort of personal mechanism. David Landes has suggested, for example, that in spite of legal prohibitions until 1825, there were at least 2,000 skilled British workers on the Continent providing indispensable assistance in the adoption of the newly developed techniques.

> The best of the British technicians to go abroad were usually entrepreneurs in their own right, or eventually became industrialists with the assistance of continental associates or government subventions. Many of them came to be leaders of their respective trades: one thinks of the Waddingtons (cotton), Job Dixon (machine-building), and James Jackson (steel) in France; James Cockerill (machine construction) and William Mulvany (mining) in Germany; Thomas Wilson (cotton) in Holland; Norman Douglas (cotton) and Edward Thomas (iron and engineering) in the Austrian empire; above all, John Cockerill in Belgium, an aggressive, shrewd businessman of supple ethical standards, who took all man-

tory 11, no. 4 (Fall 1951): 347–60. A. R. Hall has concluded: "It seems fairly clear that in most cases in the 16th century—and indeed long afterwards—the diffusion of technology was chiefly effected by persuading skilled workers to emigrate to regions where their skills were not yet plentiful" (A. R. Hall, "Early Modern Technology, to 1600," in *Technology in Western Civilization*, ed. Melvin Kranzberg and Carroll W. Pursell, Jr. [New York, 1967], 1:85). For a similar conclusion with respect to mining and metallurgy in an earlier period, see Bertrand Gille, "Technological Developments in Europe: 1100 to 1400," in *The Evolution of Science*, ed. Guy S. Metraux and François Crouzet (New York, 1963), p. 201.

[6] These episodes are described at length in the introductory essay in Rosenberg, ed., *The American System of Manufactures.*

ufacturing as his province and with the assistance of first the Dutch and then the Belgium governments made a career of exploiting the innovations of others.[7]

These episodes suggest an important point with respect to the transfer of technology. The notion of a production function as a "set of blueprints" comes off very badly if it is taken to mean a body of techniques which is available independently of the human inputs who utilize it. The point has been well expressed by Svennilson:

> It would be far too crude to assume, as often seems to be the case, that there is a common fund of technical knowledge, which is available to anybody to use by applying his individual skill. We must take into account that only a part, and mainly the broad lines, of technical knowledge is codified by non-personal means of intellectual communication or communicated by teaching outside the production process itself. The technical knowledge of persons who have been trained in actual operations has a wider scope, especially as regards the application of more broad knowledge. The "common fund," thus, covers only part of the technical knowledge to which individuals or groups of persons apply their personal skill.[8]

If, with Svennilson, we define "know-how" as "the capacity to use technical knowledge," then it is apparent that such know-how is essential for the successful utilization of the technical information incorporated in the economist's production function. And such know-how has been transmitted personally in the past simply because it is a noncodified kind of skill which has been largely acquired by direct exposure to and participation in the work process.[9] A somewhat amusing, yet apposite,

[7] David Landes, *The Unbound Prometheus* (Cambridge, 1969), pp. 148–49. Landes adds: "Perhaps the greatest contribution of these immigrants was not what they did but what they taught. . . . The growing technological independence of the Continent resulted largely from man-to-man transmission of skills on the job" (p. 150). For a more detailed discussion of this subject, see W. O. Henderson, *Britain and Industrial Europe, 1750–1870* (Liverpool, 1954).

[8] I. Svennilson, in *Economic Development with Special Reference to East Asia*, ed. K. Berrill (New York, 1964), p. 407.

[9] Some of the recent work of Derek Price lends indirect support to this position. Price argues that there exist powerful professional incentives for the scientist to publish the results of his work quickly and authoritatively, whereas the incentive structure of the working technologist is one which (with some exceptions) encourages concealment and a reluctance to publish. Science is *papyrocentric*, whereas technology is *papyrophobic* (see Derek J. De Solla Price, "Is Technology Historically Independent of Science? A Study in Statistical Historiography," *Technology and Culture* 6, no. 4 [Fall 1965]: 553–68).

instance of the role of know-how has recently been cited by Lucius Ellsworth, who points out that although we possess a complete set of the appropriate craft tools, no one in the United States today *knows how* to use coopers' tools for making wooden barrels—a once-honored and widely practiced craft.[10]

The extent of specialization in the production of machinery was of critical importance in the development of industrial skills. To appreciate this importance we must think in dynamic terms as well as in terms of static allocative efficiency. For there is a crucial learning process involved in machinery production, and a high degree of specialization is conducive both to a more effective learning process and to a more effective *application* of that which is learned. This highly developed facility in the designing and production of specialized machinery is, perhaps, the most important single characteristic of a well-developed capital goods industry and constitutes an external economy of enormous importance to other sectors of the economy.[11]

It is a common practice to look upon industrialization as involving not only growing specialization but also growing complexity and differentiation. While this is certainly true in the sense that there takes place a proliferation of new skills, facilities, commodities, and services, it also overlooks some very important facts. The most important for present purposes is that industrialization was characterized by the introduction of a relatively small number of broadly similar productive processes to a large number of industries.[12] This follows from the familiar fact that industrialization in the 19th century involved the growing adoption of a metal-using technology employing decentralized sources of power.

The use of machinery in the cutting of metal into precise shapes involves, to begin with, a relatively small number of operations (and therefore machine types): turning, boring, drilling, milling, planing, grinding, polishing, etc. Moreover, all machines performing such operations confront a similar collection of technical problems, dealing with such matters as power transmission, control devices, feed mechanisms, friction reduction, and a broad array of problems connected

[10] Lucius F. Ellsworth, "A Directory of Artifact Collections," in *Technology in Early America*, ed. Brooke Hindle (Chapel Hill, N.C., 1966), p. 96.

[11] This paragraph and the three following are taken, with minor modification, from Nathan Rosenberg, "Technological Change in the Machine Tool Industry, 1840–1910."

[12] For more recent evidence on such trends from the point of view of the input-output structure of the economy, see Anne P. Carter, "The Economics of Technological Change," *Scientific American* 214, no. 4 (April 1966): 25–31.

with the properties of metals (such as ability to withstand stresses and heat resistance). It is because these processes and problems became common to the production of a wide range of disparate commodities that industries which were apparently unrelated from the point of view of the nature and uses of the final product became very closely related (technologically convergent) on a technological basis—for example, firearms, sewing machines, and bicycles.

Because of this technological convergence, the machine tool industry, in particular, played a unique role both in the initial solution of technological problems and in the rapid transmission and application of newly learned techniques to other uses. I have found it useful to look at the machine tool industry in the 19th and early 20th centuries as a center for the acquisition and diffusion of new skills and techniques in a "machinofacture" type of economy. Its chief importance lay in its strategic role in the learning process associated with industrialization. This role was a dual one: (1) new skills and techniques were developed or perfected there in response to the demands of specific customers, and (2) once they were acquired, the machine tool industry was the main transmission center for the transfer of new skills and techniques to the entire machine-using sector of the economy.

The questions which currently trouble me are: How do we institutionalize such activities now? What current substitutes are available for the performance of the vital activities of the production and diffusion of industrial skills—activities which were largely performed by the capital goods industries in the 19th century? The experience of successfully industrializing countries in the 19th century indicates that the learning experiences in the design and use of machinery were vital sources of technological dynamism, flexibility, and vitality. Countries which rely upon the importation of a foreign technology are thereby largely cut off from this experience, but other institutional alternatives may well be available. The contemporary rise of the multinational firm in such areas as chemicals, chemical process plants, plastics, and electronic capital goods may reflect, more than anything else, an attempt to fill this gap: to provide institutional and organizational mechanisms which will facilitate the transfer of know-how. This occurs largely by institutional innovations which make the services of the personnel with the appropriate knowledge and skills readily available, on short notice, anywhere in the world.[13]

[13] In a study of the world market in chemical processing plants, Freeman has pointed out: "The main European 'design-engineering' offices of the large contractors typically employ from 300 to 700 total office-based staff. They are concentrated mainly in London (where half a dozen of them have such offices) and to a lesser

III

Anglo-American experience with the introduction and diffusion of technology in the 19th century points strongly to the importance of the composition of consumer demand and the malleability of public tastes. The willingness of the public to accept a homogeneous final product was a decisive factor in the transition from a highly labor-intensive handicraft technology to one involving a sequence of highly specialized machines. Across a whole range of commodities we find evidence that British consumers imposed their tastes on the producer in a manner which seriously constrained him with respect to the exploitation of machine technology. English observers often noted with some astonishment that American products were designed to accommodate, not the consumer, but the machine. Lloyd noted, for example, of the American cutlery trade, that "where mechanical devices cannot be adjusted to the production of the traditional product, the product must be modified to the demands of the machine. Hence, the standard American table knife is a rigid, metal shape, handle and blade forged in one piece, the whole being finished by electroplating–an implement eminently suited to factory production."[14] Even with respect to an object so ostensibly utilitarian as a gun, the British civilian market was dominated by peculiarities of taste which essentially precluded machine techniques. English civilians were in the habit of having their guns made to order, like their clothing, and treating their gunsmith in much the same manner as their tailor. In fact, we hear a Colonel Hawkins instructing his upper-class readers in the 1830s that "the length, bend and casting of a stock [gunstock] must, of course, be fitted to the shooter, who should have his measure for them as carefully entered in a gunmaker's books, as that for a suit of clothes on those of his tailor."[15]

---

extent in The Hague, Paris, and Frankfurt. London is the principal European base of the American contractors and the largest world centre for chemical 'design-engineering' work. But the large American contractors show great ingenuity and flexibility in switching work from one office to another anywhere in the world, in order to achieve an optimum loading of their total resources. The London office, even though employing 500 or fewer, is able to take on several very large contracts because it is part of a world-wide organization and can call upon process design and know-how from the parent company, as well as specialists to deal with crises or bottlenecks. A particular individual, a scale model, or a set of drawings may be flown out at a few hours' notice" (C. Freeman, "Chemical Process Plant: Innovation and the World Market," *National Institute Economic Review* [August 1968], p. 37).

[14] G. I. H. Lloyd, *The Cutlery Trades* (London, 1913), pp. 394–95.

[15] As quoted in *A Treatise on the Progressive Improvement and Present State of the Manufactures in Metal* (London, 1833), 2:105.

On the other hand, Siegfried Giedion has pointed out that "in the United States, where department stores had slowly been developing since the 'forties, ready-made clothes, in contrast to Europe, were produced from the start."[16] Giedion's point is doubly interesting in the present context because, in fact, the United States, although beginning industrialization later than England, took an early lead in the manufacture of both guns and ready-made clothing (as well as in new techniques of merchandising).

There is, it seems to me, an intimate interrelationship between composition of demand and homogeneity of product, on the one hand, and the range of technological possibilities open to society on the other. Certainly the producer must have some minimum degree of freedom to design his product in order to make it suitable for mass production techniques. There is very little doubt that, in this respect, the American producer had several degrees of freedom more than did his English counterpart. This difference had a great deal to do with the fact that mass production technology essentially originated in the United States rather than England, with early American leadership in products which were particularly well suited for mass production—firearms, sewing machines, typewriters, and later automobiles—and with the slow acceptance of many of these techniques in England.

But I am anxious not to leave the impression that Anglo-American differences can all be explained in terms of differences in consumer tastes, although I do think these were important. They seem to have been part of a much more widespread phenomenon in Britain as compared with the United States of "customer initiative" as opposed to "producer initiative." If we examine the relation between producers of capital equipment and their purchasers in both countries, we also find analogous differences. That is, in America the producer of capital goods took the initiative in matters of machine design and successfully suppressed variations in product design which served no clearly defined purpose. He brought about, in other words, a high degree of standardization in the machinery, which very much simplified his own production problems and in turn reduced the price of capital goods. Producer initiative was a very important factor in developing patterns of efficient specialization in American capital goods production.

In England, on the other hand, the capital goods producer remained, to a surprising degree, what Landes has aptly called a "custom tailor working in metal." The explanation for this difference in roles is not at all clear. In the case of many kinds of capital goods, the variations in machine design were quite unrelated to any functional justification.

[16] Siegfried Giedion, *Mechanization Takes Command* (New York, 1948), p. 90.

Buyers of machinery were in the habit of drawing up blueprints with highly detailed specifications which the machine producer had to agree to provide as a condition of fulfillment of the contract. Many of the specifications were often nothing more than the result of historical accident. In England the initiative in matters of machine design was emphatically in the hands of the buyer, and this had serious implications for the machine-producing sector both as a transmission center for new techniques and as a producer of low-cost machinery.[17]

The problems encountered by the machinery-producing sector were further intensified by the role of that strange British institution, the consulting engineer. These engineers were imbued with a professional tradition which often led to an obsession with technical perfection in a purely engineering sense, and they imposed their own tastes and idiosyncrasies upon product design. In America, by contrast, the engineer and engineering skills were more effectively subordinated to business discipline and commercial criteria and did not dominate them.[18] The result was to perpetuate, in Great Britain, a preoccupation with purely technical aspects of the final *product* rather than with the productive *process*. This is reflected in the history of the automobile, where, of course, British engineers developed an early reputation for high-priced, high-quality engineering. In 1912, when Henry Ford was preparing to demonstrate to the world the great possibilities of standardized, high volume production, an influential British trade journal commented: "It is highly to the credit of our English makers that they choose rather to maintain their reputation for high grade work than cheapen that reputation by the use of the inferior material and workmanship they would be obliged to employ to compete with American manufacturers of cheap cars."[19]

[17] At the turn of the 20th century, some evidence of change began to appear. A British observer in 1902 made the following highly revealing comments: "Some few years ago no machine-tool maker ventured to offer advice to a manufacturing engineer as to the tools he might buy or the methods he should follow. He would have been told to mind his own business. Today a good deal of help and advice is often requested and the tool maker and user, as they should be, are often in consultation as to tools, methods and organization. Machine tool making in America has been, I take it, considered a reputable business for some years, whereas here for a time it certainly was but little considered" (William H. Booth, "An English View of American Tools," *American Machinist*, no. 45, November 6, 1902, p. 1580).

[18] See, for example, Daniel H. Calhoun, *The American Civil Engineer: Origin and Conflict* (Cambridge, Mass., 1960), esp. chap. 6.

[19] *Autocar*, September 21, 1912, as quoted in S. B. Saul, "The Motor Industry in Britain to 1914," *Business History* 5, no. 1 (December 1962): 41. The belief long persisted in British industry that high quality was incompatible with mass produc-

The earlier design of locomotives was a classic case of a rampant and pointless individualism. By 1850, after the early period of experimentation (when there was an important justification for variations in design),[20] the basic features of a locomotive stabilized.[21] But innumerable minor variations of this basic design persisted. "Probably five distinct classes of locomotives would afford a variety sufficiently accommodating to suit the varied traffic of railways," an expert wrote in 1855, "whereas I suppose the varieties of locomotives in actual operation in this country and elsewhere are very nearly five hundred."[22] There is much authoritative evidence that, by comparison with American practice, a needless proliferation of designs and specifications prevailed in Great Britain throughout a broad range of iron and steel products and engineering works generally.[23]

This absence of standardization vastly complicated the process of adopting the technology and organization of mass production. It seriously inhibited the growth of specialization by firms which was so characteristic of the American machinery industry in the second half of the 19th century. For example, British railroad companies built their rolling stock in their own workshops. Private locomotive producers

---

tion. Much evidence on this point for the late 1920s may be found in Committee on Industry and Trade, *Survey of Metal Industries*, pt. 4 (London, 1928), pp. 227–28, 220–21, and passim.

[20] See Burton H. Klein, "A Radical Proposal for R. and D.," *Fortune* 57, no. 5 (May 1958): 112 ff., for a statement of the case for the desirability of parallel research efforts under conditions of uncertainty.

[21] It is interesting to note that, as early as the 1840s, American engineers were playing a vital role in bringing the railroad to Russia. See the account of the work of Major George Washington Whistler in Albert Parry, *Whistler's Father* (Indianapolis, 1939). Of Major Whistler, the awed muzhiks are supposed to have exclaimed: "How smart the foreign gentleman, who has harnessed the samovar and made it run." The book is unfortunately thin on technological history. It does, however, provide a minimal account of Whistler's role in directing the construction of the Saint Petersburg to Moscow railway and of his establishment of a factory at Alexandrovsky for building locomotives and other railroad equipment.

[22] J. H. Clapham, *An Economic History of Modern Britain*, 3 vols. (Cambridge, 1938), 2:76. English orientation toward export markets doubtless increased the difficulties of standardization of locomotive design (see *Survey of Metal Industries*, pp. 168–72). Nevertheless, the high degree of diversification had developed early in the history of the industry when it was certainly preoccupied with the needs of domestic users.

[23] J. Stephen Jeans, ed., *American Industrial Conditions and Competition*, Reports of the Commissioners Appointed by the British Iron Trade Association to Enquire into the Iron, Steel and Allied Industries of the United States (London, 1902), esp. pp. 255–59.

worked essentially for the export trade, and therefore large numbers of firms in Britain were engaged in small-batch production of a wide range of railroad equipment. During the 1920s one locomotive manufacturer was making equipment for twenty-four different gauges, varying from 18 inches to 5 feet 6 inches, and locomotive sizes ranging from 6 to 60 tons.[24] Nothing comparable to the Baldwin Locomotive Company of Philadelphia ever developed, under these circumstances, in Britain. And what was characteristic of British locomotive production was broadly true of all British engineering, where general engineering firms predominated—that is, firms which undertook to produce a range of products which would have seemed incredibly diverse by American standards. British engineering and equipment-producing firms were typically engaged in "batch" or "small-order" production. At the same time, of course, the absence of standardization further increased capital costs by raising the costs of repair and necessitating larger inventories.

Similarly, one of the most characteristic features of automobile production in Britain was the extent to which each firm produced its own components. In part this was due to the failure to standardize, but cause and effect are very hard to disentangle in these matters. There is much evidence suggesting that, in the early days of the automobile industry, British firms found it impossible to make satisfactory arrangements for the production of components by other firms and therefore fell back upon the expedient of producing such components themselves. When William Morris (later Lord Nuffield) made up his mind to purchase the main components of his automobiles from specialist producers (1913), he found that they could not meet his requirements, and he finally had to import components from America.[25]

All this was a very serious handicap because it meant that individual firms were responsible for producing a range of products which was beyond their organizational and technical competence.[26] From the per-

[24] *Survey of Metal Industries*, p. 172.

[25] H. J. Habakkuk, *American and British Technology in the 19th Century* (Cambridge, 1962), p. 203.

[26] A good deal of valuable material on Anglo-American differences in engineering methods at the beginning of the 20th century will be found in the paper by H. F. L. Orcutt, "Modern Machine Methods," and the subsequent discussion, in *Proceedings of the Institution of Mechanical Engineers* (1902), pts. 1–2, pp. 9–112. The greater degree of specialization of both American firms and American machinery as compared with their English counterparts received much attention. J. R. Richardson, who operated a general engineering workshop, justified his unwillingness to purchase American molding machines in the following way: "It was not that English engineers did not understand American methods, but that Americans did not as a rule

spective of American industrialization, where highly elaborate patterns of interfirm specialization developed, this was one of the most striking features of British industry. Not only did railways produce their own locomotives and automobile companies their own components, but users of capital equipment such as machine tools often made the tools themselves—for example, sewing machine companies and textile machinery firms—long after such production had been taken over by specialist firms in the United States.[27]

IV

Although I have already suggested several factors which may account for differences in the degree of specialization in the United States and Britain, I am far from convinced that such factors as differences in taste, standardization, and producer initiative constitute a complete explanation. To put the point in its most general way: American firms showed a much greater talent than British firms for coordinating successfully their relationships with other firms upon whom they were dependent for the supply of essential inputs. American firms learned to integrate their own operations with those of their main suppliers in a way which enabled them to confine their own productive operations to a limited number of specializations. Clearly the British failure to subcontract more efficiently was closely related to the factors which we have already discussed. But the failure seems also to have included an

---

understand the conditions which obtained in large engineering works in England having a big general practice. There must be a large run of work. Even the most enthusiastic Americans had told him that a large quantity was not needed, that it could be done perfectly well with a dozen, but very often a dozen was a large quantity. Not only was it necessary to have on his catalogue 500 different types and sizes of steam-engines, but an infinite variety of mining and general machinery; and in addition his firm was expected to do anything required, and had to do it even if it only had to be done once" (pp. 72–73). The important unanswered—indeed, unposed—question is why the general engineering firm persisted as it did in England (see S. B. Saul, "The American Impact upon British Industry," *Business History* 3 [1960]: 21–27).

[27] America's superiority in the contriving of highly specialized machinery was clearly recognized as early as the 1850s. A parliamentary committee of distinguished engineers which visited America in 1853 stated: "As regards the class of machinery usually employed by engineers and machine makers, they are upon the whole behind those of England, but in the adaptation of special apparatus to a single operation in almost all branches of industry, the Americans display an amount of ingenuity, combined with undaunted energy, which as a nation we would do well to imitate, if we mean to hold our present position in the great market of the world" (*Report of the Committee on the Machinery of the U.S.*, as reprinted in Rosenberg, ed., *The American System of Manufactures*, pp. 128–29).

inability of British firms to rely more confidently upon the fulfillment of contracts with the degree of precision in engineering specifications which is so important to the assembly of components into a complex final product. F. W. Lanchester, in looking back upon his early experiences in the auto industry, where he attempted to persuade craftsmen to work to standard instructions so as to achieve interchangeability, commented: "In those days, when a body-builder was asked to work to drawings, gauges or templates, he gave a sullen look such as one might expect from a Royal Academician if asked to colour an engineering drawing."[28] The difficulties of successful subcontracting of high-precision components in the face of such attitudes can be readily imagined.

Furthermore, I wonder if the failure might not also involve the absence of some more subtle managerial or administrative talents which may have been available in greater abundance in the United States. For example, after discussing vertical integration as a possible solution to the problems of cooperation and communication among chemical firms, contractors, and component makers, Freeman has pointed out:

> The trend in the United States has been in the opposite direction. The oil and chemical firms make increasing use of specialist contractors and specialist component-makers. The contractors tend to divest themselves of manufacturing and fabricating activities. Studies of the aircraft industry and large weapon systems have found that prime United States contractors make far more use of specialist sub-contractors and external economies than European aircraft firms. When the prime contractor (whether in aircraft or chemical plant) uses large numbers of specialist sub-contractors, this means that he must acquire considerable skills in systems management and must be capable of enforcing very high technical standards on sub-contractors.[29]

[28] As quoted in Habakkuk, p. 203.

[29] Freeman, p. 49. Earlier, Freeman had stated: "We found in our case studies that a number of successful process innovations arose from intimate technical cooperation between chemical firms and contractors or chemical firms and component makers, or all three. This is particularly evident in the United States; much less so in Europe" (p. 48). In an article on the plastics industry, Freeman states, in connection with the machinery for making plastics: "Generally speaking a technically advanced and progressive machine industry will principally benefit the country in which it is located, because material-suppliers, machine-makers and fabricators can more easily co-operate there in experiment, development and design. This tripartite co-operation appears to be closer and more satisfactory in Germany than in Britain" (C. Freeman, "The Plastics Industry: A Comparative Study of Research and Innovation," *National Institute Economic Review* [November 1963], pp. 42–43; see also p. 46).

In another study, of electronic capital goods, Freeman emphasizes that the innovative process is now highly dependent upon a successful collaboration between producers of the final product and specialist makers of components.

> The development process in the electronic capital goods industry consists largely of devising methods of assembling components in new ways, or incorporating new components to make a new design, or developing new components to meet new design requirements. (This is not quite so simple as it may sound. There are more than a hundred different components in a TV set, more than 100 thousand components in a large computer, and more than a million in a big electronic telephone exchange.) Consequently, there must be close collaboration between end-product makers and component makers.[30]

There seem to be some kinds of talents involved here which economists have not yet identified very precisely but which may be important determinants of success in the utilization of certain kinds of technology. The Japanese, it is worth noting, have been extremely successful (especially since the 1920s) at subcontracting, and this ability is almost certainly closely connected with the important role which small firms continue to play in Japanese industry.[31] The Japanese, as Strassmann points out, have been highly successful in the "inter-firm coordination of schedules and standards in subcontracts."[32] On the other hand, the Russians have been singularly unsuccessful in introducing subcontracting arrangements. This failure has been closely connected with many of the organizational failures of Russian central planning.[33]

V

The existence of a well-developed domestic capital goods industry is important to the transmission of technology for a reason which is not yet sufficiently explicit. I have discussed the role of these industries so far in terms of their *capacities*, but we must recognize also their *motivations*. It is the producers of capital goods who have the financial incentive and therefore provide the pressures (marketing, demonstra-

[30] C. Freeman, "Research and Development in Electronic Capital Goods," *National Institute Economic Review* (November 1965), p. 63.

[31] Paul Strassmann, *Technological Change and Economic Development* (Ithaca, N.Y., 1968), p. 168.

[32] Ibid., p. 274.

[33] David Granick, *Soviet Metal-fabricating and Economic Development* (Madison, Wis., 1967), chap. 5.

tion) to persuade firms to adopt the innovation (which they produce). Creating a capital goods industry is, in effect, a major way of *institutionalizing* internal pressures for the adoption of new technology. In America the producers of capital goods have always played a major role in persuading and educating machinery users about the superiority and feasibility of new techniques.

This is an extremely important activity in overcoming the inevitable combination of inertia, ignorance, and genuine uncertainty which surrounds an untried product. The introduction of the diesel locomotive by General Motors is a classic case in point. In the United States both the railroad companies and the locomotive producers were extremely skeptical of the diesel engine and resisted its introduction. It took great promotional effort on the part of GM, which developed the diesel, to induce the railroads even to consider and experiment with the innovation. This kind of promotional activity, on the part of capital goods industries with a strong personal motive to gain acceptance for their product, seems to have been a critical factor in the American experience. Here again it has to be asked: what mechanisms or institutions can be substituted for the motivational pressure, provided in the past by domestic capital goods producers, on the part of poor countries which rely on distant foreign producers for their capital equipment?

In general the relationship between machinery producers and their customers was more successful in the United States than in England or elsewhere in Europe. In the confrontations that took place between these two groups, there seems to have been an interchange of information and communication of needs to which the machinery producers responded in a highly creative way. They learned to deal with the requirements of their customers at the same time that the machinery user learned to rely heavily on the judgment and initiative of the machinery supplier. It was, in part, the relative harmony and mutual confidence of these relationships which made it possible for machinery makers to eliminate customer preferences that were technically nonessential or irrelevant and therefore to design more highly standardized machinery. This was a process which involved an intimate knowledge of customer activities and needs and which presupposed frequent face-to-face confrontations and exchange of information (at which the British were much less successful than the Americans).[34]

[34] Strassmann makes the interesting point that "face-to-face contact to establish trust appears to be a need throughout the technical transfer network. Lack of full trust invariably means some concealment of information which impairs perception of needs at subsequent stations. Moreover, a sense of insecurity may blur the trans-

Here the extent of machinery standardization intrudes itself again as a significant variable. I think it is reasonable to argue that, once the technical characteristics of a new machine have become stabilized, the amount of initiative which it is possible for capital goods suppliers to exert with their customers will depend on the extent of standardization. With extreme heterogeneity (British railroads) the role of the capital goods producer tends to become passive and adaptive rather than active and initiatory. Initiative will then reside with the user of equipment, and it will be difficult for anyone to anticipate and cater to his needs (tailor syndrome). This may have a profound effect not only on the ability of the capital goods industries to generate innovations but also on the efficiency with which they are able to produce their machines. This certainly seems to have been an important factor in accounting for differences in performance between British and American capital goods industries.

In the 19th century, a major source of capital saving for the economy as a whole came through improvements in the efficiency of capital goods production. The important analytical point is that any cost reduction in the capital goods sector—whether it is labor saving or capital saving in its immediate factor-proportion bias—is a capital-saving innovation to the economy as a whole. Such a capital-saving stage has been an inevitable but later stage of the sequence by which industrial economies have accommodated themselves to an innovation. When a new machine was introduced, there existed, by definition, no established system of organization for producing the machine. As Marx once stated: "There were mules [spinning mules] and steam-engines before there were any laborers whose exclusive occupation it was to make mules and steam-engines; just as men wore clothes before there were such people as tailors."[35] The fact of the matter is that a very high fraction of new inventions, new products, or new processes, once conceived, are

---

mitter's vision as well, making him incapable of stating his problems, even on paper, and of reaching decisions. To see faces is to see confusion, satisfaction, or antagonism; to gauge whether status is being, or should be, granted or withheld. The conversational setting from time to time allows the release of tensions through small talk, jokes, and laughter. The urge to conceal as well as the fear of concealment weakens" (Strassmann, pp. 32–33). For a highly suggestive treatment of the spatial diffusion of innovations which turns, in large measure, on the role of person-to-person contacts and personal communications between pairs of individuals, see Torsten Hagerstrand, "Quantitative Techniques for Analysis of the Spread of Information and Technology," in *Education and Economic Development*, ed. C. Arnold Anderson and Mary Jean Bowman (Chicago, 1965), chap. 12.

[35] Karl Marx, *Capital* (New York, 1936), p. 417.

of no economic relevance until the capital goods industries have successfully solved the technical and mechanical problems or developed the new machines which the inventions require. This creative process on the part of the capital goods industries has been badly slighted, and a recognition of this process may help us understand the widely-observed failure of poor countries to develop techniques with the factor-saving bias which they need. In the currently developed economies, the growing efficiency of the capital goods sector was a major source of capital saving for the economy, *both* because a large proportion of all innovations originated in that sector and because it developed a capacity for producing capital goods at lower and lower cost.[36]

From this perspective, some of my concerns about the prospects for poor countries which rely on the importation of foreign capital equipment are obvious. Of course it is of enormous benefit to them to be able to import this equipment, even where the equipment is not optimally factor-biased. But if new techniques are regularly transferred from industrial countries, how will the learning process in the design and the production of capital goods take place? Reliance on borrowed technology perpetuates a posture of dependency and passivity. It deprives a country of the development of precisely those skills which are needed if she is to design and construct capital goods that are properly adapted to her own needs. What, then, are the prospects for underdeveloped countries ever becoming efficient producers of capital goods and, in particular, developing a technology with factor-saving biases more appropriate to their own factor endowments? In the past, as I have argued, the appropriate skills were acquired through an intimate association between the user and the producer of capital goods. In the absence of these experiences, what substitute mechanisms or institutions can be established to provide the necessary skills?

The problem takes on additional importance when we adopt a more realistic conception of the nature of technical change than we have inherited from Schumpeter. Schumpeter accustomed economists to thinking of technical change as involving major breaks, giant discontinuities or disruptions with the past. This rather melodramatic conception fitted in well with his charismatic approach to entrepreneurship. But technological change is also (and perhaps even more importantly) a continuous stream of innumerable minor adjustments, modifications, and adaptations by skilled personnel, and the technical vitality of an economy employing a machine technology is critically affected by its

[36] See Nathan Rosenberg, "Capital Goods, Technology and Economic Growth," *Oxford Economic Papers*, n.s. 15, no. 3 (November 1963): 217–27.

capacity to make these adaptations.[37] The necessary skills, in the past, were developed and diffused in large measure by the capital goods sector. The skills are, inevitably, embodied in the human agent and not in the machine, and unless these skills are somehow made available, the prospects for the viability of a machine technology may not be very good.

As part of our Schumpeterian heritage, we still look upon the transformation of an "invention" into an "innovation" as the work of entrepreneurs. But from a *technological* perspective, it is much more the work of the capital goods industry. This is most obviously the case where the invention is a cost-reducing *process* and does not have to be marketed to final consumers. But in making new products and processes practicable, there is a long adjustment process during which the invention is improved, bugs ironed out, the technique modified to suit the specific needs of users, and the "tooling up" and numerous adaptations made so that the new product (process) can not only be produced but can be produced at low cost. The idea that an invention reaches a stage of commercial profitability *first* and is then "introduced" is, as a matter of fact, simple-minded. It is during a (frequently protracted) shakedown period in early introduction that it becomes obviously worthwhile to *bother* making the improvements. Improvements in the *production* of a new product occur *during* the commercial introduction.

Alternatively put, there has been a tendency to think of a long precommercial period when an invention is treated as somehow shaped and modified by exogenous factors until it is ready for commercial introduction. This is not only unrealistic; it is a view which has also been responsible for the neglect of the critical role of capital goods firms in the innovation process.

We are now coming to realize that modern technology has a long

[37] In the shaping of technology, many more kinds of adjustments are involved than the factor-proportions adjustment, with which economists are typically preoccupied. It may be possible, for example, to incorporate certain kinds of features into a machine which will compensate for (or simply bypass) deficiencies on the part of the labor force. If workers cannot be relied upon to perform appropriate maintenance procedures on expensive machinery, machines can be designed which *require* less maintenance. This occurred in the United States early in the 20th century when it was found that immigrant labor could not be relied upon to attend to the numerous separate lubrication points of their machinery, thus causing frequent breakdowns. This problem was solved by redesigning machine tools to incorporate the centralized, self-acting lubrication system of the automobile. Since this system went into operation automatically when the machine was activated, reliance upon the worker was, in this respect, bypassed (S. Einstein, "Machine-Tool Milestones, Past and Future," *Mechanical Engineering* 52, no. 11 [November 1930]: 961).

umbilical cord. Innumerable unsuccessful foreign aid projects in the past twenty years—including Russian-sponsored as well as American-sponsored projects—have confirmed that when modern technology is carried to points remote from its source, without adequate supportive services, it will often shrivel and die. This is partly because the technology emerged in a particular context, often in response to highly narrow and specific problems, such as may have been defined by a particular natural resource deposit. But, more important, the technology functions well only when it is maintained and nourished by an environment offering it a range of services which are essential to its continued operation. These would include the ability to diagnose correctly the causes of machine breakdown or other sources of inferior performance, the availability of facilities and personnel to perform repair work and to provide routine maintenance and repairs, and the provision of spare parts. In the absence of the kinds of skills produced and embodied in a capital goods industry, repair and maintenance costs in the use of machinery are likely to remain high. A major reason for a domestic capital goods industry, therefore, is that the ability to utilize complex machinery effectively—whatever the country of origin of the machinery—depends upon the kinds of skills which such an industry uniquely makes available.

Moreover, physical proximity between the producer and user of machinery seems to have been indispensable in the past for reasons which we do not really understand but which seem to be rooted basically in the problem of communications. Successful technological change seems to involve a kind of interaction that can best be provided by direct, personal contact. Successful instances of technological change in the past have involved a subtle and complex network of contacts and communication between people, a sharing of interests in similar problems, and a direct confrontation between the user of a machine, who appreciates problems in connection with its use, and the producer of machinery, who is thoroughly versed in problems of machinery production and who is alert to possibilities of reducing machinery (and therefore capital) costs.

VI

With all of the difficulties attaching to the transfer of industrial technology, such technology is nevertheless much easier to transfer than agricultural technology because industrial technology is at least very much self-contained. It tends to be a relatively closed system. Even its dependence upon human inputs can be reduced by making the technology more capital-intensive. But where ecological relationships are involved, as they are in agriculture, everything is quite different. Here there are important interactions between the human enterprise and specific features of the natural environment. Here natural phenomena participate in a much more active way, and productive activity must be more highly responsive to even minor variations and peculiarities of the environment. Agricultural activity is much more closely enmeshed with the natural environment, and this, in turn, has important consequences. It means that solutions to agricultural problems and the ability to make appropriate adaptations to local conditions will hinge upon kinds of knowledge which are not ordinarily possessed by the cultivator: biology, botany, biochemistry, genetics, etc. Major breakthroughs in agriculture are likely to come from the application to local circumstances of scientific disciplines which do not ordinarily flourish in poor countries. I would therefore argue that a major goal of institution building in poor countries should be to equip them with the kind of research abilities and facilities that will enable them to produce the new knowledge required for their own peculiar agricultural resources.

Agriculture's close involvement with nature has another important consequence for the acceptance of new techniques: because of the importance of even minor variations in rainfall, sunshine, soil content, topography, plant diseases, etc., there is a much higher degree of uncertainty concerning the application of new agricultural techniques. Institution building in agricultural environments will have to take this element of uncertainty prominently into account. Institutional arrangements which serve to reduce this uncertainty or to protect the cultivator against a loss that could be personally disastrous may prove to be extremely important in speeding the adoption of new techniques.

In examining the sources of productivity improvement in both industry and agriculture, we must pay attention to the fact that the performance of individual industries will frequently depend not only on resources within that industry but on the availability and the effectiveness of industries which stand in an important complementary relationship with it. The technological inputs (including knowledge) which crucially affect the success of industry *A* are produced by industries *B*, *C*, and *D*. Much of the discussion of the prospects and possibilities for technical improvement in poor countries has suffered from ignoring such interindustry relationships. We have to take account of these relationships and learn how to exploit them.

This point is particularly pertinent to the behavior of agriculture. The history of countries with highly productive agricultural sectors indicates clearly that the major sources of improvements in agricultural productivity have been generated *outside* the agricultural sector. Unless an economy is well equipped with these complementary sources, agriculture is not likely to experience rapid improvements in efficiency. In the American experience the sources of productivity growth in agriculture have come from the machinery-producing sector, which developed a mechanical technology appropriate to agriculture; from research at agricultural experiment stations and other educational institutions; from the fertilizer industry; and increasingly in the 20th century from the study of genetics and from chemistry (including soil chemistry). In fact, in American agriculture the industries which supply agriculture with certain of its inputs have played a role in the introduction of new technologies entirely analogous to that of the capital goods industries in relation to the manufacturing sector.[38] It is the absence of these complementary sectors and resources which is often so damaging to underdeveloped agriculture.[39] If, for example, the fertilizer industry can produce fertilizer only at a very high cost, then all the extension work and exhortation in the world will not induce the peasant to use it—he will reject its use on perfectly rational cost-benefit grounds. What is needed is research oriented toward reducing the cost of the inputs which agriculture receives from the industrial sector. An extension service will not perform an important social function until it has valuable information to extend.[40]

In broad terms, the history of Japanese agricultural development

[38] As Vernon Ruttan has pointed out: "The ratio of purchased inputs to total output has risen steadily over the last several decades in American agriculture. The farm supply industries have played an important role in channeling technological advances into agriculture. Adoption of new technology has usually increased agriculture's dependence on inputs produced in the nonfarm sectors of the economy and has lessened agriculture's dependence upon land inputs" (Vernon Ruttan, "Research on the Economics of Technological Change in American Agriculture," *Journal of Farm Economics* 42, no. 4 [November 1960]: 740).

[39] For an absorbing account of the Soviet Union's heavy reliance upon American technical personnel in her attempt to achieve a rapid mechanization of agriculture, see Dana G. Dalrymple, "The American Tractor Comes to Soviet Agriculture: The Transfer of a Technology," *Technology and Culture* 6, no. 2 (Spring 1964): 191–214. The nature and extent of early Soviet dependence upon the importation of foreign skills is explored in sector-by-sector detail in Anthony C. Sutton, *Western Technology and Soviet Economic Development, 1917 to 1930* (Stanford, Calif., 1968).

[40] On the other hand, the "feedback" role of extension services may also have an important effect upon the productivity of agricultural research. Colleges of agriculture and agricultural experiment stations have frequently suffered from the absence of direct extension-work responsibilities because this has isolated them from a valuable flow of feedback originating at the farm level. (I am grateful to Professor Vernon Ruttan for this observation.)

seems to confirm the importance of these complementary relationships. Japan represents perhaps the most remarkable success story of the 20th century—a story that began 100 years ago. The great improvements in Japanese agricultural productivity, which provided the basis for the growth of the rest of the economy, rested firmly upon inputs provided from outside the agricultural sector. The growth of her agricultural productivity was due to a long series of dynamic interactions between the needs of the farm sector, on the one hand, and the combined ingenuity of industry and the educational establishments to supply these needs, on the other. The outcome of these interactions included the provision of low-cost commercial fertilizer, new farm implements, high-quality education and extension work, and, at a later stage, the beginnings of mechanization. Although it is relatively easy to point out the things which were done well in Japan's spectacular success story, the really critical question is how to incorporate such functions in a set of institutions which will perform successfully in a drastically different environment.[41]

VII

The exact institutional arrangements which will make for success will always be difficult to specify a priori. I would like to suggest that this is so mainly because success or failure always depends on the environment in which any particular institution is immersed—that is, on the complex of *other* institutions as well as widely shared values and traditions which critically affect its operation.

If history teaches us anything in this regard, it is that a wide diversity of institutional forms have proved to be successful under differing conditions. Similarly, history indicates also the very variable experience of the *same* institution when placed in different contexts. Thus, while on the American scene it has been customary to celebrate the virtues of the family farm, the family farm in France has been a favorite whipping boy, held responsible for the continuation of outmoded farm practices, for excessively small farm units, and for an insufficient use

[41] Nor should one ignore social changes which can accelerate the rate of diffusion of existing "best-practice" techniques. Thomas C. Smith (*Agrarian Origins of Modern Japan* [Stanford, Calif., 1959], esp. chap. 7) has pointed to the numerous improvements in farming methods which had been developed under the Tokugawa Shogunate. However, the spread of these techniques had been seriously inhibited by feudal restrictions upon travel and communication. One of the important achievements of the Meiji Restoration was the elimination of barriers to the spread of the best techniques. Consequently, agricultural productivity was raised by narrowing the gap between "best-practice" and "average-practice" techniques. See also Saburo Yamada, "Changes in Output and in Conventional and Nonconventional Inputs in Japanese Agriculture since 1880," *Food Research Institute Studies* 7, no. 3 (1967): 371–413.

of capital. And, while the American owner-operated farm has indeed been a remarkable success, it has been critically dependent for this success upon a wide range of organizational arrangements which absolutely defy simple ideological classification.[42]

The kinds of institutions which will function successfully in a particular environment and in the pursuit of particular objectives is something which it is not easy to generalize about a priori.[43] Certainly this is confirmed by our own recent history and experience, which suggests that a wide diversity of organizational tactics have proven to be successful. Consider the changing mix of government participation in various economic activities in recent years. Not only do we have public production of the TVA variety, but we also have public regulation and control of private utilities, public subsidies to private enterprise, public highways financed on a user-cost basis, public and private institutions of higher education living happily together (or almost so), public support of medical research which is financed partly from private sources, and public sponsorship of specialized activities contracted to private organizations—the Atomic Energy Commission and even the RAND Corporation. If we look specifically at the realm of technology, we find that varying mixes of the public and private sectors have been responsible for a broad range of technological breakthroughs in areas such as the development of atomic energy and its application to peaceful uses, medical research, jet propulsion, electronics, and agriculture. I point this out not to suggest that there is anything optimal about the public-private mix in the American economy but rather to indicate the wide range of institutional devices and combinations which have proved to be successful. Indeed, one of the most spectacularly successful kinds of institutions in our "capitalist" economy in recent years has been the nonprofit corporation, which has been a major source of both new knowledge and new technology.

[42] "Organization for research and development in the farm supply industries varies widely. At one extreme are such industries as hybrid seed corn, where a substantial share of research and development has been conducted at USDA and land-grant college experiment stations. At the other extreme is the farm equipment industry where public contributions have apparently been considerably smaller relative to the contributions by private industry. A third variation is represented by the fertilizer industry where the TVA has provided what is in effect an industry research institute for the fertilizer industry with a research program ranging from fundamental chemical and biological research to applied engineering and economics" (Ruttan, p. 741).

[43] Similarly, the identification of obstacles to economic development is a hazardous enterprise. For a valuable and provocative discussion, see Albert O. Hirschman, "Obstacles to Development: A Classification and a Quasi-vanishing Act," *Economic Development and Cultural Change* 13, no. 4, pt. 1 (July 1965): 385–93.

Clearly there is no single "best-way-of-doing-things" to which we have rigidly adhered in all sectors of our economic life. If the American experience suggests anything, it is that there has been no single institutional formula for success. It is reasonable to expect that the success formulas will be both different and equally variable in poor countries. Without question we ought to attach the highest research priority to finding out what combinations are likely to prove successful in the special environments of individual countries. This is an area where history can provide some guidance and valuable insights but certainly no authoritative answers.

The Weapons of the West

MILITARY TECHNOLOGY AND MODERNIZATION IN 19TH-CENTURY CHINA AND JAPAN

BARTON C. HACKER

China and Japan diverged sharply in their responses to the West during the 19th century. At century's end China, still clinging to tradition, threatened to dissolve in chaos, while Japan, having traded its Confucian classics for Western technical manuals, seemed ready to assume the status of Great Power. The contrast appears all the more striking because the two societies were geographically so close and shared so many elements of a common heritage. Appearances aside, however, the differences between China and Japan were profound and perhaps nowhere more clearly attested than in the ways each reacted to Western pressures. Like other non-Western societies, China and Japan felt those pressures most decisively in military terms. For 19th-century East Asians, as for 18th-century Russians or 20th-century regimes in every developing country, weapons were the West's greatest attraction.[1] Military technology has normally been the leading edge of modernization.[2]

DR. HACKER assumed the post of historian at the University of California Lawrence Livermore National Laboratory in 1992, continuing the pattern of research positions alternating with academic teaching that has marked his career throughout. His books and articles have won several awards, including SHOT's 1993 Usher Prize for *An Annotated Index to . . . "Technology and Culture" 1959–1984* (Chicago, 1991). At present, he is working not only on nuclear weapons history but also on *Ordering Society: A World History of Military Institutions* (Westview Press).

[1]The importance of military concerns in the reforms of Peter the Great and Russian Westernization are suggested in Alexander Gerschenkon, *Europe in the Russian Mirror: Four Lectures in Economic History* (Cambridge, 1970), pp. 69–96; Vasili Klyuchevsky, *Peter the Great,* trans. Liliana Archibald (New York, 1961), pp. 57–111; and Melvin C. Wren, *The Western Impact on Tsarist Russia* (Chicago, 1971), pp. 21–67. The scope of the armaments trade may be the clearest indicator of the current attraction of Western military hardware (see George Thayer, *The War Business: The International Trade in Armaments* [New York, 1969]; and John Stanley and Maurice Pearton, *The International Trade in Arms* [New York and Washington, 1972]).

[2]"Modernization" and "Westernization" are words of protean meaning. I here use "modernization" for the process of adapting or creating social institutions to meet the demands of rapid technological change, "Westernization" for the form of modernization that adopts Western-style institutions and practices toward the same ends. These are essentially practical definitions dictated by the needs of this essay (for fuller discussions, see C. E. Black, *The Dynamics of Modernization: A Study in Comparative History* [New

Over the past two decades, American historians have devoted much effort to understanding and explaining the nature of the West's impact on China and Japan. This body of literature has a good deal to say about the significance of military technology and the role of military institutions in modernization, although only a few authors address military concerns directly. In this essay, I propose to review the recent literature with two related questions in mind: How did the attractiveness of Western weapons stimulate modernization? and, What are the implications of the very different fates met by Western military technology in 19th-century China and Japan?

I

The modern history of contacts between Western Europe and East Asia began in the 16th century, as intrepid Portuguese seamen fought their way into the Indian Ocean and beyond. They first reached China in 1514, Japan in 1543. Even then, the military superiority of the West, at least afloat, was striking. The gun-bearing sailing vessel as an engine of war far outclassed anything Westerners met all across the south coasts of Asia.[3] Both China and Japan were quick to see and exploit the value of Western firearms. In Japan, troops equipped along Western lines played a key role in the establishment of the Tokugawa Shogunate by 1600. China, too, readily adopted the West's superior firearms in the struggles that culminated in 1644 with the victorious Manchus installed as the new Ch'ing dynasty.[4]

York, 1966]; Marion J. Levy, Jr., *Modernization and the Structure of Societies: A Setting for International Affairs* [Princeton, N.J., 1966]; and Peter Berger, Brigitte Berger, and Hansfried Kellner, *The Homeless Mind: Modernization and Consciousness* [New York, 1974]). Scholars have paid little attention to the role of military technology in modernization, as suggested by the absence of any reference to military technology in John Brode, *The Process of Modernization: An Annotated Bibliography on the Sociocultural Aspects of Development* (Cambridge, Mass., 1969). The military role in modernization is the subject of a growing but largely distinct literature focused chiefly on politics and civil-military relations, without much concern for technology: Charles Kuhlman, *The Military in the Developing Countries: A General Bibliography* (Bloomington, Ind., 1961); and Arthur D. Larson, *Civil-Military Relations and Militarism: A Classified Bibliography Covering the United States and Other Nations of the World* (Manhattan, Kans., 1971). A partial exception is the recent study by economist Gavin Kennedy, *The Military in the Third World* (New York, 1974).

[3]Carlo M. Cipolla plausibly argues that this advantage explains the early success of European empire building in *Guns, Sails and Empires: Technological Innovation and the Early Phases of European Expansion, 1400–1700* (New York, 1966). Cipolla is stronger on guns than ships (see also Romola and R. C. Anderson, *The Sailing-Ship: Six Thousand Years of History* [New York, 1926], pp. 116–39; and J. H. Parry, *The Age of Reconnaissance* [New York, 1963], pp. 67–98, 130–45).

[4]C. R. Boxer, "Notes on Early European Military Influence in Japan (1543–1853),"

Strong and vigorous central governments in both Japan and China soon checked major European influence. Europeans were far too few, their military advantage too small beyond the range of shipborne guns, to challenge the new regimes. Both countries imposed severe limits on trade with the West and all but excluded Westerners from their domains. Rarely before had either country enjoyed so peaceful and prosperous a time as they did in the 18th century. During those happy years, the "Great Tradition" of East Asia far more deeply affected the West than Western scientific and technical curiosities did the East.[5]

Until the 15th century or so, even the course of technical borrowing had been more often westward than eastward.[6] That began to change after 1500. The West embarked on a remarkable self-transformation that shattered the rough parity in technology, economy, and polity then existing among civilized communities all across Eurasia. The sources of the West's modernization may ultimately be traced to the Scientific and Industrial Revolutions, but there was also a military revolution of decisive importance.[7] When the West once again ap-

---

Transactions of the Asiatic Society of Japan 8, 2d ser. (1931): 68–93; Delmer M. Brown, "The Impact of Firearms on Japanese Warfare (1543–1598)," *Far Eastern Quarterly* 7 (1948): 236–53; Boxer, "Portuguese Military Expeditions in Aid of the Mings against the Manchus, 1621–1647," *T'ien-Hsia Monthly* 7 (1938): 24–36; and Paul Pelliot, "La Date de l'apparition en Chine des canons Fo-lang-ki," *T'oung Pao* 38 (1948): 199–207.

[5]The best introduction to modern East Asian history is John K. Fairbank, Edwin C. Reischauer, and Albert M. Craig, *East Asia: The Modern Transformation* (Boston, 1965); it is the second volume of *A History of East Asian Civilization,* the first being *East Asia: The Great Tradition* (Boston, 1961) (see also Donald Lach, *Asia in the Making of Europe,* 2 vols. [Chicago, 1965–70]).

[6]The work of Joseph Needham and his collaborators, chiefly the multivolume and still incomplete *Science and Civilisation in China* (Cambridge, 1954–), points up China's technical mastery in a wide range of fields; the planned section on military technology, however, has yet to be published. In a stimulating essay on Chinese development, Mark Elvin notices both China's early technological advances and the role of its military institutions (*The Pattern of the Chinese Past* [Stanford, Calif., 1973]). China's most notable military invention was gunpowder; on that development, see the well-documented survey of Chinese military technology by Herbert Franke, "Siege and Defense of Towns in Medieval China," in *Chinese Ways in Warfare,* ed. Frank A. Kierman, Jr., and John K. Fairbank (Cambridge, Mass., 1974), pp. 151–201; and J. R. Partington, *A History of Greek Fire and Gunpowder* (Cambridge, 1960). Chinese and other Asian sources of Western technology have been traced in several works by Lynn White, jr., notably *Medieval Technology and Social Change* (New York, 1964), which also suggests the West's early military preoccupations: gunpowder may have been Chinese, but the gun was Western (p. 98).

[7]The significance and implications of the military revolution have not been so much stressed by scholars as other factors in the West's self-transformation. Lewis Mumford suggestively discusses the military roots of Western technological, scientific, and indus-

proached East Asia early in the 19th century, its military edge was far greater than it had been two centuries earlier.

II

As the 19th century opened, however, both China and Japan were showing signs of change that had little or nothing to do with the intrusive West. In China, the Ch'ing dynasty was slipping into decline. Corruption, oppressive taxation, unrest in the countryside, and all the other ills that Confucian scholars had long regarded as the symptoms of dynastic decay loomed on every hand. The presence of the West, its mounting pressure in the 1820s and 1830s, seemed to Chinese eyes more irritant than threat, for danger to the Middle Kingdom was something that came from the inner frontiers of Asia, not from the sea.

Strains in Japanese society were of a different order. The long Tokugawa peace had brought growing prosperity as productivity increased and commerce thrived. China, too, had made such gains, but Japan did not dilute them as China did with explosive population growth. Still, the realities of a society becoming urban and market oriented strained the fabric of a basically feudal political system. Deprived of their true social function by the Tokugawa peace, Japan's warrior elite had become administrators and bureaucrats. There were far too many of them, and some, like the lords of Satsuma and Chōshū, had never been fully reconciled to Tokugawa rule. The shogunate was in no danger of imminent collapse, but changes were coming.[8]

---

trial development in *Technics and Civilization* (New York, 1934), pp. 81–96; W. W. Rostow points out the economic importance of military concerns in *How It All Began: Origins of the Modern Economy* (New York, 1975), chap. 2; Samuel E. Finer addresses their political significance in "State- and Nation-Building in Europe: The Role of the Military," in *The Formation of National States in Western Europe,* ed. Charles Tilly (Princeton, N.J., 1975), pp. 84–163; and Robert K. Merton their impact on the Scientific Revolution in *Science, Technology and Society in Seventeenth Century England* (New York, 1970), chaps. 7–10. For an informed and carefully documented survey of its consequences in a number of areas, see Michael Roberts, "The Military Revolution, 1560–1660," in *Essays in Swedish History* (Minneapolis, 1967), pp. 195–225.

[8]Marion J. Levy, Jr., "Contrasting Factors in the Modernization of China and Japan," in *Economic Growth: Brazil, India, Japan,* ed. Simon Kuznets, Wilbert E. Moore, and Joseph J. Spengler (Durham, N.C., 1955), pp. 496–536; K. C. Liu, "Nineteenth-Century China: The Disintegration of the Old Order and the Impact of the West," in *China in Crisis,* ed. P. T. Ho and T. Tsou, 2 vols. (Chicago, 1968), 1:93–178; Philip A. Kuhn, *Rebellion and Its Enemies in Late Imperial China: Militarization and Social Structure, 1796–1864* (Cambridge, Mass., 1970); Frederic Wakeman, Jr., "High Ch'ing: 1683–1839," in *Modern East Asia: Essay in Interpretation,* ed. James B. Crowley (New York,

In the first half of the 19th century, East Asian military technology was much what it had been two centuries before. Gunpowder weapons were limited to matchlocks and cannons patterned on 16th- and 17th-century European models. Such weapons were, in any case, auxiliary, in the same class as lance and pike. The Manchu bannerman trusted chiefly to his bow, the samurai to his swords. The East's naval technology had declined from 17th-century levels. Tokugawa Japan turned its back on the sea, shutting off what had been a vigorous maritime enterprise with an enforced ban on sea travel. The land-oriented Manchus had never displayed much naval skill, and what water forces it had were strictly adjuncts to its armies.[9]

In sharp contrast, the West had applied itself to military matters with enormous enthusiasm and success. By the 19th century, Western military technology was beginning to draw on a maturing science to advance at an ever-accelerating rate. The first half of the century witnessed the opening stages of a revolutionary transformation of Western military and naval technology. The smoothbore flintlock musket that equipped European armies in 1800, itself a long step beyond the matchlocks still used in the East, had given way by mid-century to the caplock rifled musket. Artillery had become more mobile, and breech-loading rifled ordnance firing explosive shells had begun to replace the older muzzle-loading smoothbores both on land and at sea. Shell guns were but one aspect of a naval revolution whose most important feature, steam propulsion, was already well established by mid-century.[10]

1970), pp. 1–28; and John Whitney Hall, "Tokugawa Japan: 1800–1853," ibid., pp. 62–94.

[9]Although many of the works cited below discuss Ch'ing and Tokugawa military institutions in the context of the changes they underwent as the 19th century advanced, little has been written (in Western languages, at least) on those institutions in their own right (see, however, T. F. Wade, "The Army of the Chinese Empire: Its Two Great Divisions, the Bannerman or National Guard, and the Green Standard or Provincial Troops; Their Organization, Locations, Pay, Condition, & c.," *Chinese Repository* 20 [1851]: 250–80, 300–340, 363–422; and the chapter, "Tokugawa Military Organization," in Conrad D. Totman, *Politics in the Tokugawa Bakufu, 1600–1843* [Cambridge, Mass., 1967], pp. 43–63).

[10]There is no satisfactory general history of military technology, but Theodore Ropp, *War in the Modern World,* rev. ed. (New York, 1962), pays much attention to technology and provides a wealth of references. Thomas A. Palmer discusses major changes in the middle years of the 19th century in "Military Technology," in *Technology in Western Civilization,* ed. Melvin Kranzberg and Carroll W. Pursell, Jr., 2 vols. (New York, 1967), 1:489–502. On firearms, see W. Y. Carman, *A History of Firearms: From Earliest Times to 1914* (London, 1955). The best study of the naval revolution is James Phinney Baxter, *The Introduction of the Ironclad Warship* (Cambridge, Mass., 1933).

The new weapons of war were much in evidence in the first significant 19th-century confrontation between China and the West, the so-called Opium War of 1839–42. Small British forces easily brushed aside Chinese defenses to impose the first of the unequal treaties that came to symbolize Western imperialism in East Asia. The humiliation of this defeat triggered scattered Chinese efforts to emulate Western arms. A few, mostly obscure, officials urged the imperial court to acquire Western ships and guns in self-defense. Nothing much came of these efforts, largely because other matters seemed far more pressing in Peking than Western trade demands, even when backed up by such undeniably superior military force.[11]

China's internal problems had assumed catastrophic proportions. The classic signs of dynastic decline included rebellion, and Ch'ing suffered more than its share during the first half of the 19th century. But these were merest prelude to the vast upheaveal known as the Taiping Rebellion, breaking out in 1851 and not finally suppressed until 1864. Perhaps the greatest rebellion in Chinese history, it flared in almost every corner of the country, costing millions of lives and causing untold damage. And yet it was only one of four major uprisings, and countless local insurrections, that convulsed China between 1851 and 1873, leaving imperial officials little time to concern themselves with the Western nuisance on the coasts. But the same problems that distracted official attention for the growing threat of Western imperialism enhanced the value of Western arms in Chinese eyes. The central government and its regional agents purchased Western firearms and ordnance, as well as warships, and took the first steps toward setting up Western-style arsenals and shipyards. Forces loyal to Ch'ing increasingly outclassed their internal foes in equipment and organization, a key factor in their ultimate success in putting down the mid-century rebellions.[12]

[11]By far the most important advocate of Western arms was Lin Tse-hsü, the imperial commissioner chiefly responsible for precipitating the war (see H. P. Chang, *Commissioner Lin and the Opium War* [Cambridge, Mass., 1964]). On the early efforts by Lin and others to foster the use of Western arms, see Gideon Chen, *Lin Tse-hsü: Promoter of the Adoption of Western Means of Maritime Defense in China* (Peking, 1934); and S. Y. Teng and John K. Fairbank, *China's Response to the West: A Documentary Survey, 1839–1923* (Cambridge, Mass., 1954), pp. 23–36 (see also John K. Fairbank, "China's Response to the West: Problems and Suggestions," *Cahiers d'histoire mondiale* 3 [1956]: 381–406). John Selby describes China's 19th-century wars with Europeans in *The Paper Dragon: An Account of the China Wars, 1840–1900* (London, 1968).

[12]Liu, p. 120; Paul A. Cohen, "Ch'ing China: Confrontation with the West, 1850–1900," in Crowley (n. 8 above), pp. 29–61; Immanuel C. Y. Hsü, "The Taiping Revolution and the Nien and Moslem Rebellions," in *The Rise of Modern China* (New York, 1970) (see also Franz Michael and C. L. Chang, *The Taiping Rebellion: History and Documents,* 3 vols. [Seattle, 1966]; and S. Y. Teng, *The Nien Army and Their Guerilla Warfare* [Paris and The Hague, 1961]).

Interest in Western arms had a longer history in Japan, where news of England's easy triumph in the Opium War merely added to an already widespread concern. Knowing that Western ships and guns could not be resisted by sword-wielding samurai, the Tokugawa government bowed to American demands in 1853–54 that its ports be opened to foreign trade. Efforts to acquire the new weapons, already set in motion at both national and local levels, now intensified. Japan faced nothing like the cataclysmic problems that China did, the issue being largely confined to dealing with the Western threat to Japanese sovereignty. As the Tokugawa regime failed to expel the barbarians, those feudal lords who had long been restive under Tokugawa rule were able to mount a successful challenge. Both sides purchased and built Western arms to equip new military units organized along Western lines. But geography favored the dissidents, particularly in the southwestern domains of Satsuma and Chōshū, who had easier access to Western weapons and training. Their better-armed and better-managed forces defeated the numerically larger Tokugawa armies and overthrew the regime.[13]

In the initial confrontation between West and East during the middle decades of the 19th century, Japan showed more concern for the Western threat than China did. National and provincial officials in both lands, however, were alert to the value of Western arms in meeting their own problems. Using Western ships and guns aroused no significant resistance in either country, whatever the reaction might be to other aspects of Western culture. Through mid-century, there was little to distinguish the responses of China and Japan. Japan's military traditions might have made its leaders more alive to the implications of a new military technology, but China's scholar-officials had never been loath to borrow military technology from barbarians and so found no reason to reject the new weapons of the West. The crucial issue, and the point from which Chinese and Japanese re-

[13]The early and persistent Japanese interest in Western military technology has not usually been stressed by scholars, who have tended to subsume military technology under Western science or industry (see R. P. Dore, *Education in Tokugawa Japan* [Berkeley and Los Angeles, 1965], pp. 166–75; J. Numata, "Acceptance and Rejection of Elements of European Culture in Japan," *Cahiers d'histoire mondiale* 3 [1956]: 231–53; K. Yabuuti, "The Pre-History of Modern Science in Japan: The Importation of Western Science during the Tokugawa Period," ibid., 9 [1965]: 208–32; A. Seiho, "The Western Influence on Japanese Military Science, Shipbuilding, and Navigation," *Monumenta Nipponica* 19 [1964]: 118–46; and Thomas C. Smith, "The Introduction of Western Industry to Japan during the Last Years of the Tokugawa Period," *Harvard Journal of Asiatic Studies* 11 [1948]: 130–52). On the mid-19th-century struggle between shogun and lords, see Albert M. Craig, *Chōshū in the Meiji Restoration* (Cambridge, Mass., 1961); Roger F. Hackett, *Yamagato Aritomo in the Rise of Modern Japan, 1838–1922* (Cambridge, Mass, 1971); and Marius B. Jansen, *Sakamoto Ryōma and the Meiji Restoration* (Princeton, N.J., 1961).

sponses sharply diverged in the 1860s and later, was how much of Western culture was attached to the hardware. China and Japan found different answers.

III

The military problem that the West posed to China and Japan during the late 19th century was compounded by the still-rapid rate of change in Western military technology. By 1900, muskets had given way to breech-loading repeating rifles, smoothbore artillery had been completely replaced by longer-ranged and quicker-firing rifled ordnance, and machine guns were coming into wide use. Changes so great and so rapid left even Western armies confused, although just how deeply remained to be shown by the World War of 1914–18. For the would-be borrower in the East the problems were magnified. This may have been even more true of the continuing naval revolution, in which the pace of change was so rapid that the ships of one decade were all but worthless in the next. Screw-propelled steamships wrapped in ever-thicker armor and armed with ever-larger guns presented baffling problems to Western navies, and even harder problems to their Eastern students.[14]

During the 1860s, both China and Japan entered new historical epochs that have been dubbed restorations: the T'ung-chih Restoration in China, the Meiji Restoration in Japan. The same word, however, masks profoundly different experiences.[15]

The T'ung-chih Restoration began in 1862 and had largely run its course by the end of the decade. It was a traditional response to what was seen as a traditional problem. Several times before in Chinese

[14]In addition to the works cited in n. 10, see Edward L. Katzenbach, Jr., "The Mechanization of War, 1880–1919," in Kranzberg and Pursell (n. 10 above), 2:548–61; Brian Bond, "War and Peace: Mechanized Warfare and the Growth of Pacifism," in *The Nineteenth Century: The Contradictions of Progress,* ed. Asa Briggs (New York, 1970), pp. 186–214; Marcus Cunliffe, "War: Conceptions and Reality," in *The Age of Expansion, 1848–1917* (Springfield, Mass., 1974). For some of the difficulties of Eastern borrowers, see Ernst L. Presseisen, *Before Agression: Europeans Prepare the Japanese Army* (Tucson, Ariz., 1965); John Curtis Perry, "Great Britain and the Emergence of Japan as a Naval Power," *Monumenta Nipponica* 21 (1966): 305–21; and John L. Rawlinson, *China's Struggle for Naval Development, 1839–1895* (Cambridge, Mass., 1967).

[15]The word is, in fact, only the same in the West. The Chinese term is *chung-hsing,* literally "resurgence in the middle [of the dynastic cycle]," with connotations of a temporary halt in dynastic decline. The Japanese *ishin* (Chinese *wei-hsin*) means literally "renovation" and refers to the kind of general reform and modernization associated with Meiji; there was, of course, no place in the Japanese setting for the concept of dynastic cycle (see Mary Clabaugh Wright, *The Last Stand of Chinese Conservatism: The T'ung Chih Restoration, 1862–1874* [New York, 1966], p. 45). I am indebted to N. Sivin for clarifying this point.

history dynastic decline had been checked when a gifted ruler had overseen a return to Confucian virtue and the revival of honest government. No such paragon blessed mid-19th-century China, but a group of particularly able high officials acting for the emperor achieved results close enough to the historic pattern to justify the analogy. During the T'ung-chih period China enjoyed a remarkable renewal of administrative vigor and responsible efforts to solve the country's overwhelming problems. Associated with the restoration was a movement pursued under the slogan of "self-strengthening." Just as an older China had borrowed the superior military techniques of barbarian tribesmen to maintain itself against their raids, the new China of the 1860s could adopt Western arms to fend off Western pressure. The self-strengthening movement outlasted the restoration and saw the expanded purchase of Western weapons, as well as the founding of shipyards and arsenals to make them and schools to provide training in the new techniques.[16]

Self-strengthening produced results: the quelling of rebellion, the spectacular recovery of Chinese Turkestan in the 1870s, a sharp reverse inflicted on French forces in Indochina in 1885. As a viable policy, however, self-strengthening foundered on the outcome of the Sino-Japanese War of 1894–95. The humiliating defeat at the hands of upstart Japan was not, however, a product of technological backwardness. China's weapons were reasonably up-to-date, in some respects better than Japan's. But China lacked the military organization to coordinate, direct, and control its scattered forces.[17] In a deeper

[16]The best study is Wright's *Last Stand of Chinese Conservatism.* Gideon Chen has studied two leaders of self-strengthening: *Tseng Kuo-fan: Pioneer Promoter of the Steamship in China* (Peking, 1935), and *Tso Tsung T'ang: Pioneer Promoter of the Modern Dockyard and Woolen Mill in China* (Peking, 1938). The foremost advocate and practitioner of self-strengthening was Li Hung-chang, the subject of a book by Stanley Spector, *Li Hung-chang and the Huai Army: A Study in Nineteenth-Century Chinese Regionalism* (Seattle, Wash., 1964); and a series of articles by K. C. Liu: "The Confucian as Patriot and Pragmatist: Li Hung-chang's Formative Years, 1823–1866," *Harvard Journal of Asiatic Studies* 30 (1970): 5–45; "Li Hung-chang in Chihli: The Emergence of a Policy, 1870–1875," in *Approaches to Modern Chinese History,* ed. Albert Feuerwerker, Rhoads Murphey, and Mary C. Wright (Berkeley and Los Angeles, 1967), pp. 68–104; and "Self-Strengthening and Reform" (paper presented to American Historical Association, New York, December 1968) (see also K. H. Kim, *Japanese Perspectives on China's Early Modernization: The Self-strengthening Movement, 1860–1895, a Bibliographical Survey* [Ann Arbor, Mich., 1974]).

[17]John L. Rawlinson, "China's Failure to Coordinate Her Modern Fleets in the Late Nineteenth Century," in Feurwerker, Murphey, and Wright, pp. 105–32; Stephen Roberts, "Imperial Chinese Steam Navy," *Warships International,* vol. 11, no. 1 (1974); Ralph L. Powell, *The Rise of Chinese Military Power, 1895–1912* (Princeton, N.J., 1955); and Y. Hatano, "The New Armies," in *China in Revolution: The First Phase, 1900–1913,* ed. Mary C. Wright (New Haven, Conn., 1968), pp. 365–82.

sense, China's defeat was rooted in a fundamental miscalculation. Self-strengthening assumed that China could defend its traditional society against the West with Western weapons, that the West's military technology could be detached from Western culture as a whole. But could it? Some Chinese officials came to see, reluctantly, the unbreakable chain that led from firearms and ships to coal mines, iron foundries, and railroads; from military technology to industrialization; from the weapons of the West to Westernization. But that meant the end of traditional society, and few were willing to push the logic to its end and urge so drastic a course before the close of the 19th century.[18]

Such foot-dragging (or prudence) was almost unseen in Japan. The Meiji Restoration of 1868 was so named from the presumed return to the emperor of his former power, usurped in recent centuries by the shogun. The rhetoric of imperial rule and a return to time-honored forms disguised far-reaching changes. Younger samurai had played key roles in toppling the Tokugawa regime. Deeply impressed by the West's military technology, they assumed their new government posts determined to sustain Japan's independence with Western weapons. But they accepted, as their Chinese counterparts did not, the price of that technology, which involved not only a complete revamping of the military system but also large-scale industrialization and all it implied.[19]

The oligarchy that now ruled Japan in the name of the emperor recognized the economic underpinnings of Western military power. Arsenals and shipyards, technical schools, and foreign advisers were

[18]Adrian Arthur Bennett, *John Fryer: The Introduction of Western Science and Technology into Nineteenth-Century China* (Cambridge, Mass., 1967); Knight Biggerstaff, *The Earliest Modern Government Schools in China* (Ithaca, N.Y., 1961); Albert Feurwerker, *China's Early Industrialization: Sheng Hsuan-huai (1844–1916) and Mandarin Enterprise* (Cambridge, Mass., 1958); and P. Huard and W. Ming, "Le Développement de la technologie dans la Chine du XIX siècle," *Cahiers d'histoire mondiale* 7 (1962): 68–85 (see also Genevieve C. Dean, *Science and Technology in the Development of Modern China: An Annotated Bibliography* [London, 1974]).

[19]The best overall study is W. J. Beasley, *The Meiji Restoration* (Stanford, Calif., 1973). On Meiji military affairs, see James B. Crowley, "From Closed Door to Empire: The Formation of the Meiji Military Establishment," in *Modern Japanese Leadership: Transition and Change,* ed. Bernard S. Silberman and H. D. Harootunian (Tucson, Ariz., 1966), pp. 261–89; S. Fukushima, "The Building of a National Army," *Developing Economies* 3 (1965): 516–40; Roger F. Hackett, "The Meiji Leaders and Modernization: The Case of Yamagata Aritomo," in *Changing Japanese Attitudes toward Modernization,* ed. Marius B. Jansen (Princeton, N.J., 1965), pp. 243–83, and "The Military: Japan," in *Political Modernization in Japan and Turkey,* ed. Robert E. Ward and Dankwart A. Rustow (Princeton, N.J., 1964), pp. 328–51; and Hyman Kublin, "The 'Modern' Army of Early Meiji Japan," *Far Eastern Quarterly* 9 (1949): 20–41.

only the first steps, to be followed by full industrialization. And perhaps more. Western science, Western morals, even Western dress, might be a crucial part of the recipe. The best way to be certain that no significant detail was overlooked was to embark on thoroughgoing Westernization. That logic was not uniquely Japanese. The beginnings of China's industrialization and Westernization can be traced to the same late-19th-century military concerns. But while that logic was deeply resented and strongly resisted in China, Japan's new rulers embraced it eagerly and threw the government boldly into the process.[20]

The clear superiority of Western military techniques provided the incentive for modernization in both Japan and China, although an incentive more willingly accepted in Japan, at least before the 20th century. This close association between modernization and military technology may account for the process so often leading to imperialism and war, as it did for Japan and for so many others. Japan's war with China for hegemony over Korea in 1894–95 revealed the proficiency of its new armed forces, amply confirmed a decade later when they defeated Russia and raised Japan to the status of a Great Power.

China, in the meantime, was left prostrate. Self-strengthening had failed, and the disastrous Boxer uprising at the turn of the century confronted China with a choice that could no longer be evaded. After Japan joined Western forces to suppress the Boxers, China had either to transform itself or perish. The answer was first reform, then revolution. From 1900 on, a new spirit moved China to cast off its Confucian heritage and to embark on the road toward nationalism and modernization.[21]

[20]David S. Landes, "Japan and Europe: Contrasts in Industrialization," in *The State and Economic Enterprise in Japan: Essays in the Political Economy of Growth,* ed. William W. Lockwood (Princeton, N.J., 1965), pp. 93–182; and Marius B. Jansen, "The Meiji State: 1868–1912," in Crowley (n. 8 above), pp. 95–121 (see also Johannes Hirschmeier, *The Origins of Entrepreneurship in Meiji Japan* [Cambridge, Mass., 1964]; and Thomas C. Smith, *Political Change and Industrial Development in Japan: Government Enterprise, 1868–1880* [Stanford, Calif., 1955]).

[21]I. Akira, "Imperialism in East Asia," in Crowley (n. 8 above), pp. 122–50; Irving Louis Horowitz, "The Military as a Subculture: Militarization, Modernization, and Mobilization: Third World Development Patterns Reexamined," in *Protagonists of Change: Subcultures in Development and Revolution,* ed. Abdul A. Said (Englewood Cliffs, N.J., 1971), pp. 41–51; Ike Nobutaka, "War and Modernization," in *Political Development in Modern Japan,* ed. Robert E. Ward (Princeton, N.J., 1968), pp. 189–211; William W. Lockwood, "Economic and Political Modernization: Japan," in Ward and Rustow (n. 18 above), pp. 117–45; and Mary C. Wright, "The Rising Tide of Change," in *China in Revolution* (n. 16 above), pp. 1–63.

IV

The emphasis on military technology in this sketchy survey of the Western impact on China and Japan during the 19th century should not be miscontrued. Although military technology has tended to be undervalued as a factor in Chinese and Japanese responses to the West, it was far from being the whole story. Both societies were in the midst of major changes when the West appeared on the scene early in the 19th century, changes to which neither the West nor its weapons was a party. Technological change is itself a species of social change that both affects and is affected by other forms of change. How the introduction of novel technologies interacted with changing societies is a better question than what impact technology, military or other, had on those societies. Because technological innovations tend to be among the most visible of social changes, they are sometimes assigned a causal significance that may really belong to other factors less easily seen.

The crucial point is that Western military technology was widely viewed in both China and Japan as a desirable innovation. Western ships and guns clearly served pressing practical needs. But toward the end of the 19th century, Western military technology could scarcely be disentangled from its complex interrelationships with an evolving industrial and military order. Weapons might indeed merely be purchased, but that entailed dependence on Western suppliers. Even then, the proper use of such weapons demanded, at the very least, changes in military organization with indeterminate but far-reaching consequences. To avoid dependence demanded even greater changes, industrialization being an obvious, if not easy-to-take, step. No one could then, or for that matter can now, say with assurance what else might be required even to take that step. To avoid overlooking any essential ingredient, the surest path seemed to be a wholesale adoption of Western institutions. Westernization, in other words, was the final link in the chain of logic that began with the lure of Western weapons.

Events in neither China nor Japan suggested any real resistance to that lure. Because China's sovereignty had been so badly impaired by the end of the 19th century, while Japan's had survived largely intact, what China did achieve has not been much stressed. If, as most Chinese believed, the central problems were to restore internal order and preserve an ancient heritage, then China's response to Western technology makes sense through much of the 19th century. Reorganized armies equipped with Western arms shattered the mid-century rebellions and helped to sustain the old order long after it might otherwise have crumbled. China was more than ready to seize the chance for using Western weapons to deal with its own problems, in which, for the most part, the West's role was peripheral. Other aspects of Western technology, and of Western culture in general, met no such urgent need and found a cooler reception.

This was precisely the point from which Chinese and Japanese responses diverged. China tried to do no more than adopt the weapons themselves, while rejecting most of the military and social systems in which the technology was enmeshed. Japan acted as if the superiority of Western arms were an integral part of Western military institutions and Western civilization, as if to adopt Western weapons demanded Westernization. Japan was more concerned than China with the West as a political and military threat, and so less willing to depend on the West for its weapons. Both saw that the West could only be opposed with its own arms, but before 1900 only Japan was willing to pay the price of that opposition. The West's unmatched military technology was one product of its long-time military preoccupation. Military institutions played central roles in the West's own modernization. That they should be equally central to the modernization of non-Western countries is scarcely surprising. The path of modernization has typically led from military technology through industrialization to Westernization. Military concerns moved first Japan, then China, to take that path, and many others have trod it as well.

The "Industrial Revolution" in the Home: Household Technology and Social Change in the 20th Century

RUTH SCHWARTZ COWAN

When we think about the interaction between technology and society, we tend to think in fairly grandiose terms: massive computers invading the workplace, railroad tracks cutting through vast wildernesses, armies of woman and children toiling in the mills. These grand visions have blinded us to an important and rather peculiar technological revolution which has been going on right under our noses: the technological revolution in the home. This revolution has transformed the conduct of our daily lives, but in somewhat unexpected ways. The industrialization of the home was a process very different from the industrialization of other means of production, and the impact of that process was neither what we have been led to believe it was nor what students of the other industrial revolutions would have been led to predict.

* * *

Some years ago sociologists of the functionalist school formulated an explanation of the impact of industrial technology on the modern family. Although that explanation was not empirically verified, it has become almost universally accepted.[1] Despite some differences in emphasis, the basic tenets of the traditional interpretation can be roughly summarized as follows:

Before industrialization the family was the basic social unit. Most families were rural, large, and self-sustaining; they produced and processed almost everything that was needed for their own support and for trading in the marketplace, while at the same time perform-

DR. COWAN, professor of history at the State University of New York at Stony Brook, is the author of *A Social History of American Technology* (1997) and *More Work for Mother: The Ironies of Household Technology from the Open Hearth to the Microwave* (1983).

[1]For some classic statements of the standard view, see W. F. Ogburn and M. F. Nimkoff, *Technology and the Changing Family* (Cambridge, Mass., 1955); Robert F. Winch, *The Modern Family* (New York, 1952); and William J. Goode, *The Family* (Englewood Cliffs, N.J., 1964).

ing a host of other functions ranging from mutual protection to entertainment. In these preindustrial families women (adult women, that is) had a lot to do, and their time was almost entirely absorbed by household tasks. Under industrialization the family is much less important. The household is no longer the focus of production; production for the marketplace and production for sustenance have been removed to other locations. Families are smaller and they are urban rather than rural. The number of social functions they perform is much reduced, until almost all that remains is consumption, socialization of small children, and tension management. As their functions diminished, families became atomized; the social bonds that had held them together were loosened. In these postindustrial families women have very little to do, and the tasks with which they fill their time have lost the social utility that they once possessed. Modern women are in trouble, the analysis goes, because modern families are in trouble; and modern families are in trouble because industrial technology has either eliminated or eased almost all their former functions, but modern ideologies have not kept pace with the change. The results of this time lag are several: some women suffer from role anxiety, others land in the divorce courts, some enter the labor market, and others take to burning their brassieres and demanding liberation.

This sociological analysis is a cultural artifact of vast importance. Many Americans believe that it is true and act upon that belief in various ways: some hope to reestablish family solidarity by relearning lost productive crafts—baking bread, tending a vegetable garden—others dismiss the women's liberation movement as "simply a bunch of affluent housewives who have nothing better to do with their time." As disparate as they may seem, these reactions have a common ideological source—the standard sociological analysis of the impact of technological change on family life.

As a theory this functionalist approach has much to recommend it, but at present we have very little evidence to back it up. Family history is an infant discipline, and what evidence it has produced in recent years does not lend credence to the standard view.[2] Phillippe Ariès has shown, for example, that in France the ideal of the small nuclear family predates industrialization by more than a century.[3] Historical demographers working on data from English and French families have been surprised to find that most families were quite small and

[2]This point is made by Peter Laslett in "The Comparative History of Household and Family," in *The American Family in Social Historical Perspective,* ed. Michael Gordon (New York, 1973), pp. 28–29.

[3]Phillippe Ariès, *Centuries of Childhood: A Social History of Family Life* (New York, 1960).

that several generations did not ordinarily reside together; the extended family, which is supposed to have been the rule in preindustrial societies, did not occur in colonial New England either.[4] Rural English families routinely employed domestic servants, and even very small English villages had their butchers and bakers and candlestick makers; all these persons must have eased some of the chores that would otherwise have been the housewife's burden.[5] Preindustrial housewives no doubt had much with which to occupy their time, but we may have reason to wonder whether there was quite as much pressure on them as sociological orthodoxy has led us to suppose. The large rural family that was sufficient unto itself back there on the prairies may have been limited to the prairies—or it may never have existed at all (except, that is, in the reveries of sociologists).

Even if all the empirical evidence were to mesh with the functionalist theory, the theory would still have problems, because its logical structure is rather weak. Comparing the average farm family in 1750 (assuming that you knew what that family was like) with the average urban family in 1950 in order to discover the significant social changes that had occurred is an exercise rather like comparing apples with oranges; the differences between the fruits may have nothing to do with the differences in their evolution. Transferring the analogy to the case at hand, what we really need to know is the difference, say, between an urban laboring family of 1750 and an urban laboring family 100 and then 200 years later, or the difference between the rural nonfarm middle classes in all three centuries, or the difference between the urban rich yesterday and today. Surely in each of these cases the analyses will look very different from what we have been led to expect. As a guess we might find that for the urban laboring families the changes have been precisely the opposite of what the model predicted; that is, that their family structure is much firmer today than it was in centuries past. Similarly, for the rural nonfarm middle class the results might be equally surprising; we might find that married women of that class rarely did any housework at all in 1890 because they had farm girls as servants, whereas in 1950 they bore the full brunt of the work themselves. I could go on, but the point is, I hope, clear: in order to verify or falsify the functionalist theory, it will be necessary to know more than we presently do about the impact of industrialization on families of similar classes and geographical locations.

* * *

[4]See Laslett, pp. 20–24; and Philip J. Greven, "Family Structure in Seventeenth Century Andover, Massachusetts," *William and Mary Quarterly* 23 (1966): 234–56.

[5]Peter Laslett, *The World We Have Lost* (New York, 1965), passim.

With this problem in mind I have, for the purposes of this initial study, deliberately limited myself to one kind of technological change affecting one aspect of family life in only one of the many social classes of families that might have been considered. What happened, I asked, to middle-class American women when the implements with which they did their everyday household work changed? Did the technological change in household appliances have any effect upon the structure of American households, or upon the ideologies that governed the behavior of American women, or upon the functions that families needed to perform? Middle-class American women were defined as actual or potential readers of the better-quality women's magazines, such as the *Ladies' Home Journal, American Home, Parents' Magazine, Good Housekeeping,* and *McCall's*.[6] Nonfictional material (articles and advertisements) in those magazines was used as a partial indicator of some of the technological and social changes that were occurring.

The *Ladies' Home Journal* has been in continuous publication since 1886. A casual survey of the nonfiction in the *Journal* yields the immediate impression that that decade between the end of World War I and the beginning of the depression witnessed the most drastic changes in patterns of household work. Statistical data bear out this impression. Before 1918, for example, illustrations of homes lit by gaslight could still be found in the *Journal;* by 1928 gaslight had disappeared. In 1917 only one-quarter (24.3 percent) of the dwellings in the United States had been electrified, but by 1920 this figure had doubled (47.4 percent—for rural nonfarm and urban dwellings), and by 1930 it had risen to four-fifths percent).[7] If electrification had meant simply the change from gas or oil lamps to electric lights, the changes in the housewife's routines might not have been very great (except for eliminating the chore of cleaning and filling oil lamps);

[6]For purposes of historical inquiry, this definition of middle-class status corresponds to a sociological reality, although it is not, admittedly, very rigorous. Our contemporary experience confirms that there are class differences reflected in magazines, and this situation seems to have existed in the past as well. On this issue see Robert S. Lynd and Helen M. Lynd, *Middletown: A Study in Contemporary American Culture* (New York, 1929), pp. 240–44, where the marked difference in magazines subscribed to by the business-class wives as opposed to the working-class wives is discussed; Salme Steinberg, "Reformer in the Marketplace: E. W. Bok and *The Ladies Home Journal*" (Ph.D. diss., Johns Hopkins University, 1973), where the conscious attempt of the publisher to attract a middle-class audience is discussed; and Lee Rainwater et al., *Workingman's Wife* (New York, 1959), which was commissioned by the publisher of working-class women's magazines in an attempt to understand the attitudinal differences betweeen working-class and middle-class women.

[7]*Historical Statistics of the United States, Colonial Times to 1957* (Washington, D.C., 1960), p. 510.

but changes in lighting were the least of the changes that electrification implied. Small electric appliances followed quickly on the heels of the electric light, and some of those augured much more profound changes in the housewife's routine.

Ironing, for example, had traditionally been one of the most dreadful household chores, especially in warm weather when the kitchen stove had to be kept hot for the better part of the day; irons were heavy and they had to be returned to the stove frequently to be reheated. Electric irons eased a good part of this burden.[8] They were relatively inexpensive and very quickly replaced their predecessors; advertisements for electric irons first began to appear in the ladies' magazines after the war, and by the end of the decade the old flatiron had disappeared; by 1929 a survey of 100 Ford employees revealed that ninety-eight of them had the new electric irons in their homes.[9]

Data on the diffusion of electric washing machines are somewhat harder to come by; but it is clear from the advertisements in the magazines, particularly advertisements for laundry soap, that by the middle of the 1920s those machines could be found in a significant number of homes. The washing machine is depicted just about as frequently as the laundry tub by the middle of the 1920s; in 1929, forty-nine out of those 100 Ford workers had the machines in their homes. The washing machines did not drastically reduce the time that had to be spent on household laundry, as they did not go through their cycles automatically and did not spin dry; the housewife had to stand guard, stopping and starting the machine at appropriate times, adding soap, sometimes attaching the drain pipes, and putting the clothes through the wringer manually. The machines did, however, reduce a good part of the drudgery that once had been associated with washday, and this was a matter of no small consequence.[10] Soap powders appeared on the market in the early 1920s, thus eliminating the need to scrape and boil bars of laundry soap.[11] By the end of the

[8]The gas iron, which was available to women whose homes were supplied with natural gas, was an earlier improvement on the old-fashioned flatiron, but this kind of iron is so rarely mentioned in the sources that I used for this survey that I am unable to determine the extent of its diffusion.

[9]Hazel Kyrk, *Economic Problems of the Family* (New York, 1933), p. 368, reporting a study in *Monthly Labor Review* 30 (1930): 1209–52.

[10]Although this point seems intuitively obvious, there is some evidence that it may not be true. Studies of energy expenditure during housework have indicated that by far the greatest effort is expended in hauling and lifting the wet wash, tasks which were not eliminated by the introduction of washing machines. In addition, if the introduction of the machines served to increase the total amount of wash that was done by the housewife, this would tend to cancel the energy-saving effects of the machines themselves.

[11]Rinso was the first granulated soap; it came on the market in 1918. Lux Flakes had been available since 1906; however it was not intended to be a general laundry product

1920s Blue Monday must have been considerably less blue for some housewives—and probably considerably less "Monday," for with an electric iron, a washing machine, and a hot water heater, there was no reason to limit the washing to just one day of the week.

Like the routines of washing the laundry, the routines of personal hygiene must have been transformed for many households during the 1920s—the years of the bathroom mania.[12] More and more bathrooms were built in older homes, and new homes began to include them as a matter of course. Before the war most bathroom fixtures (tubs, sinks, and toilets) were made out of porcelain by hand; each bathroom was custom-made for the house in which it was installed. After the war industrialization descended upon the bathroom industry; cast iron enamelware went into mass production and fittings were standardized. In 1921 the dollar value of the production of enameled sanitary fixtures was $2.4 million, the same as it had been in 1915. By 1923, just two years later, that figure had doubled to $4.8 million; it rose again, to $5.1 million, in 1925.[13] The first recessed, double-shell cast iron enameled bathtub was put on the market in the early 1920s. A decade later the standard American bathroom had achieved its standard American form: the recessed tub, plus tiled floors and walls, brass plumbing, a single-unit toilet, an enameled sink, and a medicine chest, all set into a small room which was very often 5 feet square.[14] The bathroom evolved more quickly than any other room of the house; its standardized form was accomplished in just over a decade.

Along with bathrooms came modernized systems for heating hot water: 61 percent of the homes in Zanesville, Ohio, had indoor plumbing with centrally heated water by 1926, and 83 percent of the homes valued over $2,000 in Muncie, Indiana, had hot and cold running

---

but rather one for laundering delicate fabrics. "Lever Brothers," *Fortune* 26 (November 1940): 95.

[12]I take this account, and the term, from Lynd and Lynd, p. 97. Obviously, there were many American homes that had bathrooms before the 1920s, particularly urban row houses, and I have found no way of determining whether the increases of the 1920s were more marked than in previous decades. The rural situation was quite different from the urban; the President's Conference on Home Building and Home Ownership reported that in the late 1920s, 71 percent of the urban families surveyed had bathrooms, but only 33 percent of the rural families did (John M. Gries and James Ford, eds., *Homemaking, Home Furnishing and Information Services,* President's Conference on Home Building and Home Ownership, vol. 10 [Washington, D.C., 1932], p. 13).

[13]The data above come from Siegfried Giedion, *Mechanization Takes Command* (New York, 1948), pp. 685–703.

[14]For a description of the standard bathroom see Helen Sprackling, "The Modern Bathroom," *Parents' Magazine* 8 (February 1933): 25.

water by 1935.[15] These figures may not be typical of small American cities (or even large American cities) at those times, but they do jibe with the impression that one gets from the magazines: after 1918 references to hot water heated on the kitchen range, either for laundering or for bathing, become increasingly difficult to find.

Similarly, during the 1920s many homes were outfitted with central heating; in Muncie most of the homes of the business class had basement heating in 1924; by 1935 Federal Emergency Relief Administration data for the city indicated that only 22.4 percent of the dwellings valued over $2,000 were still heated by a kitchen stove.[16] What all these changes meant in terms of new habits for the average housewife is somewhat hard to calculate; changes there must have been, but it is difficult to know whether those changes produced an overall saving of labor and/or time. Some chores were eliminated—hauling water, heating water on the stove, maintaining the kitchen fire—but other chores were added—most notably the chore of keeping yet another room scrupulously clean.

It is not, however, difficult to be certain about the changing habits that were associated with the new American kitchen—a kitchen from which the coal stove had disappeared. In Muncie in 1924, cooking with gas was done in two out of three homes; in 1935 only 5 percent of the homes valued over $2,000 still had coal or wood stoves for cooking.[17] After 1918 advertisements for coal and wood stoves disappeared from the *Ladies' Home Journal;* stove manufacturers purveyed only their gas, oil, or electric models. Articles giving advice to homemakers on how to deal with the trials and tribulations of starting, stoking, and maintaining a coal or a wood fire also disappeared. Thus it seems a safe assumption that most middle-class homes had switched to the new method of cooking by the time the depression began. The change in routine that was predicated on the change from coal or wood to gas or oil was profound; aside from the elimination of such chores as loading the fuel and removing the ashes, the new stoves were much easier to light, maintain, and regulate (even when they did not have thermostats, as the earliest models did not).[18] Kitchens were, in addition, much easier to clean when they did not have coal dust regularly tracked through them; one writer in the *Ladies'*

[15] *Zanesville, Ohio and Thirty-six Other American Cities* (New York, 1927), p. 65. Also see Robert S. Lynd and Helen M. Lynd, *Middletown in Transition* (New York, 1936), p. 537. Middletown is Muncie, Indiana.

[16] Lynd and Lynd, *Middletown,* p. 96, and *Middletown in Transition,* p. 539.

[17] Lynd and Lynd, *Middletown,* p. 98, and *Middletown in Transition,* p. 562.

[18] On the advantages of the new stoves, see *Boston Cooking School Cookbook* (Boston, 1916), pp. 15–20; and Russell Lynes, *The Domesticated Americans* (New York, 1957), pp. 119–20.

Home Journal estimated that kitchen cleaning was reduced by one-half when coal stoves were eliminated.[19]

Along with new stoves came new foodstuffs and new dietary habits. Canned foods had been on the market since the middle of the 19th century, but they did not become an appreciable part of the standard middle-class diet until the 1920s—if the recipes given in cookbooks and in women's magazines are a reliable guide. By 1918 the variety of foods available in cans had been considerably expanded from the peas, corn, and succotash of the 19th century; an American housewife with sufficient means could have purchased almost any fruit or vegetable and quite a surprising array of ready-made meals in a can —from Heinz's spaghetti in meat sauce to Purity Cross's lobster à la Newburg. By the middle of the 1920s home canning was becoming a lost art. Canning recipes were relegated to the back pages of the women's magazines; the business-class wives of Muncie reported that, while their mothers had once spent the better part of the summer and fall canning, they themselves rarely put up anything, except an occasional jelly or batch of tomatoes.[20] In part this was also due to changes in the technology of marketing food; increased use of refrigerated railroad cars during this period meant that fresh fruits and vegetables were in the markets all year round at reasonable prices.[21] By the early 1920s convenience foods were also appearing on American tables: cold breakfast cereals, pancake mixes, bouillon cubes, and packaged desserts could be found. Wartime shortages accustomed Americans to eating much lighter meals than they had previously been wont to do; and as fewer family members were taking all their meals at home (businessmen started to eat lunch in restaurants downtown, and factories and schools began installing cafeterias), there was simply less cooking to be done, and what there was of it was easier to do.[22]

* * *

Many of the changes just described—from hand power to electric power, from coal and wood to gas and oil as fuels for cooking, from one-room heating to central heating, from pumping water to running water—are enormous technological changes. Changes of a similar dimension, either in the fundamental technology of an industry, in the diffusion of that technology, or in the routines of workers, would have long since been labeled an "industrial revolution." The change from the laundry tub to the washing machine is no less profound than

[19]"How to Save Coal While Cooking," *Ladies' Home Journal* 25 (January 1908): 44.
[20]Lynd and Lynd, *Middletown*, p. 156.
[21]Ibid.; see also "Safeway Stores," *Fortune* 26 (October 1940): 60.
[22]Lynd and Lynd, *Middletown*, pp. 134–35 and 153–54.

the change from the hand loom to the power loom; the change from pumping water to turning on a water faucet is no less destructive of traditional habits than the change from manual to electric calculating. It seems odd to speak of an "industrial revolution" connected with housework, odd because we are talking about the technology of such homely things, and odd because we are not accustomed to thinking of housewives as a labor force or of housework as an economic commodity—but despite this oddity, I think the term is altogether appropriate.

In this case other questions come immediately to mind, questions that we do not hesitate to ask, say, about textile workers in Britain in the early 19th century, but we have never thought to ask about housewives in America in the 20th century. What happened to this particular work force when the technology of its work was revolutionized? Did structural changes occur? Were new jobs created for which new skills were required? Can we discern new ideologies that influenced the behavior of the workers?

The answer to all of these questions, surprisingly enough, seems to be yes. There were marked structural changes in the work force, changes that increased the work load and the job description of the workers that remained. New jobs were created for which new skills were required; these jobs were not physically burdensome, but they may have taken up as much time as the jobs they had replaced. New ideologies were also created, ideologies which reinforced new behavioral patterns, patterns that we might not have been led to expect if we had followed the sociologists' model to the letter. Middle-class housewives, the women who must have first felt the impact of the new household technology, were not flocking into the divorce courts or the labor market or the forums of political protest in the years immediately after the revolution in their work. What they were doing was sterilizing baby bottles, shepherding their children to dancing classes and music lessons, planning nutritious meals, shopping for new clothes, studying child psychology, and hand stitching color-coordinated curtains—all of which chores (and others like them) the standard sociological model has apparently not provided for.

The significant change in the structure of the household labor force was the disappearance of paid and unpaid servants (unmarried daughters, maiden aunts, and grandparents fall in the latter category) as household workers—and the imposition of the entire job on the housewife herself. Leaving aside for a moment the question of which was cause and which effect (did the disappearance of the servant create a demand for the new technology, or did the new technology make the servant obsolete?), the phenomenon itself is relatively easy to document. Before World War I, when illustrators in the women's magazines depicted women doing housework, the women were very often servants. When the lady of the house was drawn, she was often the person being served, or she was supervising the serving, or she was adding an elegant finishing touch to the work. Nursemaids diapered babies, seamstresses pinned up hems, waitresses served meals, laundresses did the wash, and cooks did the cooking. By the end of the 1920s the servants had disappeared from those illustrations; all those jobs were being done by housewives—elegantly manicured and coiffed, to be sure, but housewives nonetheless (compare figs. 1 and 2).

If we are tempted to suppose that illustrations in advertisements are not a reliable indicator of structural changes of this sort, we can corroborate the changes in other ways. Apparently, the illustrators really did know whereof they drew. Statistically the number of persons throughout the country employed in household service dropped from 1,851,000 in 1910 to 1,411,000 in 1920, while the number of households enumerated in the census rose from 20.3 million to 24.4 million.[23] In Indiana the ratio of households to servants increased from 13.5/1 in 1890 to 30.5/1 in 1920, and in the country as a whole the number of paid domestic servants per 1,000 population dropped from 98.9 in 1900 to 58.0 in 1920.[24] The business-class housewives of Muncie reported that they employed approximately one-half as many woman-hours of domestic service as their mothers had done.[25]

In case we are tempted to doubt these statistics (and indeed statistics about household labor are particularly unreliable, as the labor is often transient, part-time, or simply unreported), we can turn to articles on the servant problem, the disappearance of unpaid family workers, the design of kitchens, or to architectural drawings for houses. All of this evidence reiterates the same point: qualified servants were difficult to find; their wages had risen and their numbers fallen; houses were being designed without maid's rooms; daughters and unmarried aunts were finding jobs downtown; kitchens were being designed for housewives, not for servants.[26] The first home with a

[23]*Historical Statistics,* pp. 16 and 77.

[24]For Indiana data, see Lynd and Lynd, *Middletown,* p. 169. For national data, see D. L. Kaplan and M. Claire Casey, *Occupational Trends in the United States, 1900–1950,* U.S. Bureau of the Census Working Paper no. 5 (Washington, D.C., 1958), table 6. The extreme drop in numbers of servants between 1910 and 1920 also lends credence to the notion that this demographic factor stimulated the industrial revolution in housework.

[25]Lynd and Lynd, *Middletown,* p. 169.

[26]On the disappearance of maiden aunts, unmarried daughters, and grandparents, see Lynd and Lynd, *Middletown,* pp. 25, 99, and 110; Edward Bok, "Editorial," *American Home* 1 (October 1928): 15; "How to Buy Life Insurance," *Ladies' Home Journal* 45

The Ladies' Home Journal for April, 1918 107

The things you'd never put in the family laundry

YES, it's beginning to look dusky around the edges of the cuff and along the roll of the collar. Your precious new Georgette—you'd never dream of putting it in with the general laundry.

Your silk underwear, silk stockings, white satin collars—how they discolor, or yellow—how the threads break and grow weak when they are washed in the family laundry.

If only you knew the heartless laundress would not rub the life and the newness out of your dainty things!

You cannot afford to have your nicest things go so fast.

You, yourself, with a fraction of the energy you once spent hating the laundress, can now gently rinse the dirt out of your filmiest things—take them from the pure Lux suds soft and gleaming and new!

The secret? No ruinous rubbing of a cake of soap on fine fabrics! No rubbing again to get the soap and the dirt out. Just the gentle, tender cleansing with pure Lux suds that frail things must have to keep them unhurt.

No ruinous rubbing of fine fabrics

Lux is the most modern form of soap. There is nothing else like it. Lux comes in wonderful, delicate white flakes—pure and transparent. They dissolve instantly in hot water. You whisk them into the richest, sudsiest lather that loosens all the dirt—leaves the finest silk fabric clean and new—not a fiber roughened or torn or weakened in any way.

Write for free booklet and simple Lux directions for laundering. Learn how easy it is to launder perfectly the most delicate fabrics.

Be sure to get your package of Lux today. Your grocer, druggist or department store has it—Lever Bros. Co., Dept. A1, Cambridge, Mass.

These things need never be spoiled by washing. Try washing them the Lux way.

To wash silk blouses

FIG. 1.—The housewife as manager. (*Ladies' Home Journal,* April 1918. Courtesy of Lever Brothers Co.)

LADIES' HOME JOURNAL

NEW! *Beads of Soap*

banish washday drudgery

Super Suds dissolves quicker . . . works faster . . . rinses out easier

HERE is a way to wash clothes that's faster . . . easier . . . better. Saves one rinsing. Gets clothes whiter. A revolutionary discovery that brings you soap in the form of tissue-thin beads—the fastest-working form of soap ever made.

First bar soap—then chips . . . now Super Suds

Years ago women had only bar soap. How hard it was to rub the clothes with the soap and to rinse out those clinging soap particles.

Next came chips. Many women changed to this form of soap because it could be *stirred* into a cleansing solution. But clothes and dishes had to be rinsed very carefully in order to get rid of the undissolved soap.

Now comes Super Suds, and women are changing to it by thousands in preference to all other forms of soap because it dissolves twice as fast as any soap they have ever seen before.

Super Suds is not a chip . . . not a powder . . . but a remarkable new form of soap in tiny hollow beads, so thin that they burst into suds the instant they touch water.

Four times as thin as chips, Super Suds is the thinnest soap made.

In this new soap women have discovered two distinct advantages.

First, Super Suds is so thin it dissolves instantly . . . saves time and trouble.

Second, Super Suds dissolves completely . . . no undissolved soap to leave spots on clothes or film on dishes. Women like Super Suds because it does the work faster and better than other forms of soap.

Super Suds is simply wonderful for dishes. It makes them sparkle and glisten like jewels, and yet you never even touch a dishtowel. Just give them a quick hot rinse—and let them drain!

Already thousands of progressive women have been delighted with Super Suds. Won't you try it today? Just say "Super Suds" and your grocer will hand you the biggest box of soap you have ever seen for ten cents!

An Octagon Soap Product. Every box of Super Suds carries a premium coupon, our discount to you!

The BIGGEST box of soap on the market for 10¢

FIG. 2.—The housewife as laundress. (*Ladies' Home Journal*, August 1928. Courtesy of Colgate-Palmolive-Peet.)

kitchen that was not an entirely separate room was designed by Frank Lloyd Wright in 1934.[27] In 1937 Emily Post invented a new character for her etiquette books: Mrs. Three-in-One, the woman who is her own cook, waitress, and hostess.[28] There must have been many new Mrs. Three-in-Ones abroad in the land during the 1920s.

As the number of household assistants declined, the number of household tasks increased. The middle-class housewife was expected to demonstrate competence at several tasks that previously had not been in her purview or had not existed at all. Child care is the most obvious example. The average housewife had fewer children than her mother had had, but she was expected to do things for her children that her mother would never have dreamed of doing: to prepare their special infant formulas, sterilize their bottles, weigh them every day, see to it that they ate nutritionally balanced meals, keep them isolated and confined when they had even the slightest illness, consult with their teachers frequently, and chauffeur them to dancing lessons, music lessons, and evening parties.[29] There was very little Freudianism in this new attitude toward child care: mothers were not spending more time and effort on their children because they feared the psychological trauma of separation, but because competent nursemaids could not be found, and the new theories of child care required constant attention from well-informed persons—persons who were willing and able to read about the latest discoveries in nutrition, in the control of contagious diseases, or in the techniques of behavioral psychology. These persons simply had to be their mothers.

Consumption of economic goods provides another example of the housewife's expanded job description; like child care, the new tasks associated with consumption were not necessarily physically burdensome, but they were time consuming, and they required the acquisi-

---

(March 1928): 35. The house plans appeared every month in *American Home,* which began publication in 1928. On kitchen design, see Giedion, pp. 603–21; "Editorial," *Ladies' Home Journal* 45 (April 1928): 36; advertisement for Hoosier kitchen cabinets, *Ladies' Home Journal* 45 (April 1928): 117. Articles on servant problems include "The Vanishing Servant Girl," *Ladies Home Journal* 35 (May 1918): 48; "Housework, Then and Now," *American Home* 8 (June 1932): 128; "The Servant Problem," *Fortune* 24 (March 1938): 80–84; and *Report of the YWCA Commission on Domestic Service* (Los Angeles, 1915).

[27]Giedion, p. 619. Wright's new kitchen was installed in the Malcolm Willey House, Minneapolis.

[28]Emily Post, *Etiquette: The Blue Book of Social Usage,* 5th ed. rev. (New York, 1937), p. 823.

[29]This analysis is based upon various child-care articles that appeared during the period in the *Ladies' Home Journal, American Home,* and *Parents' Magazine.* See also Lynd and Lynd, *Middletown,* chap. 11.

tion of new skills.[30] Home economists and the editors of women's magazines tried to teach housewives to spend their money wisely. The present generation of housewives, it was argued, had been reared by mothers who did not ordinarily shop for things like clothing, bed linens, or towels; consequently modern housewives did not know how to shop and would have to be taught. Furthermore, their mothers had not been accustomed to the wide variety of goods that were now available in the modern marketplace; the new housewives had to be taught not just to be consumers, but to be informed consumers.[31] Several contemporary observers believed that shopping and shopping wisely were occupying increasing amounts of housewives' time.[32]

Several of these contemporary observers also believed that standards of household care changed during the decade of the 1920s.[33] The discovery of the "household germ" led to almost fetishistic concern about the cleanliness of the home. The amount and frequency of laundering probably increased, as bed linen and underwear were changed more often, children's clothes were made increasingly out of washable fabrics, and men's shirts no longer had replaceable collars and cuffs.[34] Unfortunately all these changes in standards are difficult to document, being changes in the things that people regard as so insignificant as to be unworthy of comment; the improvement in standards seems a likely possibility, but not something that can be proved.

In any event we do have various time studies which demonstrate somewhat surprisingly that housewives with conveniences were spending just as much time on household duties as were housewives without them—or, to put it another way, housework, like so many

[30]John Kenneth Galbraith has remarked upon the advent of woman as consumer in *Economics and the Public Purpose* (Boston, 1973), pp. 29–37.

[31]There was a sharp reduction in the number of patterns for home sewing offered by the women's magazines during the 1920s; the patterns were replaced by articles on "what is available in the shops this season." On consumer education see, for example, "How to Buy Towels," *Ladies' Home Journal* 45 (February 1928): 134; "Buying Table Linen," *Ladies' Home Journal* 45 (March 1928): 43; and "When the Bride Goes Shopping," *American Home* 1 (January 1928): 370.

[32]See, for example, Lynd and Lynd, *Middletown,* pp. 176 and 196; and Margaret G. Reid, *Economics of Household Production* (New York, 1934), chap. 13.

[33]See Reid, pp. 64–68; and Kyrk, p. 98.

[34]See advertisement for Cleanliness Institute—"Self-respect thrives on soap and water," *Ladies' Home Journal* 45 (February 1928): 107. On changing bed linen, see "When the Bride Goes Shopping," *American Home* 1 (January 1928): 370. On laundering children's clothes, see, "Making a Layette," *Ladies' Home Journal* 45 (January 1928): 20; and Josephine Baker, "The Youngest Generation," *Ladies' Home Journal* 45 (March 1928): 185.

other types of work, expands to fill the time available.[35] A study comparing the time spent per week in housework by 288 farm families and 154 town families in Oregon in 1928 revealed 61 hours spent by farm wives and 63.4 hours by town wives; in 1929 a U.S. Department of Agriculture study of families in various states produced almost identical results.[36] Surely if the standard sociological model were valid, housewives in towns, where presumably the benefits of specialization and electrification were most likely to be available, should have been spending far less time at their work than their rural sisters. However, just after World War II economists at Bryn Mawr College reported the same phenomenon: 60.55 hours spent by farm housewives, 78.35 hours by women in small cities, 80.57 hours by women in large ones—precisely the reverse of the results that were expected.[37] A recent survey of time studies conducted between 1920 and 1970 concludes that the time spent on housework by nonemployed housewives has remained remarkably constant throughout the period.[38] All these results point in the same direction: mechanization of the household meant that time expended on some jobs decreased, but also that new jobs were substituted, and in some cases—notably laundering—time expenditures for old jobs increased because of higher standards. The advantages of mechanization may be somewhat more dubious than they seem at first glance.

* * *

As the job of the housewife changed, the connected ideologies also changed; there was a clearly perceptible difference in the attitudes that women brought to housework before and after World War I.[39]

[35]This point is also discussed at length in my paper "What Did Labor-saving Devices Really Save?" (unpublished).

[36]As reported in Lyrk, p. 51.

[37]Bryn Mawr College Department of Social Economy, *Women During the War and After* (Philadelphia, 1945); and Ethel Goldwater, "Woman's Place," *Commentary* 4 (December 1947): 578–85.

[38]JoAnn Vanek, "Keeping Busy: Time Spent in Housework, United States, 1920–1970" (Ph.D. diss., University of Michigan, 1973). Vanek reports an average of 53 hours per week over the whole period. This figure is significantly lower than the figures reported above, because each time study of housework has been done on a different basis, including different activities under the aegis of housework, and using different methods of reporting time expenditures; the Bryn Mawr and Oregon studies are useful for the comparative figures that they report internally, but they cannot easily be compared with each other.

[39]This analysis is based upon my reading of the middle-class women's magazines between 1918 and 1930. For detailed documentation see my paper "Two Washes in the Morning and a Bridge Party at Night: The American Housewife between the Wars," *Women's Studies* (in press). It is quite possible that the appearance of guilt as a strong

Before the war the trials of doing housework in a servantless home were discussed and they were regarded as just that—trials, necessary chores that had to be got through until a qualified servant could be found. After the war, housework changed: it was no longer a trial and a chore, but something quite different—an emotional "trip." Laundering was not just laundering, but an expression of love; the housewife who truly loved her family would protect them from the embarrassment of tattletale gray. Feeding the family was not just feeding the family, but a way to express the housewife's artistic inclinations and a way to encourage feelings of family loyalty and affection. Diapering the baby was not just diapering, but a time to build the baby's sense of security and love for the mother. Cleaning the bathroom sink was not just cleaning, but an exercise of protective maternal instincts, providing a way for the housewife to keep her family safe from disease. Tasks of this emotional magnitude could not possibly be delegated to servants, even assuming that qualified servants could be found.

Women who failed at these new household tasks were bound to feel guilt about their failure. If I had to choose one word to characterize the temper of the women's magazines during the 1920s, it would be "guilt." Readers of the better-quality women's magazines are portrayed as feeling guilty a good lot of the time, and when they are not guilty they are embarrassed: guilty if their infants have not gained enough weight, embarrassed if their drains are clogged, guilty if their children go to school in soiled clothes, guilty if all the germs behind the bathroom sink are not eradicated, guilty if they fail to notice the first signs of an oncoming cold, embarrassed if accused of having body odor, guilty if their sons go to school without good breakfasts, guilty if their daughters are unpopular because of old-fashioned, or unironed, or—heaven forbid—dirty dresses (see figs. 3 and 4). In earlier times women were made to feel guilty if they abandoned their children or were too free with their affections. In the years after World War I, American women were made to feel guilty about sending their children to school in scuffed shoes. Between the two kinds of guilt there is a world of difference.

* * *

Let us return for a moment to the sociological model with which this essay began. The model predicts that changing patterns of

---

element in advertising is more the result of new techniques developed by the advertising industry than the result of attitudinal changes in the audience—a possibility that I had not considered when doing the initial research for this paper. See A. Michael McMahon, "An American Courtship: Psychologists and Advertising Theory in the Progressive Era," *American Studies* 13 (1972): 5–18.

LADIES' HOME JOURNAL 95

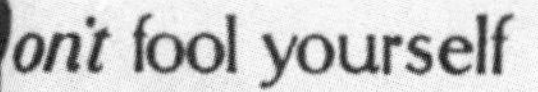

Since halitosis never announces itself, to the victim, you simply cannot know when you have it.

Are you unpopular *with your own children?*

Make sure that you don't have halitosis. It is inexcusable. And unnecessary.

Some mothers blame everything but halitosis (unpleasant breath) when children show resentment or lack of affection.

More often than you would imagine, however, halitosis is at fault. Children are quick to resent it.

Intelligent people realize that as a result of modern habits halitosis is very common and that anyone may have it, *and yet be ignorant of the fact.*

Realizing this, they eliminate any risk of offending, by the systematic use of Listerine in the mouth. Every morning. Every night. And between times when necessary, especially before meeting people. Keep a bottle handy in home and office for this purpose.

Listerine checks halitosis instantly. Being antiseptic, it strikes at its commonest cause—fermentation in the oral cavity. Then, being a powerful deodorant, it destroys the odors themselves.

If you have any doubt of Listerine's power to deodorize, make this test: Rub a slice of onion on your hand. Then apply Listerine clear. Immediately, every trace of onion odor is gone. Lambert Pharmacal Company, St. Louis, Mo.

INTENDED FOR MEN
but try and keep the women away

Have you tried the new LISTERINE SHAVING CREAM? Keeps your face marvelously cool long after shaving. Incidentally, women think rather highly of it as a shampoo.

LISTERINE
—the safe antiseptic

FIG. 3.—Sources of housewifely guilt: the good mother smells sweet. (*Ladies' Home Journal,* August 1928. Courtesy of Warner-Lambert, Inc.)

FIG. 4.—Sources of housewifely guilt: the good mother must be beautiful. (*Ladies' Home Journal,* July 1928. Courtesy of Colgate-Palmolive-Peet.)

household work will be correlated with at least two striking indicators of social change: the divorce rate and the rate of married women's labor force participation. That correlation may indeed exist, but it certainly is not reflected in the women's magazines of the 1920s and 1930s: divorce and full-time paid employment were not part of the life-style or the life pattern of the middle-class housewife as she was idealized in her magazines.

There were social changes attendant upon the introduction of modern technology into the home, but they were not the changes that the traditional functionalist model predicts; on this point a close analysis of the statistical data corroborates the impression conveyed in the magazines. The divorce rate was indeed rising during the years between the wars, but it was not rising nearly so fast for the middle and upper classes (who had, presumably, easier access to the new technology) as it was for the lower classes. By almost every gauge of socioeconomic status—income, prestige of husband's work, education—the divorce rate is higher for persons lower on the socioeconomic scale—and this is a phenomenon that has been constant over time.[40]

The supposed connection between improved household technology and married women's labor force participation seems just as dubious, and on the same grounds. The single socioeconomic factor which correlates most strongly (in cross-sectional studies) with married women's employment is husband's income, and the correlation is strongly negative; the higher his income, the less likely it will be that she is working.[41] Women's labor force participation increased during the 1920s but this increase was due to the influx of single women into the force. Married women's participation increased slightly during those years, but that increase was largely in factory labor —precisely the kind of work that middle-class women (who were, again, much more likely to have labor-saving devices at home) were least likely to do.[42] If there were a necessary connection between the improvement of household technology and either of these two social indicators, we would expect the data to be precisely the reverse of what in fact has occurred: women in the higher social classes should have fewer func-

[40]For a summary of the literature on differential divorce rates, see Winch, p. 706; and William J. Goode, *After Divorce* (New York, 1956) p. 44. The earliest papers demonstrating this differential rate appeared in 1927, 1935, and 1939.

[41]For a summary of the literature on married women's labor force participation, see Juanita Kreps, *Sex in the Marketplace: American Women at Work* (Baltimore, 1971), pp. 19–24.

[42]Valerie Kincaid Oppenheimer, *The Female Labor Force in the United States*, Population Monograph Series, no. 5 (Berkeley, 1970), pp. 1–15; and Lynd and Lynd, *Middletown*, pp. 124–27.

tions at home and should therefore be more (rather than less) likely to seek paid employment or divorce.

Thus for middle-class American housewives between the wars, the social changes that we can document are not the social changes that the functionalist model predicts; rather than changes in divorce or patterns of paid employment, we find changes in the structure of the work force, in its skills, and in its ideology. These social changes were concomitant with a series of technological changes in the equipment that was used to do the work. What is the relationship between these two series of phenomena? Is it possible to demonstrate causality or the direction of that causality? Was the decline in the number of households employing servants a cause or an effect of the mechanization of those households? Both are, after all, equally possible. The declining supply of household servants, as well as their rising wages, may have stimulated a demand for new appliances at the same time that the acquisition of new appliances may have made householders less inclined to employ the laborers who were on the market. Are there any techniques available to the historian to help us answer these questions?

* * *

In order to establish causality, we need to find a connecting link between the two sets of phenomena, a mechanism that, in real life, could have made the causality work. In this case a connecting link, an intervening agent between the social and the technological changes, comes immediately to mind: the advertiser—by which term I mean a combination of the manufacturer of the new goods, the advertising agent who promoted the goods, and the periodical that published the promotion. All the new devices and new foodstuffs that were being offered to American households were being manufactured and marketed by large companies which had considerable amounts of capital invested in their production: General Electric, Procter & Gamble, General Foods, Lever Brothers, Frigidaire, Campbell's, Del Monte, American Can, Atlantic & Pacific Tea—these were all well-established firms by the time the household revolution began, and they were all in a position to pay for national advertising campaigns to promote their new products and services. And pay they did; one reason for the expanding size and number of women's magazines in the 1920s was, no doubt, the expansion in revenues from available advertisers.[43]

Those national advertising campaigns were likely to have been powerful stimulators of the social changes that occurred in the

[43]On the expanding size, number, and influence of women's magazines during the 1920s, see Lynd and Lynd, *Middletown*, pp. 150 and 240–44.

household labor force; the advertisers probably did not initiate the changes, but they certainly encouraged them. Most of the advertising campaigns manifestly worked, so they must have touched upon areas of real concern for American housewives. Appliance ads specifically suggested that the acquisition of one gadget or another would make it possible to fire the maid, spend more time with the children, or have the afternoon free for shopping.[44] Similarly, many advertisements played upon the embarrassment and guilt which were now associated with household work. Ralston, Cream of Wheat, and Ovaltine were not themselves responsible for the compulsive practice of weighing infants and children repeatedly (after every meal for newborns, every day in infancy, every week later on), but the manufacturers certainly did not stint on capitalizing upon the guilt that women apparently felt if their offspring did not gain the required amounts of weight.[45] And yet again, many of the earliest attempts to spread "wise" consumer practices were undertaken by large corporations and the magazines that desired their advertising: mail-order shopping guides, "product-testing" services, pseudoinformative pamphlets, and other such promotional devices were all techniques for urging the housewife to buy new things under the guise of training her in her role as skilled consumer.[46]

Thus the advertisers could well be called the "ideologues" of the 1920s, encouraging certain very specific social changes—as ideologues are wont to do. Not surprisingly, the changes that occurred were precisely the ones that would gladden the hearts and fatten the purses of the advertisers; fewer household servants meant a greater demand for labor and timesaving devices; more household tasks for women meant more and more specialized products that they would need to buy; more guilt and embarrassment about their failure to succeed at their work meant a greater likelihood that they would buy the products that were intended to minimize that failure. Happy,

[44]See, for example, the advertising campaigns of General Electric and Hotpoint from 1918 through the rest of the decade of the 1920s; both campaigns stressed the likelihood that electric appliances would become a thrifty replacement for domestic servants.

[45]The practice of carefully observing children's weight was initiated by medical authorities, national and local governments, and social welfare agencies, as part of the campaign to improve child health which began about the time of World War I.

[46]These practices were ubiquitous, *American Home,* for example, which was published by Doubleday, assisted its advertisers by publishing a list of informative pamphlets that readers could obtain; devoting half a page to an index of its advertisers; specifically naming manufacturer's and list prices in articles about products and services; allotting almost one-quarter of the magazine to a mail-order shopping guide which was not (at least ostensibly) paid advertisement; and as part of its editorial policy, urging its readers to buy new goods.

full-time housewives in intact families spend a lot of money to maintain their households; divorced women and working women do not. The advertisers may not have created the image of the ideal American housewife that dominated the 1920s—the woman who cheerfully and skillfully set about making everyone in her family perfectly happy and perfectly healthy—but they certainly helped to perpetuate it.

The role of the advertiser as connecting link between social change and technological change is at this juncture simply a hypothesis, with nothing much more to recommend it than an argument from plausibility. Further research may serve to test the hypothesis, but testing it may not settle the question of which was cause and which effect—if that question can ever be settled definitively in historical work. What seems most likely in this case, as in so many others, is that cause and effect are not separable, that there is a dynamic interaction between the social changes that married women were experiencing and the technological changes that were occurring in their homes. Viewed this way, the disappearance of competent servants becomes one of the factors that stimulated the mechanization of homes, and this mechanization of homes becomes a factor (though by no means the only one) in the disappearance of servants. Similarly, the emotionalization of housework becomes both cause and effect of the mechanization of that work; and the expansion of time spent on new tasks becomes both cause and effect of the introduction of time-saving devices. For example the social pressure to spend more time in child care may have led to a decision to purchase the devices; once purchased, the devices could indeed have been used to save time— although often they were not.

* * *

If one holds the question of causality in abeyance, the example of household work still has some useful lessons to teach about the general problem of technology and social change. The standard sociological model for the impact of modern technology on family life clearly needs some revision: at least for middle-class nonrural American families in the 20th century, the social changes were not the ones that the standard model predicts. In these families the functions of at least one member, the housewife, have increased rather than decreased; and the dissolution of family life has not in fact occurred.

Our standard notions about what happens to a work force under the pressure of technological change may also need revision. When industries become mechanized and rationalized, we expect certain general changes in the work force to occur: its structure becomes more highly differentiated, individual workers become more specialized, managerial functions increase, and the emotional context of the work disappears. On all four counts our expectations are reversed with regard to household work. The work force became less rather than more differentiated as domestic servants, unmarried daughters, maiden aunts, and grandparents left the household and as chores which had once been performed by commercial agencies (laundries, delivery services, milkmen) were delegated to the housewife. The individual workers also became less specialized; the new housewife was now responsible for every aspect of life in her household, from scrubbing the bathroom floor to keeping abreast of the latest literature in child psychology.

The housewife is just about the only unspecialized worker left in America—a veritable jane-of-all-trades at a time when the jacks-of-all-trades have disappeared. As her work became generalized the housewife was also proletarianized: formerly she was ideally the manager of several other subordinate workers; now she was idealized as the manager and the worker combined. Her managerial functions have not entirely disappeared, but they have certainly diminished and have been replaced by simple manual labor; the middle-class, fairly well educated housewife ceased to be a personnel manager and became, instead, a chauffeur, charwoman, and short-order cook. The implications of this phenomenon, the proletarianization of a work force that had previously seen itself as predominantly managerial, deserve to be explored at greater length than is possible here, because I suspect that they will explain certain aspects of the women's liberation movement of the 1960s and 1970s which have previously eluded explanation: why, for example, the movement's greatest strength lies in social and economic groups who seem, on the surface at least, to need it least—women who are white, well-educated, and middle-class.

Finally, instead of desensitizing the emotions that were connected with household work, the industrial revolution in the home seems to have heightened the emotional context of the work, until a woman's sense of self-worth became a function of her success at arranging bits of fruit to form a clown's face in a gelatin salad. That pervasive social illness, which Betty Friedan characterized as "the problem that has no name," arose not among workers who found that their labor brought no emotional satisfaction, but among workers who found that their work was invested with emotional weight far out of proportion to its own inherent value: "How long," a friend of mine is fond of asking, "can we continue to believe that we will have orgasms while waxing the kitchen floor?"

The Development of the Diesel Engine

LYNWOOD BRYANT

In 1912, when the diesel engine was about twenty years old, just coming of age after a prolonged infancy and a painful adolescence, it was the subject of a celebrated controversy, in which the inventor, Rudolf Diesel (1858–1913), and two distinguished professors of engineering discussed the very topic that concerns us in this symposium: the distinctions among invention, development, and innovation as parts of the total process of technological evolution. These three words are commonly used rather loosely, and I do not need to make very sharp distinctions among them because my main point is that in the real world the three processes to which these words refer are not sharply separated. But let me begin by saying that I am thinking of an invention as the appearance of an idea in someone's mind, an event in intellectual history; of development as the conversion of an idea into some kind of workable reality, such as an engine that runs; and of innovation as the introduction of the developed invention into the economy as a useful, salable product.

The 1912 controversy began when Diesel heard that Adolph Nägel of Dresden was planning a book on the history of the diesel engine. Diesel was naturally nervous about what Nägel would say, for he was a sensitive and proud man, and there had been some troublesome uncertainties and misunderstandings about the invention of his engine and the validity of his patent. So the inventor prepared his own account of the origin of his engine and presented it at a meeting of the German Society of Naval Architects in November 1912. In the discussion period following the paper, two professors launched an attack on Diesel that raised disturbing questions about his professional integrity. Their main point was that the engine that emerged from the development process was not the same as the engine that Diesel invented, and that credit for it should go to the practical engineers who

MR. BRYANT is professor of history, emeritus, at the Massachusetts Institute of Technology. He is indebted to Friedrich Sass, Eugen Diesel, Georg Strössner, and Kurt Schnauffer for many helpful conversations about Diesel and his engine; to Irmgard Denkinger for her kindness and her guidance in the MAN Werkarchiv in Augsburg; and to A. R. Rogowski of M.I.T. for ten years of patient explanation of the mysteries of engines.

developed it. Diesel, they said, was a promoter, a mere businessman, not an inventor.[1] This was the unkindest cut of all, for Diesel was an engineer with scientific pretensions. He thought of himself as the James Watt of the 20th century.

Four books grew out of this controversy, one by Diesel and three by his critics.[2] They were all published shortly before or shortly after Diesel's mysterious death,[3] and they all include theoretical discussion of the nature of invention and development, and argument about Diesel's work. So the Diesel story is well documented in the sense that much has been written about it, but the literature is mostly polemical or promotional, so that there are still many uncertainties about what actually happened.

I propose to go over this story once more, and use it as a case study in the process of development. I shall try to indicate what different kinds of activity were going on at different times in the course of the early evolution of this engine, and then make some general remarks about the nature of development and how it is related to invention on the one hand and innovation on the other.

My main point is that these processes may be conceptually distinct, but when you look closely at what is going on, it is hard to say when one leaves off and another begins. It may be better to regard them not

[1]Diesel's paper appeared under the title "Die Entstehung des Dieselmotors" in *Jahrbuch der Schiffbautechnischen Gesellschaft* 14 (1913): 267–355, with a transcript of the following discussion on pp. 355–67. Marine engineers were especially interested in diesel power at this time. The major powers had recently adopted it for submarines, and the first motor ship had crossed the Atlantic in 1911.

[2]The four books were: Rudolf Diesel, *Die Entstehung des Dieselmotors* (Berlin, 1913), which is the 1912 paper, slightly revised, without the following discussion; P. Meyer, *Beiträge zur Geschichte des Dieselmotors* (Berlin, 1913); J. Lüders, *Der Dieselmythus: Quellenmässige Geschichte der Entstehung des heutigen Oelmotors* (Berlin, 1913); and A. Riedler, *Dieselmotoren: Beiträge zur Kenntnis der Hochdruckmotoren* (Berlin, 1914). The inventor's son, Eugen Diesel, wrote two books and many articles about his father, which tend to be a little colorful and obscure, but they are fundamentally sound, I think, and the best source for biographical information. They also include technical information about the development of the engine. The standard biography is *Diesel: der Mensch, das Werk, das Schicksal* (Stuttgart, 1937). This book was reissued a number of times, at least once with sensitive passages slightly revised. Another important biographical work of Eugen Diesel's is *Jahrhundertwende: Gesehen im Schicksal meines Vaters* (Stuttgart, 1949).

[3]Diesel disappeared from a channel steamer in the night of September 29–30, 1913. His body was recovered ten days later by a pilot boat at the mouth of the Scheldt River and returned to the sea, but unmistakably identified by articles taken from the body. Lurid rumors appeared at once and persist to this day, for example the legend that Diesel was executed by the German secret service because he was about to betray submarine secrets to the British. All known facts are consistent with the conclusion that he was a suicide. He was in financial trouble, and deeply disturbed by these attacks on his integrity.

as chronological stages but as different kinds of forces operating in an evolutionary process, or different types of human interest and activity, all more or less involved in all stages of technological progress.

First Period: Invention, 1890–93

The first period begins with the conception in Diesel's mind, which occurred in 1890 or 1891, and ends with the first attempt at a real engine in 1893. In this period the engine is a concept, not a reality, and its history is intellectual history, the story of how a man got an idea and was guided by this idea.[4]

The system of ideas that constitutes the diesel engine includes a dozen important concepts and techniques, all of which appeared before 1890. But one key idea Diesel claimed as novel: the idea that combustion in an internal-combustion engine could be made to take place at constant temperature, that is, isothermally. As Diesel conceived it, such an engine (on a down stroke of the piston) would draw in a charge of plain air, and then (on an up stroke) compress it to a very high pressure and temperature. Then just as the piston started down again, fuel would be introduced gradually into the hot, expanding air in such a way that the tendency of the temperature to rise with combustion would be exactly counterbalanced by the tendency of the temperature to fall as the air expanded.

If this way of achieving isothermal combustion works, thought Diesel, then an approximation to the ideal engine embodying the Carnot cycle is possible. Such an engine would be much more efficient than any existing heat engine because it would operate through a wider range of temperature, and, with isothermal combustion, all the

[4]The best summary of this piece of intellectual history is in Friedrich Sass, *Geschichte des deutschen Verbrennungsmotorenbaues von 1860 bis 1918* (Berlin, 1962). The last third of this monumental work is an excellent history of the diesel engine. It rests on several years of research by Kurt Schnauffer, which is recorded in many volumes of typescript now in the library of the Deutsches Museum in Munich. Schnauffer's own account of the invention is "Die Erfindung Rudolf Diesels," *Zeitschrift des Vereines deutscher Ingenieure* 100 (1958): 308–20 (hereafter cited as *ZVDI*). All the books cited in note 2 discuss the invention. Paul Meyer, the most objective and temperate of these contemporary critics of Diesel's work, wrote his book in 1913 on the basis of some experience with Diesel—he worked with him for a while beginning in 1898—but without access to Diesel's papers. Thirty years later, during the Second World War, he went through the Diesel papers and wrote a new history of the invention of the diesel. This work was sponsored by the Verein deutscher Ingenieure (VDI) but never published. It remains a manuscript entitled "Die Geschichte des Dieselmotors" in the Diesel papers in the Deutsches Museum. Meyer later published two important articles based on this work: "Aus der Enstehungsgeschichte des Dieselmotors," *Schweizerische Bauzeitung* 66 (1948): 485–87; and "War der Dieselmotor jemals durch Patente geschützt?" *Schweizerische Bauzeitung* 67 (1949): 309–10.

heat added would be converted to work on the piston. In a Carnot heat engine the heat to be converted into work has to be added at the highest temperature of the cycle and it has to be added without raising the temperature further. The practical difficulty in realizing this ideal in a combustion engine is that it is hard to see how combustion can take place without a rise in temperature. This was the essence of Diesel's invention: he conceived a way of burning fuel in an engine without raising the temperature.[5]

Diesel began by formulating the idea carefully and working out the details of a possible engine. He prepared a manuscript, with supporting calculations and illustrations, and sent it out for criticism to several experienced engineers and industrialists. One or two critics said encouraging things, but mostly they regarded the engine as impractical because of the extreme pressures and temperatures required. Diesel went back to the drawing board to prepare a more modest proposal for a more realistic engine, and with this revision as a supplement, sent the manuscript to Springer to be published.

Diesel had also formulated his ideas in the form of a patent application, which went through a similar process of negotiation and revision. The patent was issued and the book published early in 1893.[6]

Meanwhile Diesel had offered his engine, or his idea for an engine, to a number of machine-building firms (it was clearly much too large a project to be developed by a single man in a home workshop) and was rejected by all of them, but as a result of a second approach with more modest claims, the Maschinenfabrik Augsburg agreed to help Diesel develop his engine. This was a large firm led for almost fifty years by Heinrich Buz, an industrial statesman who plays a role in the Diesel

[5]My article "Rudolf Diesel and His Rational Engine," *Scientific American* 221 (August 1969): 108–18, tries to explain the Carnot cycle and its relation to the diesel engine. The famous Carnot cycle was first described in Sadi Carnot, *Réflexions sur la puissance motrice du feu* (Paris, 1824).

[6]The German patent is number 67,207, dated February 28, 1892. (Another patent, number 82,168, dated November 30, 1893, is for a modified process and was applied for after the experimental work began.) The book is *Theorie und Konstruktion eines rationellen Wärmemotors zum Ersatz der Dampfmaschinen und der heute bekannten Verbrennungsmotoren* (Berlin, 1893). An English translation by Bryan Donkin (London, 1894) omits some material on applications of the engine that appears in the German original. The idea of achieving this kind of isothermal combustion in an engine appears in Otto Köhler, *Theorie der Gasmotoren* (Leipzig, 1887), and the engine that Köhler hypothesizes is similar in a number of interesting ways to the engine that Diesel describes in his book and his patent. This prior publication of Diesel's key idea was a disturbing threat to the validity of the patent. Diesel said that Köhler's book was unknown to him at the time. But Köhler's idea was picked up and published in F. Grashof, *Theorie der Kraftmaschinen* (Hamburg and Leipzig, 1890). This is the third volume of Grashof's *Theoretische Maschinenlehre.* It is hard to see how Diesel could have overlooked this book by a distinguished authority in his field.

story something like the role of Boulton in the Watt story.[7] After the patent was issued, the firm of Krupp agreed to join the Maschinenfabrik Augsburg in this project, to share the expenses of development, and to pay Diesel a salary of 30,000 marks a year while he worked on the development.[8] It was a good bargain for Diesel.

When Buz called for drawings, Diesel already had them prepared, for a two-cylinder, 50-horsepower engine. Buz proposed a more modest start, with a one-cylinder, 25-horsepower unit. There was no model, no preliminary work with components or processes. Diesel was so sure of himself that he moved at once to a full-size working engine. To him it was a scientific engine, built on sure thermodynamic principles.

I am labeling this first period "invention" because what is going on is primarily mental activity: thinking, calculation, argument, writing. But it is by no means a simple conception of an idea and nothing else. It includes an elaborate consideration of what is involved in realizing the idea, and a good deal of hard work in trying to sell the idea, to secure the endorsements of scientists and the financial support of business men. I notice that the compromise between idea and reality, which we usually associate with development, is there from the beginning and that the business negotiations and the balancing of economic interests, which we expect to be dominant in the later stages, have an important bearing on the early decisions to undertake development.

Second Period: Making the First Real Engine, 1893–97

The second period was a four-year struggle to get an engine that would run, a struggle that took place inside a large steam-engine factory in Augsburg. Diesel was in direct charge of the work himself, usually with one or two full-time mechanics assigned to the project.

Now what was Diesel doing during this period? We have a great deal of information about the technical work that was going on, for Diesel kept a meticulous journal in his own hand and wrote a book about it. He even kept thousands of indicator diagrams, so that you can still tell how the engine was doing on any given day, if you want to.[9]

[7]The Maschinenfabrik Augsburg merged with the Maschinenbaugesellschaft Nürnberg in 1898. The firm, which took the name Maschinenfabrik Augsburg-Nürnberg (MAN) in 1908, has a well-managed museum and library at its headquarters in Augsburg, with much interesting Diesel material.

[8]The role of the Krupp firm in the development of the diesel can be followed in the correspondence in the Diesel papers in the Deutsches Museum, and it is described in a mimeographed publication by Wilhelm Worsoe, *Die Mitarbeit der Werke Fried. Krupp an der Entstehung des Dieselmotors in den Jahren 1893–97 und an der Anfangsentwicklung 1897–99*, Erweiterte Ausgabe von 1933 (Kiel, 1940).

[9]Diesel was a self-conscious inventor who saw himself as a historical figure and kept

In his own account Diesel divides the period into six series of experiments, each lasting a few weeks or a few months. At the end of each period of work Diesel would summarize his findings in the journal and prepare drawings for the changes he wanted to try next. While waiting for a modified or a new engine—it might be several weeks or months—Diesel did a good deal of thinking and writing, corresponding with experts, searching the literature for help with troublesome problems—perhaps in compressed air technology, lubrication, or ignition devices. He was also on the road a good deal, attending meetings, taking care of his patent affairs, keeping in touch with firms already signed up, such as Krupp and Sulzer, and trying to interest others in taking licenses to make and sell the engine when it was ready.

The first engine never ran at all under its own power, but it began to show Diesel what his problems were, and it sent him back to revise his theory. The first task was to get the high pressures necessary for the new process. The first experimental engine was designed for a pressure of 44 atmospheres, but at the beginning Diesel could reach only 18. With a good deal of work on mechanical details he was able to raise this to 33 atmospheres and to assure himself that compression ignition was possible, if not easy, at this pressure. The second engine, a completely new one, ran idle for a minute in 1894. In the next two years Diesel did a great deal of work on fuels, fuel injection, carburetion, and ignition, and finally settled on a kerosene for fuel and a compressed-air system for delivering it to the combustion chamber. It was 1897 before a prototype engine was running smoothly and Diesel was ready to announce the end of the period of development.[10]

Now let me list the technical problems that Diesel had to solve in order to transform his idea for an engine into a real engine. In order to make his engine work, Diesel had to: (1) compress air to a very high pressure and temperature, (2) choose the right fuel, (3) inject the fuel into the high-pressure air, (4) control the timing of the injection and the amount of fuel injected (in very small amounts, under very high

---

careful records. Some of these he destroyed before he left Munich on the trip that ended in his death, but the journal covering development between 1893 and 1897, with supplementary notes and memoranda and a great many indicator cards, he gave to the Deutsches Museum. An indicator card is a pressure-volume diagram, a curve that shows how pressure changes with changing volume as the piston moves inside the cylinder of a heat engine. The area enclosed by the curve measures the work done inside the engine.

[10]Sass, pp. 431–81, gives a good summary of the development. Diesel, *Entstehung*, gives more details. Most popular accounts of diesel development are legendary. The most widely repeated legend is that the engine exploded and nearly killed the inventor. What actually happened was that the indicator broke under the high pressure (80 atmospheres) of the first firing of the engine.

pressure, in pulses), (5) mix the fuel with air in the very short time available, and (6) ignite the mixture. These problems may not sound very challenging, considering the experience engineers had already had with steam, hot-air, and gas engines by 1893, but they were all well beyond the state of the art at the time. Diesel certainly underrated their novelty and difficulty.

One set of problems, caused mostly by the need to achieve and contain very high pressures, was in the mechanical domain of pistons, cylinders, valves, pumps, compressors, and all the associated plumbing. Diesel had already had ten years of experience in this field (with refrigeration machinery) and the support of a large firm of engine builders with excellent resources and skills. He also had a well-developed technology of steam, air, and hydraulic systems to draw on, as well as an internal-combustion engine technology that had been developing for twenty or thirty years (to which Diesel had not paid much attention). His achievements in this field were mostly refinements and extensions of existing technologies, and I will pass over them with the remark that everybody else seems to pass over them too—they do not have the dramatic or theoretical interest that attracts writers and readers. They ought not to be passed over, because this sort of activity, unglamorous and frustrating, is at the heart of the development process. I am sure that Diesel would have preferred not to bother with this type of problem himself. I think of him as the man in the white coat in the German laboratory who prefers to leave the routine work to the man in the gray coat. But Diesel put a great deal of time and effort into the mechanical details himself.

Another set of problems was in the area of fuels and combustion, and here Diesel was operating in the dark, with many unknowns and no guidance from theory. He did not even know what it was that he needed to know, which was the principles of combustion. At the beginning he probably assumed that almost any fuel would naturally flash into flame as soon as it entered the combustion chamber if only the air were hot enough. But he found it was not so easy. Fuels behaved in mysterious ways. When he could not get a kerosene to ignite, for example, he turned to gasoline, which common sense says ought to be easier to burn. But it turned out that kerosene burned better than gasoline, no one knew why. Work in this area was wholly empirical, with the simplest cut-and-try methods. It was also experimentally difficult because success depended on so many variables interrelated in sensitive ways, and because the combustion event of interest was an extremely brief one. In a slow one-cylinder engine (say, 120 revolutions per minute) the pulses of combustion would come at one-second intervals. Each pulse might last one-fiftieth of a second, and Diesel had promised a gradual, controlled burning in this interval.

The key problem in getting a diesel engine to run, we now know, is to get a good mixture of fuel and air inside the cylinder in the very short time available. A gasoline engine has combustion in very brief pulses too, usually much briefer than a diesel, but it mixes fuel and air at leisure outside the cylinder and draws in a charge already mixed. In a diesel the mixture cannot be prepared in advance. The fuel has to be driven into the combustion chamber at the time of maximum compression, and somehow get pulverized or atomized so that each particle of fuel can find a particle of oxygen to combine with in a small fraction of a second. Successful combustion depends on the size and location of the jets of fuel and the force behind them, the shape of the spray and the degree of atomization, and the shape of the combustion chamber. Different sizes of cylinder, different speeds, and different types of fuel all require different tactics of injection. After seventy years of development there are still plenty of uncertainties in this area, and in the first decade getting an engine to run smoothly was mostly a matter of accidentally hitting on the right combination of fuel, injection mechanism, and shape of combustion chamber, and sticking to the right formula with fingers crossed. The engine might falter with the slightest change, or with no apparent change at all.

In these first four years, Diesel tried nearly everything to get reliable ignition: different fuels; different ways of mixing fuel ("external" and "internal" carburetors); and different kinds of artificial ignition (spark, flame, and hot-tube).[11] For fuel injection he began by trying to drive the fuel into the combustion chamber by some sort of plunger or pump, but after a year or more of work he concluded that this kind of injection (called solid injection) was impossible, and turned to a compressed-air technique. Eventually this turned out to be a good solution: the blast of air not only carried the fuel into the cylinder, but it also helped atomize the fuel and mix it violently with the air inside. But it was expensive: it required extra pumps and cylinders of air which added weight and plumbing complexity to the engine, and absorbed a good deal of its power. Air injection made the diesel a large and expensive engine, unable to compete in the market for small powers.

Diesel's activity during this period could be classified as either invention or development. Looking back later, Diesel regarded it as

[11]Compression ignition, now usually thought of as the defining characteristic of the diesel, was not essential to the process as Diesel viewed it (see Diesel, *Entstehung,* pp. 3–4). It was foreseen by Carnot in 1824, and attempted by a number of inventors before Diesel.

invention. An invention, he said, is not a pure idea, but rather the product of a struggle between Idea and Nature. The product is always a compromise between the Ideal and the Attainable, and working out the details of this compromise is a part of the process of invention.[12] To me this work looks more like what I would call development: the messy job of constructing something new and trying to make it work, adjusting something and trying again, changing dimensions, working out the details of piston ring or valve or gasket, coping with unexpected difficulties, retreating from one approach and trying a new one, giving up for a while and groping for guidance from books and journals and people—that sort of thing.

At the same time, every day, Diesel was deeply involved in theoretical work, for the first experience with a real engine had revealed new phenomena that had to be accounted for, and forced a revision of his original theory. The real engine had to be rationalized. An invention as described in a patent may be the embodiment of a permanent, fixed idea, but in real life ideas may not stay fixed. In this case, at least, the development modified the invention.

Diesel found that to get a practical engine he had to use much more fuel than his theory of isothermal combustion called for, and with more fuel the temperature had to go up. So he wrote a justification for using a constant-presssure instead of a constant-temperature process and tried to reconcile it with his patent. He also discovered at once that the engine had to be cooled, and the cooling required an explanation, because the patent and the book had said that cooling would not be necessary because no heat would be wasted: all the added heat would be turned into work on the piston. So in addition to keeping a running account of his work in the journal, Diesel wrote 150 pages of theoretical discussion intended for publication in the next edition of his book, which never appeared, and a curious rationalization of cooling as a requirement of thermodynamic law.[13]

For this theoretical work that accompanied the development Diesel had two incentives. One was the normal human need to explain puzzling phenomena and to publish the explanation. This need Diesel felt rather more keenly than most men, I think: he seemed to have a neurotic need to rationalize what he was doing. The other incentive was to protect his economic interests. The plan was to exploit the engine primarily through the sale of patent rights. Before develop-

[12]*Entstehung,* pp. 1, 151.

[13]The manuscript intended as a supplement to the 1893 book, entitled "Nachträge zur Brochüre: Theorie und Construction . . . ," is in the Diesel papers in the Deutches Museum. Diesel also applied for a supplementary patent (see n. 6 above). The rationalization of cooling is repeated in Diesel's 1897 paper cited in note 16.

ment began, the book had been attacked in public as describing an unworkable process.[14] The success of the whole enterprise seemed to depend on the validity of the patent. One of Diesel's sponsors, Krupp, was already uneasy about it, and other prospective licensees had expressed misgivings.[15] So Diesel felt a strong pressure to reconcile the real engine as it was developing with the original theory as expressed in the patent. The theoretical work was an essential part of the exploitation of the engine.

This early development experience also revealed certain economic characteristics of the new type of heat engine that influenced the marketing plans being developed in this period. High thermal efficiency—that is, fuel economy—was supposed to give the diesel its chief competitive advantage over steam, but its versatility was also an important selling point. Diesel thought of it as a universal heat engine, adaptable to any fuel, solid, liquid, or gas, and practical in any size, so that it could drive anything from a sewing machine to a battleship. But the extreme sensitiveness to fuel revealed in the development limited the range of applicability of the diesel, and the need for air injection made it uneconomical in small sizes. Diesel remained optimistic on these points all his life: in papers and speeches and in some experimental work he continued his plans to exploit different fuels—and to develop different sizes and applications of his engine. But for the prudent businessman the process of development revealed characteristics of the engine that limited its economic role. It has remained an oil engine to this day, and it was twenty-five years before the development of direct injection made the small diesel practical.

Third Period: Premature Innovation, 1897–1902

By early 1897 Diesel felt that the main problems were solved and it was time to turn over the final details of development to lesser talents. In the spring of 1897 a prototype engine was running smoothly on the Augsburg test stand where it was inspected by engineers and businessmen from many countries. Its performance was measured and evaluated by the eminent professor Moritz Schröter, the results were published, and the triumph was announced with appropriate ceremony in June 1897 at the annual meeting of the Society of German Engineers. The engine was formally introduced as fully developed, ready to be sold.[16]

[14]Speech of Otto Köhler at a meeting of the VDI in Cologne, April 10, 1893, reported in *ZVDI* 37 (1893): 1103–9.

[15]Diesel-Krupp correspondence in Diesel papers, Deutsches Museum.

[16]Rudolf Diesel, "Diesels rationelle Wärmemotor," *ZVDI* 31 (1897): 785–821; and M.

In 1897 Diesel devoted himself chiefly to promotion. The policy was to market the engine by selling patent rights to established machine manufacturing firms in all industrial countries. The German market was divided among three firms, and licenses were negotiated for a dozen foreign countries, usually with the leading engine makers of the country,[17] except in America, where the brewer Adolphus Busch acquired the rights for a million marks cash and organized a new firm to exploit the Diesel patents.[18] Diesel did most of the negotiating himself. The arrangement usually provided for a royalty, with a large advance payment in cash. Augsburg supplied drawings and technical advice, and the licensees agreed to exchange information on technical improvements (so development was expected to continue into the innovation period).

The licensees began at once to build prototype engines. Four of them managed to get engines running, more or less, at an exhibition of machinery in Munich in 1898.[19] The first diesel in the United States, a German import with a German engineer in attendance ran in public for a month in May–June 1898 at an exhibition of electrical machinery in Madison Square Garden, New York.[20]

Diesel himself was badly overworked during this critical period, on the verge of a nervous breakdown, trying to do everything at once. He wanted the Maschinenfabrik Augsburg to move at once into the manufacture of diesel engines in quantity, but Buz was not willing to shift his plant over to the new engine fast enough to suit Diesel. Therefore Diesel formed two new companies, with the cooperation of Buz and Augsburg bankers: one to manage his contracts and patent rights and coordinate the activities of the licensees, and one to manufacture and sell engines in a part of the German market.[21]

All these enterprises failed. No licensee was able to build an engine that would run reliably in the hands of a customer, not even with blueprints from Augsburg and borrowed Augsburg mechanics.

Schröter, "Diesels rationelle Wärmemotor," *ZVDI* 31 (1897): 845–52. Diesel's paper was widely noted in the technical press. In England a "summarized translation" by Bryan Donkin appeared in the *Engineer* 84 (1897): 364–66, and in the United States a full translation was published in *Progressive Age* (December 1 and 15, 1897, and January 1 and 15, 1898).

[17]Sass, pp. 481–82.

[18]The best account of the diesel in America is Eugen Diesel and Georg Strössner, *Kampf um eine Maschine* (Berlin, 1950). See also Richard H. Lytle, "The Introduction of Diesel Power in the United States 1897–1912," *Business History Review* 42 (1968): 115–48.

[19]Sass, p. 488.

[20]Diesel and Strössner, p. 53; Sass, p. 490.

[21]Sass, pp. 484–86, 492–93.

Diesel's own manufacturing firm took back every engine it made and went bankrupt. The diesel got a bad name that set back development by several years and made innovation difficult.

Diesel had stopped keeping the development journal himself in early 1897, when he became preoccupied with the business and the public-relations side of the enterprise, so that it is not easy to say exactly what he contributed to the continuing development. But he was clearly more concerned with distant and visionary projects than with the incremental improvement of the existing engine. He was always talking about new applications—to automobiles, locomotives, ships, and later even aircraft—and planning new types of engines burning new fuel—powdered coal, blast-furnace gas, even peanut oil. In 1898 and 1899 he found time for some experimental work himself. He actually built and operated a monstrous three-cylinder compound diesel of the type described in the patent as the preferred form for the ideal Carnot engine. It proved to be less efficient than the simpler compromise engine and was scrapped. He also attempted a coal-burning engine, and worked on a gas-burning type of special interest to Krupp.[22]

This work came at the lowest ebb of the diesel's reputation. Buz finally cut it short at the end of 1899 and began the long task of developing the 1897 engine into a reliable oil engine, and the longer task of rebuilding the diesel reputation. Diesel had his nervous breakdown and made no more important contributions to hardware development.

Meanwhile, main-line development continued in Augsburg. A two-cylinder 60-horsepower engine (a large step forward in power) was built and sold to a match factory in Kempten, which was managed by Buz's brother. This was the first realistic trial: the engine replaced a steam engine supplying power to a factory. Two experienced mechanics went along, to tend the engine during the day and to maintain and overhaul it at night. From this experience the engineers learned what kind of cooling and lubrication was necessary, and how to control the engine under a fluctuating load. Fuel injection remained the key problem. The system they were using formed a spray by forcing the oil through a mesh of brass wire. It worked well enough if they took it apart and cleaned it every night. After about a year of this experience, they dismantled the engine and shipped it

[22]Rudolf Diesel reports on development during this period in "Mitteilungen über den Dieselschen Wärmemotor," *ZVDI* 43 (1899): 36–42, 128–30. A translation appeared in *Progressive Age* (May 1 and 15, 1899). Imanuel Lauster, who was in charge of diesel development for MAN, covers this period in pp. 8–18 of a manuscript "Die Entwicklung des Dieselmotors" in the library of the *VDI* in Düsseldorf.

back to Augsburg for a complete redesign and rebuilding.[23] The rebuilt engine might be called the first successful diesel.

One of Diesel's uncharitable critics at the 1912 meeting, Alois Riedler, a distinguished professor and an experienced machine designer himself, said some interesting things in the discussion period about engine development, and development in general. He divided the process of developing the diesel (or anything else) into three stages. First, he said, is the development of a machine that works, an engine that runs (that is *gangbar*). The second stage is the development of an engine that is useful (*brauchbar*), that can carry a load, reliably, for a reasonable time. This is a quite different thing from the *gangbar* engine, he said, and it may be a greater and more costly achievement. But this is not the end of development; a third stage, the most important of all, is necessary to make the engine *marktfähig,* that is, ready to be sold, able to take its place and hold its own in the existing economic system. Diesel's 1897 engine, said Riedler, with the advantage of hindsight, was barely *gangbar*.[24]

In any case, the judgment that the 1897 engine was fully developed, ready to be sold, was a disastrous mistake. Such an error of judgment might be expected from an overoptimistic inventor, but in this case the hard-headed businessman, the practical mechanic, the professor of engineering, and many of the expert observers agreed. It is extraordinary that so many good people could be so wrong.

The lesson, I suppose, is that development takes longer than anyone expects. It needs to continue well into the innovation phase. The inventor tends to underestimate the need for development, and the developer is ready to pronounce his work finished too soon. For some kinds of products, at least, a realistic trial in the hands of the ultimate user is a necessary part of development, necessary to reveal problems that cannot be anticipated in the laboratory. The mechanical and economic performance of a complex and sensitive system like a diesel engine cannot be safely predicted from laboratory simulation or theoretical analysis, but only from prolonged experience on the job.

Fourth Period: Continuing Incremental Development, 1902–8

My last period is a time of incremental progress in technical development and of steady, slow growth in numbers of engines in use.

[23]Sass, pp. 505–11; Lauster, pp. 10–12.

[24]I am paraphrasing Riedler's analysis of the process of development offered in the discussion period following Diesel's 1912 lecture in *Jahrbuch der Schiffbautechnischen Gesellschaft* 14 (1912): 356–57 (see also Riedler [n. 2 above], pp. 4–5; and Meyer, *Beiträge zur Geschichte des Dieselmotors,* pp. 3, 16–19). Meyer was present at the 1912 lecture but did not participate in the open attack on Diesel.

Call it both development and innovation, running parallel. The period begins with one more or less reliable engine offered for sale by the Augsburg firm, and ends in 1908 when the basic patent runs out and the Augsburg plant drops steam to devote itself to the manufacture of diesels. By this time a thousand engines are in service, averaging perhaps 50 horsepower, mostly replacing steam engines in small stationary power plants.[25] The licensees that had dropped the diesel around 1900 picked it up again, one by one, and by 1908 a dozen or more firms in half a dozen countries were contributing to the development, providing a considerable diversity in form and variety of detail: good raw material for the forces of natural selection to work with.

The mechanical development of the engine in this period is the kind of continuous refinement of design that I normally associate with the word "development": improvements in detail that make for reliability and economy, slow growth in power and speed of individual units, weight reduction, and simplification of design to reduce cost of manufacture. This type of development is controlled throughout by strong economic pressures, especially strong for the diesel because it was competing with the mature and reliable steam technology. It was not a wholly new thing like an airplane or computer. It did exactly the same job as the steam engine it replaced, and its initial cost was higher. It had nothing to offer a buyer except an uncertain fuel economy, and in some applications a certain amount of convenience. Nevertheless the diesel gradually won itself a modest place in the stationary power market in the intermediate range of say 20–100 horsepower, largely on the promise of fuel economy. It was by no means the universal rational heat engine that Diesel had in mind in 1890, but it was clearly a viable species of engine in 1908.

Later Development

The Diesel story has many more chapters, in which I think I can see more or less the same pattern of the interacting forces that I am calling invention, development, and innovation. The most interesting chapters cover the development of special forms for special purposes, such as power for ships, heavy road vehicles, and locomotives. The general lesson that I see in these chapters is that each of these special applications requires what really amounts to a new species of engine, and each species has to go through its own independent evolutionary

[25]This is a conservative guess. An MAN promotional booklet in volume 15 of a collection of pamphlets entitled "Gas Engines" in the Detroit Public Library lists all MAN engines delivered and ordered to March 1909. The total is 1,298 engines amounting to 104,952 horsepower.

process. Diesel himself assumed that once the basic problems were solved, a change of application or a change of scale was a rather simple matter. He was wrong. The period of development for each of these species was surprisingly long—say twenty years—but once it was fully developed, it took over the field surprisingly fast: ships in the 1920s, trucks in the 1930s, and locomotives in the 1950s.

* * *

So I conclude that the three kinds of human behavior that we label invention, development, and innovation are going on more or less all the time in the process of technological evolution. The emergence of new ideas, which we label "invention," clearly takes place throughout the process. The refinement of design that comes with experience in the real world, which we label "development," is endless. The effort to fit a new technique into the existing economic and social structure, which we label "innovation," is a guiding force from the beginning.

In considering the creative process, we naturally think of the idea as coming before its object; we conceive of a machine as the realization of a preexisting idea. Our institutions, our language, and our patent law reenforce this impression. Our language seems to require that a machine be conceived, constructed, and operated, in that order. But in real life the idle curiosity, the tendency to play with ideas and to try new ways of doing things just for fun, which we regard as the special virtue of the inventor, is also a virtue for the developer and the innovator. The experience of development is in fact a great generator of new ideas, and it can modify the idea being developed, as it did for Diesel. The experience of innovation, the effort to find an economic role for a new machine, can reveal problems unsuspected in the laboratory, and redefine the task of the developer. The economic considerations, which are the special field of the innovator, are usually of great interest to the inventor. Economic questions may be suppressed for the moment while a developer devotes himself to the urgent task of getting the thing to run regardless of cost, but they cannot be put off for long.

I suggest that the model of some sort of dialectical interaction among these three types of forces may be a more realistic description of what is going on in technological evolution than the conception of chronological stages.

The Emergence of Basic Research in the Bell Telephone System, 1875–1915

LILLIAN HODDESON

By the 1920s, some of the largest industrial firms in America had come to regard scientific research as essential to their continuing success.[1] Directors of such major corporations as American Telephone

DR. HODDESON, associate professor in the Department of History at the University of Illinois at Urbana-Champaign, is also the historian at the Fermi National Accelerator Laboratory in Batavia, Illinois. The present article was part of a detailed investigation into the historical development of the transistor which culminated in the book, coauthored with Michael Riordan, *Crystal Fire: The Birth of the Information Age* (New York: W. W. Norton & Co., 1997). The author would like to thank Charles Gillispie for the initial encouragement of this program; Spencer Weart, Charles Weiner, Joan Warnow, Gordon Baym, George Schindler, Mort Fagen, Arthur Norberg, and three reviewers for *Technology and Culture* for valuable criticism; Carl Thurmond, Joseph Burton, Philip Platzman, Viola Graeper, Michael Majorossy, Bernard English, Mary Lynn Toto, and Young Hi Quick for generous research help throughout; and Stanley Goldberg and Gerald Tyne for useful correspondence. She would also like to acknowledge assistance from the American Institute of Physics Center for History of Physics, Princeton University (Program in History and Philosophy of Science), Bell Laboratories in Murray Hill, Rutgers University (Physics Department), the University of Illinois at Urbana-Champaign (Physics Department), and the American Council of Learned Societies.

[1] The National Research Council–National Academy of Sciences publication, *Industrial Research Laboratories of the United States* (Washington, D.C.), lists 296 research laboratories in March 1920, 1,575 in August 1933, and 1,769 in December 1938. For additional growth statistics, see *Research–a National Resource, Part II, Industrial Research: Report of the National Research Council to the National Resources Planning Board* (Washington, D.C., 1941) (hereafter cited as *Research–a National Resource*). Spencer Weart points out in "The Physics Business in America, 1919–1940: A Statistical Reconnaissance," *The Sciences in the American Context: New Perspectives*, ed. Nathan Reingold (Washington, D.C., 1979), that in 1900 only one-tenth of the Ph.D.'s graduated each year entered industry within a few years of their graduation, while by the 1920s one-fifth of a much greater number of Ph.D.'s were going into industry. General background and references on the development of American industrial research can be found in David F. Noble, *America by Design: Science, Technology, and the Rise of Corporate Capitalism* (New York, 1977); Howard R. Bartlett, "The Development of Industrial Research in the United States," *Research–a National Resource;* Kendall Birr, "*Science in American Industry*," in *Science and Society in the United States*, ed. David D. Van Tassel and Michael G. Hall (Homewood, Ill., 1966), and *Pioneering in Industrial Research: The Story of the General Electric Research Laboratory* (Washington, D.C., 1957); W. David Lewis, "Industrial Re-

and Telegraph, General Electric, DuPont, and Eastman Kodak had established in-house groups devoted to fundamental scientific studies.[2] The impact of science and technology on industry (as well as on each other) was now more direct than ever before, and reciprocally, industrial policies would play a major role in shaping both science and technology.

The emergence of the industrial research laboratory in America poses many important historical problems, including: When and why did particular firms become committed to research? How did the earliest industrial laboratories develop? What relationships existed between different laboratories? How did the social, economic, political, and technological forces acting in common upon these laboratories during their formative years affect them? Which technological or scientific problems were particularly important? Who were the most influential individuals, and what roles did they play?

The purpose of this paper is to document the development of basic

---

search and Development," *Technology in Western Civilization,* vol. 2, ed. Melvin Kranzberg and Carroll W. Pursell, Jr. (Oxford, 1967), pp. 615–34; Edward Weidlein and William Hamor, *Science in Action: A Sketch of the Value of Scientific Research in American Industries* (New York, 1931); A. P. M. Fleming, *Industrial Research in the United States of America* (London, 1917); C. E. K. Mees and J. A. Leermakers, *The Organization of Industrial Scientific Research* (1920; reprint ed., New York, 1950); and Courtney Robert Hall, *History of American Industrial Science* (New York, 1954). A convenient summary appears in chap. 1 of Leonard Reich, "Radio Electronics and the Development of Industrial Research in the Bell System" (Ph.D. thesis, Johns Hopkins University, 1977).

[2] General Electric established its research laboratory in Schenectady, N.Y., in 1900, under the Leipzig-trained MIT chemistry professor Willis Whitney; DuPont established its research laboratory in Repauno, N.J., in 1902, under the Heidelberg-trained chemist Charles Reese; and Eastman Kodak established a research laboratory in Rochester, N.Y., in 1913, under the London-trained chemist C. E. Kenneth Mees. The Bell System's research activities, discussed in the present study, were consolidated in 1907, with major responsibility delegated to the Chicago-trained MIT physicist Frank Jewett. The relationships among the directors of these early laboratories, many of whom knew one another, deserve scholarly attention. Oliver Buckley (Bell Laboratories' president from 1940 to 1951) recalls that between 1902 and 1904, Whitney (G.E.), Frank Jewett (AT&T), and William Coolidge (then at MIT and after 1905 at G.E.) attended a symposium at MIT led by A. A. Noyes and attended by Charles Cross, then head of MIT's Physics Department. Buckley suggests that, through Noyes's contacts with Reese of DuPont and Cross's contacts with the executives of American Bell and G.E., "an idea, developed in this symposium, led to a new movement in those industries that later got labeled as industrial research." Furthermore, as Buckley points out, Noyes, Reese, and Whitney had all studied in Germany "and knew something of the utilization there of young Ph.D. chemists in industry." See Buckley, "Some Implications of Expanding Industrial Research," talk delivered in December 1945 to American Institute of Chemical Engineers (Bell Laboratories Collection, Warren, N. J., hereafter cited here as BLC).

research[3] in one prominent industrial firm, the Bell Telephone System,[4] and to construct from the data a "first approximation" model of the principles that govern the relationships between science, technology, and industry in the process of establishing scientific research in such an industry. Previous historical accounts have dealt with the role of external influences, both specific, such as the possibility (realized

[3] "Fundamental" and "pure research" refer to the attempt by experimental and theoretical means to understand the physical underpinnings of phenomena. The special term "basic research" refers here to fundamental studies carried out in the context of industry, which may lead to, but do not aim primarily at, application. Applied research, on the other hand, which encompasses engineering and technology, does aim primarily at practical application. Many distinctions between "fundamental," "pure," and "basic" research have been discussed. For example, Nathan Reingold, in "American Indifference to Basic Research: A Reappraisal," *Nineteenth Century American Science,* ed. G. Daniels (Evanston, Ill., 1972), p. 45, associates "pure research" with a "psychological motivation" and employs the concept of "basic research" only when merit is a factor in the evaluation of the research. On the other hand, Harvey Brooks notes in "Applied Research: Definitions, Concepts, Themes," in *Applied Science and Technological Progress: A Report to the Committee on Science and Astronautics, U.S. House of Representatives, by the National Academy of Sciences* (Washington, D.C., June 1967), pp. 21–25, that the same work a performing scientist may view as basic may be seen as applied by the sponsoring agency. A significant portion of the critical literature on the science-technology relationship can be traced in *Technology and Culture (T & C),* e.g., Robert Multhauf, "The Scientist and the 'Improver' of Technology," *T&C* 1 (1959): 38–47; David Gruender, "On Distinguishing Science and Technology," *T&C* 12 (July 1971): 456–63; and almost the entire October 1976 issue, particularly Thomas Hughes, "The Science-Technology Interaction: The Case of High-Voltage Power Transmission Systems," pp. 646–59; Otto Mayr, "The Science-Technology Relationship as a Historiographic Problem," pp. 663–72; and Edwin Layton, "American Ideologies of Science and Engineering," pp. 688–70.

[4] The literature on early research in the Bell System can be divided into two categories: works that discuss specific devices or particular episodes, and works that consider the general questions of how and why research developed. An outstanding study of a particular development is James E. Brittain, "The Introduction of the Loading Coil: George A. Campbell and Michael I. Pupin," *T&C* 11 (January 1970): 36–57. The extensive volume, *A History of Engineering and Science in the Bell System: The Early Years (1875–1925),* ed. M. D. Fagen (Warren, N.J., 1975), is a reliable source of detailed information on many of Bell's early engineering developments and is recommended as an introduction to early telephone technology. Studies which address the general question of the emergence of Bell's basic research include: Reich; Stanley Goldberg, "Basic Research in the Industrial Laboratory: The Case of Bell Telephone Laboratories" (unpublished MS); N. R. Danielian, *AT&T: The Story of Industrial Conquest* (New York, 1939); and the reports of the Federal Communications Commission, *Report on Telephone Investigation (Pursuant to the Public Resolution No. 8, 74th Congress) House Document #340,* 76th Cong., 1st sess., 1939 (hereafter cited as FCC), along with the seventy-seven volumes of associated staff reports (hereafter cited as Exhibits) from the investigation of the telephone industry that began in 1935 and continued through 1937, available in the legal department at AT&T, New York. Exhibits of interest to this study are 1951-A (1879–

about the time that Bell established its first research division) that radio telephony would be developed by outside researchers,[5] and more general, such as the growth of science and technology in the larger community, the emergence of the professional engineer, and the concurrent rise of industrial capitalism in America.[6] In this study, we seek to unravel partially the knot (which other writers—see Noble—consider impossible to untangle) of technological and social concerns, and pinpoint the particular technological factors that initiated the establishment of a permanent research effort within the Bell Company.

The principal finding, discussed in the first section, is that Bell's research program had no sudden beginning, but was a steadily evolving interior growth, rooted in the 1880s, which responded to particular telephone problems including: attenuation (loss of energy of telephone signals during transmission), cross-talk (transfer of speech energy from one circuit to a parallel circuit), and interference with other electrical systems. As such problems grew more complex, the company came to recognize the usefulness of in-house scientists and by 1907 had shifted its official view of in-house research from tolerant to strongly supportive. The Bell System's permanent commitment to research came in 1907 with the consolidation of its research activities in the Western Electric Company and AT&T, and was established institutionally in 1911 with the creation of its first research branch.

In the second section, we show that the direct technological motivation for establishing in-house research came from the need to develop a repeater (or amplifier) suitable for use on transcontinental lines. However, commercially defined objectives (including the company's goal to achieve a "universal telephone system" or to maintain its dominant position in telephone communication) and social developments (such as the emergence of industrial research in America) gave rise to the technological problems that led to the research. And the commercial and social success of the repeater also strongly re-

1918) and 1951-B (1918–35), vols. 1 and 2 of *The Engineering and Research Departments of the Bell System* and *1989, Patent Structure of the Bell System.* Both FCC and Exhibits contain important documents, but the text, written at a time when Congress was interested in attacking big businesses, especially monopolies, is strongly biased against AT&T and is marred by considerable error, as pointed out in *Brief of Bell System Companies on Commissioner Walker's Proposed Report on the Telephone Investigation* (New York, 1939) and noted in the *Wall Street Journal* of April 15, 1938.

[5] Goldberg, Reich, Danielian, FCC, and Exhibits.

[6] Noble.

inforced the company's movement toward increased support of basic research.

These results can be incorporated into the following set of three principles governing the development of scientific research in the Bell System, which may be useful as a working hypothesis in studying other examples: (1) nonscientific objectives lead the company to particular technological problems; (2) technological problems, so profound or complex that the usual approaches to such problems fail, lead the company to seek deeper understanding of the underlying physical phenomena; and (3) research that is successful in technological terms reinforces the company's commitment to scientific studies in the particular area. Our detailed emphasis here is limited to the relationships between science and technology; a fully satisfactory account of the relationships between the company's larger objectives and decisions and its technological goals requires an analysis of business factors beyond the scope of this paper.

Let us first attempt to locate the start of the company's support of in-house basic research by surveying its technical activities between 1875 and 1907.

Early Technical Activities in the Bell System, 1875–1907

The start of the Bell Company was the funding and patenting agreement signed in February 1875.[7] According to that agreement, Thomas Sanders and Gardiner G. Hubbard would support Alexander Graham Bell's experiments to build a device for transmitting speech along wire in return for a share in any patents that might emerge. None of the three was a scientist. Sanders, the father of one of Bell's deaf students, was in the leather business; Hubbard, Bell's future father-in-law, was an attorney; and Bell, an amateur inventor, was a professor of vocal physiology concerned with teaching his father's system of visible speech to deaf-mutes. Although Bell became quite knowledgeable in certain scientific areas (he attended many scientific lectures, read numerous scientific texts, and interacted with scientists, including MIT's Charles Cross, Harvard's Joseph Lovering, and inventors such as Moses Farmer, who could help him to deepen

[7] This agreement, the basis of the Bell Patent Association, was the first of seven corporate steps in the growth of the Bell System that culminated with the absorption in 1899 of the American Bell Telephone Company by AT&T (Exhibit 1360-A, p. 8). Most of the references on the early history of the telephone and telephone system pertinent to this study are contained in Robert V. Bruce, *Bell: Alexander Graham Bell and the Conquest of Solitude* (Boston, 1973); Fagen; Prescott C. Mabon, *Mission Communications: The Story of Bell Laboratories* (Murray Hill, N.J., 1975), and *Events in Telephone History* (New York, 1974) (hereafter referred to as *Events*).

his understanding of scientific subjects[8]) his primary concern was to apply science, not create it.

Neither was Bell's technical assistant, Thomas Watson, a scientist. Having left school at age thirteen, and after trying clerking, bookkeeping, and carpentry, he eventually found a job as an instrument maker in Charles Williams's shop in Boston, which specialized in building electrical devices including call bells, fire alarms, telegraph keys, and batteries.[9] According to Watson, the Williams shop was "one of the largest and best fitted in the country," an environment where "individual initiative was the rule." It attracted "wild-eyed inventors, with big ideas in their heads and little money in their pockets," including Thomas Edison, Moses Farmer, and Alexander Bell.[10] Although Watson's electrical instrument work in Williams's shop brought him to read and think about scientific concepts,[11] such informal study was not basic research aimed toward new understanding of the phenomena.

The telephone experiments of Bell and Watson, which began in 1875,[12] drew heavily on the science that both men had picked up, and received continuing input from available scientists (including Cross and a group of professors at Brown University).[13] But their striking success, in March 1876, when Bell transmitted the famous first telephone message, "Mr. Watson, come here, I want you," was defined in

[8] Bell's notebooks, esp. "Electrical Experiments," Alexander Graham Bell Family Papers, Manuscript Division, Library of Congress, Washington, D.C. (hereafter referred to as AGBF). Among the scientific works Bell studied were Helmholtz's *On the Sensations of Tone* and Tyndall's books on electricity and acoustics. The scientific lectures included those on experimental mechanics by Cross, who in 1882 was responsible for establishing the first course in the United States leading to a degree in electrical engineering (*Who Was Who in America,* vol. 1, *1897–1942* [Chicago, 1942], p. 279). Bell subsequently discussed many physical subjects with Cross including tuning forks and other acoustical devices. Cross later offered Bell the use of apparatus and laboratories at MIT (Bruce, pp. 50–119, 131–32).

[9] Thomas Watson, "The Birth and Babyhood of the Telephone," address at the Third Annual Convention of the Telephone Pioneers of America, Chicago, October 17, 1913 (AT&T pamphlet no. 8–69).

[10] Ibid., pp. 3, 4; Bruce, pp. 92–94, 132–36.

[11] Watson, p. 3, mentions that he read Davis's *Manual of Magnetism,* published in 1847.

[12] Initially, Watson worked on the telephone on a half-time shop assignment. In September 1876, he contracted with Hubbard to devote his time to the harmonic telegraph and telephone in return for one-tenth interest in the Bell patents (*Events,* p. 3).

[13] *Brown Alumni Monthly* 39 (May 1939): 279–82. Brown professors, Eli Whitney Blake, John Pierce, and William F. Channing were helpful in the development of the telephone transmitter.

practical not scientific terms;[14] three days earlier Bell had obtained the basic telephone patent by which this technological success would be converted into financial terms.[15]

When the telephone industry started in May 1877, the subscribers were encouraged to make their own electrical connections and string their own wires, but for an additional fee, the proprietors would carry out these jobs for them.[16] As the telephone system expanded, the company assumed an increasing number of technical engineering roles,[17] and it would be out of these that the Bell System's basic research eventually developed.

The early telephone users faced many practical difficulties including weak transmission and noise on the line, for at first the system lacked microphone transmitters, switchboards, suitable signaling systems, and satisfactory lines and cables.[18] To help cope with such

[14] As discussed by David Hounshell, "Bell and Gray: Contrasts in Style, Politics, and Etiquette," *Proceedings of the Institute of Electrical and Electronics Engineers* 64 (September 1976): 1305–14, and "Elisha Gray and the Telephone: On the Disadvantages of Being an Expert," *T&C* 16 (April 1975): 133–62, being an amateur rather than professional inventor gave Bell certain practical advantages over his most serious competitor, Elisha Gray, who required acknowledgment from his colleagues. James Clerk Maxwell referred to the telephone as "an instance of the benefit to be derived from the cross-fertilization of the sciences" (*Scientific Papers,* ed. W. D. Niven, 2 vols. [Cambridge, 1890], 2:751).

[15] This patent (no. 174, 465) of March 7, 1876, together with a second (no. 186, 787) that Bell obtained in January 1877, covering structural improvements of his device, generated the Bell System and maintained its monopoly on telephones until these patents expired in 1893 and 1894 (although not without an eleven-year struggle with Western Union) (*Events,* pp. 3, 4; "Organization of the Western Electric Company for Training Purposes" [hereafter "Organization of Western Electric"]; see Historical Library of Western Electric Company, New York City (hereafter referred to as WEL), and Historical Library of AT&T (hereafter referred to as ATTL).

[16] See the first telephone advertisement, May 1877, signed by Gardiner Hubbard and reprinted in Watson, p. 15.

[17] The early technical reports can be found in WEL and ATTL. Many relevant sections are reprinted in Exhibits.

[18] Edison took the first important step toward solving the telephone transmitter problem in 1877 with the discovery that granulated carbon was useful as the variable resistance element. His transmitter, developed for Western Union, "talked louder" than Bell's and therefore presented a considerable threat to the Bell Company (Watson, p. 22). But in 1878 Francis Blake, a recent addition to Bell's staff, developed an improved microphone transmitter, based on principles described by David Hughes, which put the Bell Company on a more equal footing with Western Union. In 1878, Henry Hunning patented his discovery that microphone action could be improved if loosely packed carbon granules filled the space between the diaphragm and a rear disk electrode. The powerful "solid-back transmitter" was not developed until 1890, when Anthony White on Hayes's staff provided microphone action by using granular carbon in a "button" chamber (White to Hayes, October 1, 1890, p. 3, cited in Exhibit 1951–A, p. 46; P. Norton to Kenneth Mark, "A Brief Statement regarding the Origin of Research

problems, the company put Watson in charge of technical services. He remained the only member of the technical staff until Emile Berliner joined in late 1878, soon to be followed in 1879 by George Lee Anders.[19] Berliner and Anders were both highly talented and productive inventors who had worked previously on telephone devices, but neither was an academically trained scientist.[20] Nor was Thomas Lockwood, who joined the technical staff in 1879 as assistant inspector and electrician and soon began to devote an increasing portion of his time to patent matters.[21]

The documents show no clear-cut evidence of basic research in the late 1870s or early 1880s, but there are suggestions. Watson was responsible for instruments and quality of workmanship, while Berliner had "charge of *general experiments* upon the transmitter and Magneto Telephone" and was expected "to make *original investigation,* in regard to speaking telephones generally," and "*special experiments* upon forms or modifications of Transmitters belonging to the Company" (italics added).[22] In 1880, William Jacques, with a recent physics Ph.D. from Johns Hopkins (1879) and graduate studies in Berlin, Leipzig, Vienna, and Göttingen, joined the staff.[23] He was "to have made under his direction, such *special experiments* as he may be directed to make, and *report on the same*" (italics added),[24] which he did in good pro-

---

and Experimental Work in the Bell System and the Nature of the Work Conducted at the Laboratories of the Parent Bell Company in Boston, Mass.," October 29, 1930, ATTL, box 1047; memorandum for J. Carty, January 20, 1930, by Outside Plant Development Engineer, BLC; Roger B. Hill, "Technical Records of the Parent Bell Company," June 26, 1946 ATTL: and Fagen, pp. 70–83).

[19] Alexander Bell had no active connection with the company after 1877, the year in which he married and went to England.

[20] Anders, who had a patent on bells for selective ringing on two-party lines, became committed late in 1880 to full-time work on these as well as switching devices. Berliner, a recent immigrant from Germany (1870), working in 1878 as a dry-goods clerk, had invented a loose-contact carbon microphone (1876) and a microphone transmitter (1877); he was the first to propose the use of an induction coil in connection with the telephone transmitter. Watson recommended that the Bell company buy the rights to Berliner's transmitter and hire him as a technical expert (*Events,* pp. 5–6; *Who Was Who in America;* Frederick William Wile, *Emile Berliner: Maker of the Microphone* [Indianapolis, 1926], esp. pp. 69–132; and *Dictionary of American Biography,* s.v. "Emile Berliner").

[21] Hill, p. 3.

[22] "Electrical Department," a memorandum from T. N. Vail to Lockwood, February 5, 1881, Exhibit 1951-A, app. C; see also Lockwood's testimony in the Western Union Company, American Bell Telephone Company, Case 187, Fed. 425, Exhibit B, "Evidence for the Defendant," p. 628, Exhibit 1951-A, pp. 2, 3.

[23] *Who Was Who in America,* vol. 1, *1897–1942,* p. 626, lists Jacques as an "Expert" for American Bell from 1880 to 1897 and a lecturer in electrical engineering at MIT from 1887 to 1890.

[24] Vail to Lockwood, 1881, Exhibit 1951-A, app. C.

fessional style.[25] Jacques's measurements and special tests (particularly on cables and pole lines) could have led him to study phenomena in depth. However, while the possibility cannot be dismissed that Jacques was the father of Bell's basic research, references to Jacques's accomplishments in the documents studied are inconclusive, and he does not appear to have influenced Bell's policy regarding research in the 1880s.

The company established its first formal technical department, the "Electrical Department," in 1881, the year Watson resigned; an "Experimental Shop" under Jacques organized in 1883 "supplemented" the new department's work. Tasks included inspecting the manufacture of telephone instruments; testing pole lines, transmitters, and receivers; and deciding which patents to buy. Here, too, opportunities for research must have arisen, and may well have been taken, but the documents studied do not report any examples in the 1880s; rather, they stress the department's practical work, such as Thomas Doolittle's experiments, starting in 1881, which led to extensive testing of "hard-drawn" copper wires, and in 1884 demonstrated the "practicability" of long-distance telephone communication on a successful test line between Boston and New York.[26]

In 1881, the Bell System took a crucial step toward standardizing telephone instruments by gaining control of the Western Electric Company and making it the sole manufacturing firm for the Bell System.[27] Western Electric, which had evolved out of the association in 1869 between Elisha Gray and former Western Union telegraph operator Enos Barton,[28] having on its staff the inventive Gray[29] and later E. A. Hill, a Ph.D. chemist,[30] appears to have supported some experimenting and perhaps a modest amount of basic research. However, until 1907, Western Electric's research activities were distinct from the parent Bell Company.

The list of functions of the "Mechanical Department," the new name given in June 1884 to the Experimental Shop when it merged with Charles Williams's shop, strongly suggests the possibility of re-

[25] *Electrical World* (October 1887), p. 182.

[26] FCC, pp. 181–82; Hill (who reports the date of the formation of the Electrical Department as 1880); *Events,* p. 11; and p. 155 of Roger B. Hill and Thomas Shaw, "Hammond V. Hayes: 1860–1947," *Bell Telephone Magazine* 26 (Autumn 1947): 150–73.

[27] The Bell Company acquired the licenses of Williams's shop and the Gilliland Electric Company (*Events,* p. 10).

[28] Frank Lovette, "Western Electric's First 75 Years: A Chronology," *Bell Telephone Magazine* 23 (Winter 1944–45): 271–87.

[29] Hounshell, "Elisha Gray"; and "Organization of Western Electric."

[30] Hounshell; and C. E. Scribner, "History of the Engineering Department," *Western Electric News,* vol. 8, no. 9 (November 1919).

search. The list includes designing and experimenting with the cast-iron receiver, the magneto and carbon transmitters, repeating coils, carbon granules (obtained from Edison's laboratory in New Jersey), hard-drawn copper long-distance wire, long line apparatus, insulation, underground cable systems, and protection from "strong currents" such as lightning, electric railways, or electric-light currents. Members were expected to study and report on inventions for commercial use.[31]

But here again, the documents stress practical functions and results, and whether any of the studies were basic is difficult to ascertain. The Mechanical Department's first superintendent, Ezra T. Gilliland, was not a scientist but a mechanic, electrician, and engineer, chosen on the basis of his earlier work on the construction of telephones and switchboards in Indianapolis.[32]

The character of the research began to change in the second year of the Mechanical Department when Gilliland, after his resignation in 1885,[33] was replaced by the Harvard-trained research physicist Hammond Vinton Hayes, who, like Jacques, had one of the first physics Ph.D.'s in the country (and the second awarded by Harvard). Hayes had studied under Alexander Bell's former friend and scientific advisor, Charles Cross, and while his position as director of the Mechanical Department appears to have allowed him little time for scientific studies, he was both interested in and capable of carrying out basic research.[34] His early telephone contributions were practical: installing a central common battery in the offices of the American Bell Company; directing the work leading to the solid-back transmitter; and developing a thermally operated "heat coil" switch for protecting telephone apparatus against stray currents. Among the devices that received considerable attention from Hayes's department in 1885–88 were the transmitter, batteries, and the metallic circuit. Elec-

[31] Hill; "Annual Report to Stockholders by Directors of American Bell" (1884), p. 11; Gilliland to Vail, February 3 and 18, 1885; Gilliland to Devonshire, December 16, 1885, pp. 1, 2; "Report of directors of American Bell Telephone Company," year ending December 31 1886, p. 8 (all cited in Exhibit 1951-A, pp. 33–36, 52, app. G); and "Annual Report of Mechanical Department," year ending 1885, Devonshire to Hudson, December 30, 1885, ATTL, box 1008.

[32] Events, p. 8.

[33] Ibid., p. 11.

[34] After graduating from Harvard College in 1883, Hayes studied at MIT and Harvard, from which he subsequently received both his master's degree and doctorate (Hill and Shaw; and Edward L. Bowles, "Hammond Vinton Hayes: August 28, 1860–March 22, 1947, Scientist, Pioneer, and Benefactor," *Proceedings of the Institute of Radio Engineers* 36 [April 1948]: 443–45 [hereafter cited as Bowles, "Hayes"]; and Edward Bowles to Walter Gifford, attachment "Hammond Vinton Hayes," April 14, 1947, BLC [hereafter cited as Bowles, "Gifford," at.]).

tromagnetic problems of the 1880s included: transmitting current with maximum efficiency; developing switching (it was then still necessary for subscribers to "ring off" before an operator removed a connection); diminishing interference and cross-talk; replacing the single-wire grounded circuits by two-wire all-metallic circuits; and (in response to legislative concerns regarding the appearance of telephone wires) placing the wires in large cities in underground metallic circuit cables.[35]

A number of the engineering problems of the 1880s (such as attenuation, cross-talk, switching, and interference by other electric systems), while ostensibly preventing Hayes's department from pursuing basic studies, proved to be the source of Bell's soon-to-be-instituted policy of supporting basic research. Confronting these problems, which were insoluble by cut-and-try methods, required an understanding of electromagnetic principles still on the frontiers of physics, in particular, the theory of transmission lines operating at high (i.e., vocal) frequencies.

This theory was based on the principles of electromagnetic wave propagation, discussed in Maxwell's treatise on *Electricity and Magnetism,* published in Britain in 1873 but not entirely confirmed experimentally until Hertz's experiments in 1888. Only in the 1880s did Lord Rayleigh and Oliver Heaviside begin to apply Maxwell's theory to high-frequency transmission problems:[36] in 1884 Rayleigh derived a relationship between frequency, capacity, and attenuation which predicted the limit of the range of telephony for current deep-sea cables to a distance of the order of 20 miles; and between 1881 and 1887, Heaviside independently developed a more extensive theory which furthermore took into account the effects of magnetic induction in relation to distortion and attenuation.[37]

But to apply the theories of Heaviside and Rayleigh required mathematical facility beyond that of anyone on Hayes's staff in the 1880s. Hayes began to express his belief that it would be profitable to hire a number of mathematically trained scientists who would address not only the immediate engineering problems but, as Hayes wrote in 1887 to John Hudson, president of AT&T, the "many problems daily arising in the broad subject of telephony which require solution but are

[35] Bowles, "Gifford," at.; Hayes to Hudson, February 11, 1886, p. 2; March 1, 1886, p. 1; April 3, 1886, p. 1; January 3, 1886, p. 7; January 10, 1887, pp. 2–6; January 3, 1888, pp. 3, 4; December 31, 1888, p. 5; Mechanical Department Letterpress (1890), 21:375, Exhibit 1951-A, pp. 53–66, 41–42; and Hill and Shaw, pp. 157–81.

[36] An earlier theory of William Thomson (later Lord Kelvin) was applicable to the transmission of low-frequency telegraph signals but not voice transmission.

[37] W. H. Doherty, "The Bell System and the People Who Built It: The Growing Years," *Bell Laboratories Record* 46 (March 1968): 77–83.

not studied as they will not lead to any direct advantage to ourselves." He continues, "All these questions should be answered and I write to ask you to allow me to broaden the field of our work to embrace such problems."[38]

Some of the electromagnetic problems of the 1880s were solved by men such as John Carty who were without formal scientific training but had ample experience in practical telephony. Carty, who had joined the company in 1879 at age eighteen as a switchboard operator, having given up his plans to go to Harvard due to a temporary vision impairment, proved to be an exceptionally gifted telephone engineer. For example, in 1881 he showed how to apply the two-wire metallic circuit (which had considerably less interference than the single-grounded wire circuit used previously) and how to operate two or more telephone circuits connected with a "common" battery. Carty's twenty-four patents issued between 1883 and 1896 included one on the first multiple switchboard, the bridging bell (which connected many telephones in parallel, making the party line possible), and the repeating coil "phantom circuit" (which creates three voice paths over two circuits). One of his most important discoveries, published in 1889, was that, under many conditions, the disturbances responsible for cross-talk are electrostatic rather than electromagnetic; this result enabled him to develop a method for minimizing cross-talk by suitably intertwining lines at particular points called "neutral points."[39]

By the 1890s, practical problems such as implementing the solid-back transmitter (see n. 18) and replacing the existing open-wire circuits (which required extensive support structures and were frequently interfered with during storms) with underground metallic circuits so overcommitted the Mechanical Department that the carrying out of basic scientific studies in the department was de-

[38] Hayes to Hudson, October 26, 1887, Mechanical Department Letterpress, 14:497, Exhibit 1951-A, p. 44. Lockwood reflected the same view several years later, referring to "the great additions constantly being made to the sum of electrical knowledge . . . the various branches of this knowledge being so inter-related that it is not wise not to keep a lookout over the whole field while of course giving particular attention to such matters as particularly concern our business" (T. Lockwood, "Annual Report," directors of American Bell Telephone Company, December 31, 1891, p. 8, Exhibit 1951-A), p. 64.

[39] A sizable collection of Carty's early experimental papers, including his important treatment of cross-talk, "A New View of Telephone Inducting" read to the Electric Club, November 21, 1889, are included in BLC. See also Frank Jewett, "John J. Carty: Telephone Engineer," *Bell Telephone Quarterly* 16 (1937): 160–77 (hereafter cited as Jewett, "Carty: Telephone Engineer"); Frederick Lieland Rhodes, *John J. Carty: An Appreciation* (New York, 1932); Bancroft Gherardi, "The Dean of Telephone Engineers," *Bell Laboratories Record* 9 (September 1930): 4–19; *Electrical Review* (November 30, 1889); and Fagen, pp. 203–7, 236–40, 476–502.

emphasized.[40] In 1892, after discussing the work on cross-talk, attenuation, impedance, distortion, and self-induction, Hayes wrote to Hudson: "I have determined for the future to abandon this portion of the work of this department, devoting all our attention to practical development of instruments and apparatus. I think the theoretical work can be accomplished quite as well, and more economically, by collaboration with the students of the [Massachusetts] Institute of Technology and possibly of Harvard College." In accordance with this resolve, he rejected the idea of undertaking research into the methods of transmitting speech because of the more immediate responsibilities associated with designing instruments and supervising telephone manufacture.[41]

But while focusing the department on "practical" developments, Hayes also began to add a few university-trained scientists to his staff. The first, hired in 1890 from Johns Hopkins University, on Rowland's recommendation, was John Stone Stone, who applied his advanced mathematical training first to the theory of voice transmission and subsequently to amplifiers and Hertzian waves.[42]

In 1897, Hayes hired George Campbell, with a bachelor's degree in engineering from MIT in 1891, a master's degree from Harvard in 1893, a year at Göttingen studying with the mathematician Felix Klein, a year at Vienna studying with Boltzmann, and a year in Paris studying under Poincaré.[43] Campbell was indeed, as Frank Jewett, the first president of the Bell Telephone Laboratories later described him, "thoroughly scientifically trained."[44]

Campbell immediately turned to the principal problems at the turn of the century, reducing attenuation and phase distortion. These ef-

[40] Annual report, Hayes to Hudson, January 2, 1890; Hayes to Devonshire, March 9, 1893; and Hayes to Hudson, March 19, 1894; in Exhibit 1951-A, pp. 43, 47. The Mechanical Department was, in this period, one of two technical departments in the Bell System; the other, the "Engineering Department," was formed in 1891 under Joseph Davis. In 1893, the two departments merged and Davis was put in charge of both with Hayes directing the Mechanical Department. Later, in 1902, Hayes became the head of both.

[41] Hayes to Hudson, March 7, 1892, pp. 2, 3, Exhibit 1951-A, p. 47; see also Goldberg.

[42] Stone also studied the theory of high-induction, or "loaded," cables. In 1897 he received a patent on a proposed such cable made of iron and copper. Hayes agreed to support future research on this device, work George Campbell continued after Stone left the company in 1899 (Brittain, pp. 40–41; "Annual Report of the Mechanical Department," December 31, 1898, cited in Bowles, "Hayes," p. 444; Fagen, pp. 890–94; and Lloyd Espenshied to Wesley Fuller, Case 37014, November 27, 1946, concerning the draft, *Aladdin, Inc.*, by Milton Silverman, BLC).

[43] Brittain, p. 41.

[44] Jewett to Abraham Flexner, February 9, 1944, BLC.

fects set a practical limit on "unloaded" lines of some 1,200 miles, a distance achieved in 1893 with the building of the line connecting Boston and Chicago. Extending the mathematical studies of Heaviside and Vashy, reported in 1887, that showed attenuation could be substantially reduced by uniformly distributing inductance along a line, Campbell, by 1899, developed the fundamental theory of the loading coil. This major contribution to telephone transmission involved connecting induction coils into telephone lines at definite intervals determined by the wavelength of the signal.[45] In 1901, Campbell would take his doctorate in physics from Harvard with a dissertation based on his loading-coil studies. It is worth noting in passing how Campbell's work on the loading coil provides a working example of the principles, described in the introduction, of the development of scientific research in an industry: the nonscientific goal of extending long-distance service gave rise to the technological problems of attenuation and phase distortion, which then stimulated Campbell's scientific investigation. His technological success would eventually increase the company's appreciation for the usefulness and importance of in-house basic research.

The third university-trained scientist Hayes hired before the turn of the century was the Canadian, Edwin Colpitts, who some years later

[45] Another important problem that Campbell worked on in 1900–1907, besides loading, was clarifying the role of capacitance in transmission. He would later invent the electric wave filter, popularize the use of Fourier transform techniques and propagate the MKS system (Brittain; R. B. Hill, "Birth of the Loading Coil," *Bell Laboratories Record* 30 [June 1952]: 263–39; Frank B. Jewett, "Dr. George A. Campbell," *Bell System Technical Journal* 14 (October 1935): 553–57; Edwin Colpitts, "Introduction," *The Collected Papers of George Ashley Campbell, Research Engineer of the American Telephone and Telegraph Company* [New York and London, 1917]; Fagen, pp. 240–52; John Brainerd, "Some Unanswered Questions," *T&C* 11 [October 1970]: 601–3). Professor Michael Pupin of Columbia University independently achieved the loading coil. Pupin's patents, issued in 1900, were interfered with by the claims of Campbell, whose theory was somewhat more detailed than Pupin's. But with some support from Bell (for reasons not entirely apparent) Pupin was able to establish an earlier conception date and in 1904 win the patent fight. Lloyd Espenshied ("Communication: The Campbell-Pupin Loading-Coil Controversy," *T&C* 11 [October 1970]: 596–97) suggests that the Boston patent office confused Pupin's early network theory with the theory of the loading coil. Joseph Gray Jackson ("Patent Interference Proceedings and Priority of Invention," *T&C* 11 [October 1970]: 598–600) recalls the belief of his father, an examiner in the patent office, that Carty and co-workers supported Pupin because "a certain group in the Bell Telephone Company was trying to avoid the financial obligation under the Bell contract, to pay continuing royalties to Pupin." During the proceedings, Pupin sold his rights to American Bell. Campbell was, as Brittain points out, the first to load *actual* circuits. See also Colpitts; Memorandum, "The Pupin Contract," and associated letters, ATTL, box 1008; and John Mills, "The Line and the Laboratory," *Bell Telephone Quarterly* 19 (January 1940): 5–21.

would direct Bell's first division devoted specifically to research. Colpitts, hired in 1899 to replace Stone (who left in that year to carry out independent research) had a physics master's degree from Harvard taken in 1897 and two years of additional study in physics and mathematics, also at Harvard. Like Stone and Campbell before him, Colpitts was initially assigned to engineering problems, working with Campbell on developing new methods and instruments for measuring alternating currents, on how to load phantom circuits, and on the problem of reducing inductive interference caused by electrical trains and trolley cars.[46]

In 1904, at Campbell's suggestion, Hayes hired Frank Baldwin Jewett, an instructor in physics and electrical engineering at MIT,[47] as a transmission engineer. Jewett's first assignment concerned protecting telephone lines from noise and high voltages induced by nearby electric railways. Within only two years, Jewett succeeded Campbell as head of the Boston Engineering Department, allowing Campbell to return to theoretical studies.[48] By 1912, Jewett was assistant chief engineer at Western Electric; by 1921 he was vice-president and director; and in 1925 he became the first president of Bell Telephone Laboratories. Jewett's rise in the hierarchy exemplifies what would become an unwritten Bell policy of choosing research directors from the company's own technical staff. Association with leading physicists ran deep in Jewett's career. At the University of Chicago, where he took his doctorate in physics, Jewett worked for the noted experimental physicist A. A. Michelson and formed a close friendship with Robert Millikan, then a young physics instructor, who exposed Jewett to some of the recent developments in the new field of electron physics. Jewett would later draw upon these associations in helping to establish Bell's first research branch.

In the period 1900–1907, Hayes's reports continued to list work on problems dependent on electromagnetic theory, including analytical study of transmission on loaded long-distance lines, electrical interference, the phantom circuit, and the multiple switchboard,[49] while at the same time they played down science, and even original invention, by in-house staff. For example, in 1906 Hayes wrote to Frederick

[46] *Western Electric News* (April 1924), p. 40; *Western Electric Engineer* (July 1960), p. 10.

[47] See pp. 340–42 in Thomas Shaw, "The Conquest of Distance by Wire Telephony," *Bell System Technical Journal,* 23 (October 1944): 337–421. Jewett was hired with a starting salary of $1,600 per year, almost three times the standard rate for college men without postgraduate study.

[48] Ibid., p. 342.

[49] Hayes to Fish, January 28, 1903, pp. 2–6; Davis to Fish, January 14, 1904, pp. 7, 8; Hayes to Fish, January 31, 1905, p. 6; Exhibit 1951-A, pp. 83–84.

Fish, who became president of AT&T in 1901 (following Hudson's sudden death in 1900): "Every effort in the department is being exerted toward perfecting the engineering methods; no one is employed who, as an inventor, is capable of originating new apparatus of novel design. In consequence of this it will be necessary in many cases to depend upon the acquisition of inventions of outside men, leaving the adaptation of them to our own engineers and to the Western Electric Company."[50]

But, as we have seen, Stone, Colpitts, Campbell, and Jewett (all of whom Hayes had hired) and even Hayes himself were not only "capable of originating new apparatus of novel design," they had successfully done so. How may we interpret Hayes's dual position regarding science and invention during the 1890s? Why, despite the obvious success of research, such as that leading to the loading coil, did he continue in his reports to deemphasize original invention and scientific investigation, and indeed even question the logic of placing scientists like himself, Colpitts, or Jewett in the role of research director? In Hayes's words: "The very fact that any great invention at the present must in all probability come from some man of unusual scientific attainments would render a laboratory under the guidance of such men a most expensive and probably unproductive undertaking."[51]

One reasonable explanation is that in his official correspondence Hayes was responding to attitudes from above that considered science a somewhat risky investment. Facing remnants of the long-standing skepticism and conservatism that many businessmen still held toward in-house research,[52] Hayes was, perhaps, as we may infer from

[50] Reich, p. 20.

[51] Ibid., p. 14.

[52] Early barriers to the development of industrial research are discussed in John Beer and W. David Lewis, "Aspects of the Professionalization of Science," in Kenneth Lynn and the editors of *Daedalus, The Professions in America* (New York, 1965); William D. Coolidge "Organized Industrial Research," *Science* 79 (February 9, 1934): 129–31; Frank Jewett to Abraham Flexner, February 9, 1944, BLC; Jewett, "Utilizing the Results of Fundamental Research in the Communications Field," *Bell Telephone Quarterly* 11 (April 1932): 143–61; Jewett, "The Laboratory—a Potent Source of Progress in Industry" (talk to Association of Life Insurance Presidents annual convention, New York, December 2, 1938), p. 1, BLC (hereafter referred to as Jewett, "The Laboratory"); and U.S. Temporary National Economic Committee Investigation of Concentration of Economic Power Hearings, pt. 3, Patents, January 16–20, 1939 (Washington, D.C., 1939), Jewett testimony, p. 949. Industrialists would note the typically long lag time between the development of scientific ideas and their application to industrial needs, the fact that science is only sometimes useful commercially, and that scientific concepts, unlike devices, belong to the entire international scientific community. Conversely, academic scientists, such as A. A. Michelson, often attached a stigma to doing applied

Bowles, simply too unassertive and reserved a gentleman to defend his conviction of the value of in-house basic research within the company. Hayes's pivotal role in the beginning of Bell's basic research program did not derive from his influence on policy but came, rather, from his having hired by 1904 the small core of academically trained scientists who in time would create the Bell System's first research division.

As the following section will show, the company's stance toward in-house research changed drastically between 1903 and 1911, through a series of innovative steps initiated by business decisions. The company took the first of these steps in 1903 when it announced a new policy of strengthening its corporate position by developing an image of scientific and technological superiority rather than by participating in expensive patent litigation.[53] At about the same time, Bell reversed its policy of withholding its research results from publication,[54] and we find Campbell, for example, reporting in detail on his research in journals such as *Philosophical Magazine* (1903) and *Electrical World* (1904).[55] But the most dramatic and influential changes began six months before the Wall Street panic in mid-October 1907.[56]

The Transcontinental Line and the First Research Branch

The Bell System's financial position became increasingly fragile in the decade and a half after Alexander Bell's original patents expired in 1893 and 1894. Many independent telephone companies sprang up, for example, the Home Company and the Farmer's Lines. By 1900 there were over 6,000 companies, and by 1907 almost half of the telephones in the United States were non-Bell. Subscribers were becoming increasingly dissatisfied with the service. For example, they would often accidentally reach one of the other companies, and in most cities they had to pay two or more telephone bills each month. The Bell System, having developed out of many different companies,

---

research (see Weart). It is hoped that future detailed case studies will uncover sufficient evidence to pin down why by the early 1900s a number of industries, such as American Bell, were hiring scientists and why leading scientists, such as Robert Millikan, were beginning at this time to take an active interest in industrial research.

[53] "Annual Report," 1903.

[54] *Electrical World and Engineer* 43 (April 2, 1904): 633.

[55] G. Campbell, "Telephone Lines," *Philosophical Magazine* 5 (March 1903): 313–30; "The Shielded Balance," *Electrical World and Engineer* 43 (April 2, 1904): 647–49.

[56] Determination of and how the reform wave influenced Bell's policy change in this period, while beyond the scope of this paper, surely merits close examination in future studies of Bell. See, e.g., Robert H. Wiebe, *Businessmen and Reform: A Study of the Progressive Movement* (Cambridge, Mass., 1962).

was inefficiently and uneconomically organized.[57] In April 1907, AT&T, finding itself in severe financial straits, underwent a management reorganization which brought a New York banking syndicate under J. P. Morgan into control of the company. Fish was replaced as president of AT&T by the vigorous and assertive Theodore N. Vail, who had held that post twenty years earlier,[58] and company headquarters were moved to New York City.

Vail immediately embarked upon a program designed to consolidate the Bell System and absorb the nationwide telephone system into Bell. Two of his early moves strongly influenced the development of Bell's program of research. The first was to join all ongoing research and development in the Bell System—consisting of AT&T's department of research and development in Boston and Western Electric's engineering departments in New York and Chicago. Some of the technical staff from Boston continued to function as the AT&T engineering department at AT&T headquarters in New York City; the rest were formed into a single centralized engineering department housed at Western Electric in New York City, the department that would evolve into the Bell Telephone Laboratories.[59] Hayes was dismissed, and Carty, who had been Vail's associate in the 1880s, was appointed chief engineer.[60]

Carty later proved to be an eloquent spokesman for Bell's new policy of strongly supporting scientific research. He was often to profess in the decades ahead that through scientific research man could achieve "mastery of the forces of nature, the elimination of poverty and disease, the prolongation of life, the advancement of learning, the growth of right living and sound thinking and of good under-

[57] "Organization of Western Electric"; Albert Bigelow Paine, *In One Man's Life* (New York, 1921), p. 228; "How There Came to Be Only One Telephone Company in Town" (New York, 1975) (one in a series of reports on the first 100 years of the telephone); Doherty, p. 80; and John Brooks, *Telephone: The First Hundred Years* (New York, 1975), pp. 120–37.

[58] Vail had been general manager of the American Bell Company as well as president of the Metropolitan Telephone and Telegraph Company in New York. He became president of AT&T in 1885, but left the company in 1887 over financial disagreements (*Events*, p. 11). During his twenty-year absence he worked on underground cables for the Empire City Subway Company.

[59] W. S. Gifford, "The Place of the Bell Telephone Laboratories in the Bell System," *Bell Telephone Quarterly* 4 (April 1925): 89–93.

[60] According to Bowles (see p. 445), who had access to Hayes's diary, Vail informed Hayes that the engineering work of his department had been "costing far too much," and he was released with "a pitifully small retainer." The question of why the Ph.D. physicists Hayes and Jacques were less successful in the corporate environment than Carty and others with no scientific credentials deserves close study by historians (if relevant documents exist and become available).

standing among men."[61] He would be propagandizing faith in the possibility of accelerating scientific work by means of a well-directed research laboratory, which he liked to describe (as Jewett recalls) as "a sort of collective mind which, made up of experts in many fields who collaborated continually with one another, could arrive quickly at the solutions of problems so intricate in their ramifications as to require years of single-handed effort, if indeed they could be solved at all single-handed."[62] Research applied to development of the Bell System was to play in Carty's mind a most powerful social role, for he saw the telephone system as society's nervous system[63] and, as he would fervently explain, "I believe it will be found in any social organism that the degree of development reached by its telephone system will be an important indication of the progress which it has made in attaining coordination and solidarity."[64]

Carty's many writings were to be studded by the term "fundamental research," evidently one he enjoyed. And it is not therefore surprising that, starting in 1907, the engineering department's reports frequently would refer to "fundamental research," whereas earlier under Hayes's tenure this term had rarely been used.[65] This rhetorical shift suggests the process that Vail set in motion in 1907 by placing Carty, a man who in time became a forceful science advocate, at the head of the new centralized engineering department. The move was to alter the way science was publicized in the company and create a supportive environment for the institutionalization of research.

Vail's decision in late 1908–early 1909 to build a transcontinental telephone line also strongly influenced the development of research. He had set the broad goal, already referred to in the 1870s by Alexander Bell, of developing the Bell System into a communications network that would be capable of reaching "anyone—at any possible place," and publicized it widely as Bell's motto, "One policy, one system, and universal service."[66] Establishing coast-to-coast telphone service was an essential step. According to Shaw, some Pacific busi-

[61] J. Carty, "Science and Progress in the Industries," talk in June 1929, quoted in Rhodes, 257.

[62] Jewett, "Carty: Telephone Engineer," pp. 163–64.

[63] Ibid.

[64] Rhodes, pp. 247–48.

[65] For example, in 1908, "It is hoped that by comprehensive investigation of the fundamental principles controlling the operation of transmitters we can produce a more efficient device than that at present used" (Exhibit 1951-A, p. 125).

[66] T. S. Vail, "Annual Report of President of AT&T," March 1910, included in *Views on Public Questions: A Collection of Papers and Addresses of Theodore Newton Vail 1907–1917* (New York, 1917), p. 27; Ithiel de Sola Pool, ed., *The Social Impact of the Telephone* (Cambridge, Mass., 1977), pp. 130–31.

nessmen had suggested to Vail that a line between New York and San Francisco be opened at the San Francisco Panama-Pacific Exposition, which was initially scheduled for 1914 (and later postponed to 1915). Although Bell economists were skeptical that such a line would be profitable, Vail, believing in the line's importance, authorized it anyway. Furthermore, Carty had concluded that the line would be technologically feasible.[67]

But building the transcontinental line proved to be far more difficult than Carty expected; the day-to-day problems led him, as Jewett recalls, to drive "himself and his associates with a force that was untiring and unsparing. Sleep and relaxation in small doses were grudgingly accorded."[68] The technology for economically and accurately sending signals over distances of the order of 3,000 miles did not then exist, and the need to develop this technology would lead—as we shall now see—to the authorization of the company's first official research branch.

The length of telephone lines had grown dramatically over the previous three decades. Unloaded lines spanned only 2 miles (between Boston and Cambridge) in 1876 but reached 900 miles (between New York and Chicago) by 1892. Energy loss was the chief limiting factor and set the practical extent of such lines at the 1,200 miles (between Boston and Chicago) achieved in 1893. Campbell invented the device by which this limit could be exceeded in 1900, and by 1911 the practical limit of loaded lines, the 2,100-mile distance from New York to Denver, was achieved by inserting loading coils every 8 miles, doubling the line's transmission efficiency. However, since loading coils do not replenish dissipated energy, any further increase in distance required a device, a "repeater," that would amplify attenuated speech waves.[69]

In 1903–4 Herbert Shreeve, a member of Hayes's staff, developed a mechanical repeater that proved to be effective on relatively short lines (e.g., between Amesbury, Massachusetts, and Boston)[70] but was not adequate for long lines since it favored certain pitches over others and was highly distorting when used two or more in series; it was also disproportionately insensitive when the incoming signal was weak,

[67] Shaw, pp. 353, 370; Frank Jewett, "Carty—the Engineer and the Man," *Bell Laboratories Record* 9 (September 1930): 9–13.

[68] Jewett, "Carty: Telephone Engineer," p. 170.

[69] Mills; Fagen, pp. 253–58; and C. A. Smith, "Fifty Years of Telephone Repeaters," *Bell Laboratories Record* 27 (January 1949): 5.

[70] Fagen, pp. 254–55; Bancroft Gherardi and Frank Jewett, "Telephone Relays, Repeaters or Amplifiers," *Proceedings of the American Institute of Electrical Engineers* 38 (November 1919): 1255–1313; *Events,* p. 17.

and it failed entirely when connected into loaded lines.[71] Shreeve's device consisted of: a telephone receiver, through which speech current entered; a mechanical diaphragm which the current activated; and a carbon transmitter which transformed the current into amplified speech at the output. The mechanical repeater's difficulty on long-distance lines lay in its diaphragm being too sluggish to vibrate over the band of speech frequencies while driving the electrode of the carbon-button transmitter at each frequency. In 1910, Campbell suggested, in a memorandum to Bancroft Gherardi, then the plant engineer at AT&T, that the company look into the possibility of developing a repeater with a vibrating part consisting of mercury gas molecules or cathode rays (electrons). There is no evidence that this suggestion was approved.[72] However, later that year, Jewett, on whom Carty had placed technical responsibility for the transcontinental line, wrote to Gherardi that the company needed "to employ skilled physicists who are familiar with the recent advances in molecular physics and who are capable of appreciating such further advances as are continually being made." He had demonstrated in a cost study that New York to San Francisco transmission without a suitable repeater would be uneconomical, but felt that "if this repeater matter is tackled in the proper manner by suitably equipped men working with full coordination and under proper direction the desired results can be obtained at a relatively small cost."[73]

Considering that the solution of the repeater problem might lie in the area of microscopic physics, and recalling his conversations on electronic phenomena years earlier with Millikan in Chicago, Jewett contacted Millikan. As Millikan recalls in his autobiography:

> It was in the fall of 1910 that Jewett, with whom I had kept in close contact since he took his doctorate in 1902, came from New York to see me at Chicago. . . . He started our conversation as follows: "Mr. John J. Carty, my chief, and the other higher-ups in the Bell System, have decided that by 1914, when the San Francisco Fair is to be held, we must be in position, if possible, to telephone from New York to San Francisco. With the aid of

[71] Mills, p. 9; Fagen, pp. 254–55. For a few days in 1915, the Shreeve repeater was used successfully in commercial service over the transcontinental line.

[72] Espenshied to Fuller, p. 3, cites Campbell's memorandum entitled "Mercury Arc Applications," in the Boston Black Files no. 193–2. In an interview with Espenshied by J. Tebo and L . Gumm, November 1971, ATTL (hereafter cited as Espenshied interview), Espenshied recalls that while Campbell was advocating that the company research "a radically new kind of repeater concerned with gaseous conduction," he was separated from experimental work when the engineering departments moved to New York in 1907. Also see Fagen, p. 896.

[73] Jewett to Gherardi, December 6, 1910, cited in Shaw, p. 371.

> Pupinized lines we have been able thus far to talk between New York and Chicago, but to get through to San Francisco by the present methods is out of the question. The cost is completely prohibitive. We have got to develop somehow a telephone repeater that will boost up the speech currents when they become too attenuated, just as the enfeebled currents in dot-dash telegraphy are now boosted up by the repeater on a long telegraph line. You have been working ever since we were together from 1900 to 1902 on these pure electronic discharges in your very high vacua which you were developing at that time. It seems to me that these essentially inertialess electronic discharges are the best bet there is for a telephone repeater. This will be a very difficult task, that of following all the modulations of speech and boosting up their intensity without in any way distorting them. We want you to help us in this job, as follows: Let us have one or two, or even three, of the best young men who are taking their doctorates with you and are intimately familiar with your field. Let us take them into our laboratory in New York and assign them the sole task of developing the telephone repeater.[74]

Millikan sent several young Ph.D.'s to AT&T.[75] The first, Harold Arnold, reported in January 1911; Millikan praised him as "one of the ablest men whose research I have ever directed and had in classes."[76]

In-house basic research became an official commitment of the Bell System in April 1911, when a special research branch was organized within the Western Electric Engineering Department. According to the engineering department's 1911 annual report, the new division, under the direction of Colpitts, was to address the "increasing number of problems intimately associated with the development of the telephone business, which require especially exhaustive and complete laboratory investigation."[77] The report emphasizes "fundamental" studies in connection with transmitters, receivers, duplex cables, telegraphs, and particularly repeaters, which were to be investigated "on the broadest possible lines."[78] A subgroup under Colpitts studied the repeater itself, while another subgroup under Jewett

[74] Cited on p. 247–48 of Oliver Buckley, "Frank Baldwin Jewett, 1879–1949," *National Academy Biographical Memoirs* 27 (1952): 239–64.

[75] H. D. Arnold, H. J. van der Bijl, H. W. Nichols, John Mills, K. K. Darrow, and M. J. Kelly.

[76] Harold de Forest Arnold," *Bell Laboratories Record* 11, no. 12 (August 1933): 350–60.

[77] 1911 "Report of Western Electric Co. Engineering Department," p. 23, cited in Exhibit 1951-A, pp. 133–34.

[78] Ibid., pp. 23–26, cited in Exhibit 1951-A, pp. 134–36.

studied problems that loaded lines present to repeaters. It was specified, further, that "to make adequate progress" the new branch "should include in its personnel the best talent available and in its equipment the best facilities possible for the highest grade research laboratory work."[79] The department had twenty members in 1912; by 1915 it had between forty and forty-five, at least seven of whom were Ph.D. scientists: Arnold, C. R. Englund, E. C. Wente, I. B. Crandall, O. E. Buckley, H. van der Bijl, and W. Wilson.[80]

The mechanism by which technological needs of the Bell System were transformed into scientific research during the 1910s is illustrated by Jewett's (April 1911) outline of his repeater program:

> (1) A complete study of the characteristics of the existing [Shreeve] receiver-transmitter type of repeater with a view to determining whether the action of this repeater cannot be improved upon and whether modifications in the repeater element, its circuit or in the line conditions, will make it suitable for general use on loaded lines.
> (2) A study of other possible repeater ideas, particularly in the domain of molecular physics. Certain characteristics of discharge of electricity through gases and vapors seem to offer considerable possibility of obtaining a telephone amplifier that will be suitable for use on loaded or nonloaded lines and which will give the desired amplification without a great deal of distortion.
> (3) A mathematical and laboratory study of two-way repeater circuits with a view to determining the best form of repeater circuit to be used in combination with any desired repeater element and any kind of loaded line.
> (4) A mathematical and experimental investigation of loaded line characteristics in the existing plant, and a determination of what changes, if any, must be made in the construction and installation of loading coils and cables in order to make loaded lines suitable for the application of telephone repeaters.[81]

Once again, we observe a clear-cut example of the principles of the development of scientific research in industry. Out of Vail's larger nonscientific goal to create a universal telephone system grew his decision to build a transcontinental line, from which came the technological problem of developing a nonmechanical repeater, and this problem in turn contributed crucially to the start of Bell's formal commitment to in-house basic research. "Basic" industrial research

[79] Ibid.; see also Goldberg.
[80] Shaw, p. 405; Colpitts to R. T. Barrett, January 27, 1939, WEL; Fagen, p. 51.
[81] Jewett's work order no. 7655, "General Repeater Study," approved by Carty on April 1, 1911, cited in Shaw, p. 372.

was now recognized as intrinsically dual in nature, being fundamental from the point of view of the researchers while at the same time supported by the company for its possible applications. The earliest problems of the new branch, while approached scientifically, were all directed toward solving specific engineering problems. In time the department would broaden its activities to include scientific studies having no immediate, but only possible future, applications.

On joining Bell's research staff, Arnold immersed himself in detailed investigation of the repeater problem. Searching for a lighter vibrating part to replace the mechanical diaphragm in Shreeve's model, he experimented in 1911 and 1912 with mercury vapor molecules (as suggested earlier by Campbell) and developed a promising model mercury are amplifier based on work by Peter Cooper Hewitt. However, Arnold dropped this work abruptly in November 1912 because of an important company decision: to try to adapt to the repeater problem a device whose "vibrating part" consisted of electrons, the audion (or triode) invented by Lee de Forest.[82] Inside the audion's vacuum tube were a filament which emitted electrons when heated; a positively biased metal plate which attracted electrons; and a negatively biased grid, placed near the filament, which controlled the electron current between filament and plate. A signal applied to the grid could modulate the electron current and produce an amplified signal in the plate circuit.

Although invented in 1906, the audion's potential application to amplifiers remained unrecognized for about five years.[83] John Stone Stone called AT&T's attention to the audion's amplification potential in 1912 by sending Carty a copy of his recent (1912) paper in the *Journal of the Franklin Institute* discussing the audion.[84] Stone demonstrated the device at AT&T, together with de Forest, October 30 and 31, 1912.[85] At this demonstration, the audion did not appear over-

[82] Mills, p. 11; Gerald Tyne, *Saga of the Vacuum Tube* (Indianapolis, 1977), pp. 57–61. De Forest developed the audion late in 1906. He received a patent on the device in 1908.

[83] According to Espenshied, "De Forest had his good tube in hand five years before making it really amplify" which he learned how to do from Fritz Loewenstein in New York (Espenshied interview, p. 13). One factor which undoubtedly contributed to this oversight was that the wireless development was then dominated by promoters and stock jobbers who were not anxious to disseminate information about the audion (Fagen, p. 260, esp. n. 38).

[84] Fagen, p. 260; Espenshied to Fuller.

[85] Tyne, pp. 57–58, 84–86. According to Tyne, De Forest had asked Stone to act as a go-between in his efforts to sell his patent rights to AT&T for not less than $50,000, of which Stone would keep 10 percent. Earlier, in January 1912, Fritz Loewenstein had demonstrated a version of the audion in a sealed box at AT&T. He did not, however, disclose the circuitry or amplifying device.

whelmingly promising, for although it effectively rectified weak signals (and could therefore be used in detecting wireless telegraph signals), it was unstable at the higher voltages needed for effective amplification of telephone currents. At the required plate voltage, the device would be caused, as John Mills (then on the staff) put it, "to fill with blue haze, seem to choke, and then transmit no further speech until the incoming current had been greatly reduced."[86] But Arnold, on observing the audion on November 1, 1912, was immediately optimistic that it could be adapted to telephone needs and began to research the device on that very day.[87] The company also began the negotiations that by 1914 resulted in its buying the rights from de Forest to use the audion as a repeater.[88]

After explaining the cause of the blue haze (recombination, in collisions, of ionized air molecules in the tube), Arnold developed theories which dictated optimal physical constants of the circuit elements, and he redesigned the tube to make best use of the space charge of electrons inside, employing a high vacuum, an oxide-coated filament, a more precisely placed grid, and a new grid circuit.[89] Arnold's "high vacuum thermionic tube" solved the repeater problem, thus providing the pivotal technology necessary for the transcontinental line; it was first used on a commercial circuit (at Philadelphia, on a New York to Baltimore line) in October 1913.

The 3,400-mile transcontinental line, built with Arnold's repeaters initially located at Pittsburgh, Omaha, and Salt Lake City, and later also at Philadelphia, Chicago, Denver, and Winnemucca, Nevada,[90] was first spoken over by Vail, in July 1914. The achievement of the

[86] Mills, p. 13. For a first-person account of Arnold's experiments in November 1912, including study of the "blue haze," see the recollections of Arnold's assistant Paul Pierce, written November 12, 1921, "Supplemental Statement by P. H. Pierce," WEL.

[87] Tyne, pp. 84–88.

[88] Ibid., pp. 115–25. AT&T first paid de Forest $90,000 on August 7, 1914, for a nonexclusive license in the field of wireless telegraphy. Later, on March 16, 1917, he was paid $250,000 for all remaining rights except for making and selling audions to users for radio applications and granting a license for use to the Marconi Company.

[89] Shaw, pp. 375–82. Irving Langmuir of General Electric also arrived at a similar result and applied for a patent in 1913. An interference on this patent, Arnold-Langmuir no. 40, 386, led to a long contest in which the U.S. Supreme Court ultimately decided that Langmuir's patent was invalid and that the high-vacuum tube did not constitute invention; Arnold had antedated Langmuir but both were antedated by de Forest (August 1912). See Decisions of the U.S. Supreme Court—1931 on Case 630, May 21, 1931, pp. 734–45. Further facts are included in the November 1915 memorandum of Colpitts to Jewett and the letter of May 10, 1915, from Charles Neave to Thomas Lockwood, BLC.

[90] Ibid.; Shaw, p. 381; Doherty, p. 81; Espenshied, "Question of Just When We Produced the High-Vacuum Tube-Case 37014," November 26, 1946, BLC.

line was properly celebrated at its commercial opening on January 25, 1915, with well-publicized cross-continental conversations. "It appeals to the imagination to speak across the continent," exclaimed President Woodrow Wilson in Washington to President (of the Exposition) Moore in San Francisco. Alexander Bell, stationed in New York, reissued his famous command, "Mr. Watson, come here, I want you," and received Watson's reply from San Francisco, "It will take me five days to get there now!"[91] Other participants in the demonstration included: Vail at Jekyll Island, Georgia; Carty with Mayor John Mitchel in New York; and Mayor James Rolf in San Francisco. At the Exposition, fascinated listeners could sit in the Palace of Liberal Arts and "hear, at 11 A.M., 1:30, 2:30, 3:30 and 4:30 P.M., any day of the Exposition season, readings of head-lines from the New York evening papers of that date, phonograph music, and the waves of the Atlantic Ocean breaking on Rockaway Beach. At times, personal conversations across the continent were arranged for distinguished visitors; such as Governor Whitman of New York, who listened to the rebellious proclamations of his three-month-old baby boy 3,400 miles away."[92] The appeal of the transcontinental line increased the public's as well as the Bell System's appreciation of the enormous power of industrial research. And the success of the new repeater then stimulated further research into vacuum tube physics, including thermionic emission, electric discharges in gases, and the dynamics of electronic and ionic flow, by associates of Arnold (including van der Bijl, Buckley, and Wilson).

While the account given here emphasizes the role of internal technological problems in the evolution of the Bell System's research, external factors were not without influence. As already noted, the development in the larger scientific community of electromagnetic theory and later electron physics were crucial, as was the growing availability of trained physicists. Several studies[93] have particularly emphasized the role of outside work in radio telephony, or "wireless," in the same period that Bell was developing the repeater for the transcontinental line.[94] The first, by the FCC, whose investigators in 1935–37 explored the start of Bell's research program, concluded that

[91] February 22, 1916, Beinn Breagh Recorder, 19:18, AGBF; and *Telephone Review Supplement* (January 1915).

[92] *History of the Exposition*, pp. 79–80, Carty Files, Publicity Matters, BLC.

[93] In particular, Goldberg, Danielian, Reich, and Walker's FCC staff. See also Leonard Reich, "Research, Patents, and the Struggle to Control Radio: A Study of Big Business and the Uses of Industrial Research," *Business History Review* 51 (Summer 1977): 208–35.

[94] Gherardi to Thayer, May 17, 1919, Exhibit 1951-B, app. E-1.

"the threatened invasion of their [Bell's] wire telephone industry by the emerging science of 'wireless' telephony" was the principal cause of "the present broad scope of the Bell System's research."[95] But the historical sequence of Bell's radio activities in the period 1890–1914, which we now briefly summarize, does not support this interpretation, for the research broadened well before radio might have appeared "threatening."

Bell began to take an interest in radio even before Marconi's first (1894) experiments; in 1892, Hayes asked Stone to explore whether Hertzian waves and Tesla oscillations could be used to telephone ships. (Stone's failure to achieve ship-to-shore telephony in the 1890s was mainly due to lack of an adequate detector.)[96] In 1902, Hayes interested Greenleaf Pickard on his staff in extending Stone's work on radio telephony, and Pickard succeeded in voice modulating a high-frequency spark discharge. These studies were discontinued after approximately six months, at about the same time that Reginald Fessenden, at the University of Pittsburgh, was granted what was thought at Bell to be a fundamental radio patent. Early in 1907, Hayes suggested that Bell collaborate with Fessenden, who had recently (1906) succeeded in transmitting sounds over long distances.[97] But Carty, who soon replaced Hayes, argued against Bell's entry into radio in this time of financial difficulties, judging, in consultation with Campbell, that a wireless telephone system was not within imminent practical realization.[98] Carty soon recognized the possibility that non-Bell wireless research might lead outsiders to the telephone repeater before Bell could achieve it and mentioned this possibility in a 1909 memorandum urging "vigorous work upon the development of a more powerful repeater":

> One additional argument making for vigorous work upon the development of a more powerful repeater I call to your particular attention. At the present time scientists in Germany, France,

[95] FCC, pp. 185 and 187–90.

[96] Espenshied to Fuller. The Bell System's early involvement with radio is also discussed in Fagen, pp. 362–66.

[97] Fagen, pp. 895, 362–63; Hayes to Fish, 1907, quoted in Fagen. Fessenden was one of the earliest to recognize the advantages of continuous oscillating waves over the damped waves produced by the intermittent and noncoherent oscillations of the spark discharges. Pickard joined the company in 1902 with study at the Westbrook Seminary in Portland, Maine, the Lawrence Scientific School of Harvard University, and MIT, as well as wireless experience at the Blue Hill Observatory, the American Wireless Telephone and Telegraph Company, and the Federal Wireless Telephone and Telegraph Company (*The National Cyclopaedia of American Biography* [New York, 1962], pp. 180–81).

[98] Fagen, p. 363; and Espenshied to Fuller.

> and Italy, and a number of able experimenters in America are at work upon the problem of wireless telephony. . . . Whoever can supply and control the necessary telephone repeater will exert a dominating influence in the art of wireless telephony when it is developed. The lack of such a repeater for the art of wireless telephony and the number of able people at work upon that art create a situation which may result in some of those outsiders developing a telephone repeater before we have obtained one ourselves, unless we adopt vigorous measures from now on. A successful telephone repeater, therefore, would not only react most favorably upon our service where wires are used, but might put us in a position of control with respect to the art of wireless telephony should it turn out to be a factor of importance.[99]

The FCC investigators largely based their argument of the outstanding importance of the wireless threat on this passage.[100] But it is to be noted that here Carty is employing the idea of establishing control in the wireless area as an *additional* argument to strengthen his request for intensified research on the repeater; he is not arguing in particular for the support of radio research. Carty apparently recognized in 1909 the possibility—which indeed history fulfilled and now allows alternate readings—that the key to solving both the repeater and wireless problems would be the same.

Carty kept a careful eye on progress in the radio field. Lloyd Espenshied recalls that when he joined the research staff in 1910, the Bell System was not yet aggressively moving into the radio field, but nevertheless had "quietly maintained an active interest in this new branch of electric communications . . . [and had] a card catalog of developments."[101] In the crucial years 1907–10, when the Bell System was establishing its new science policy and its first research division, the company was therefore quite capable of assessing its vulnerability to "threatened invasion" by wireless. That it did not consider itself to be in any imminent danger is evidenced by the fact that it took until July 1914 for Vail and the board of directors to decide to support a comprehensive program of radio research.

The point is that there was little technological basis for a wireless threat before late 1912 since the apparatus for radio and telephony were very different. Receivers of that period required high power levels, and long-distance radio waves therefore had to be transmitted at correspondingly high powers. But the existing (low-power) tele-

[99] Ibid.; J. Carty "Additional Force Required—Engineer," April 8, 1909, BLC.
[100] FCC, pp. 185, 187–90.
[101] Espenshied to Fuller.

phone technology could not modulate high-power radio waves with a speech signal, nor did the technology for receiving and faithfully amplifying modulated waves exist. With the application of the vacuum tube as an amplifier in 1912 it became possible to modulate at low power levels and then amplify to the level needed to achieve the desired distance.[102] Between 1912 and 1914, the company had prudently taken steps to purchase the patent rights to the audion from de Forest, thus preventing any possible patent threat.

Bell's intensive research into radio, starting in 1914, was highly successful. Detailed investigation of radio techniques by R. V. L. Hartley, R. A. Heising, and others resulted in a working model of a vacuum tube system that became the basis of the Bell System's exciting transcontinental and transoceanic wireless test conversations in late 1915 between Arlington (Virginia), Montauk (New York), Wilmington (Delaware), St. Simons Island (Georgia), Darien (Canal Zone), Mare Island (California), Pearl Harbor (Hawaii), and Paris. These dramatic demonstrations not only captured the public's interest, but *in fact* brought Bell—as Vail publicly reminded Carty in a telegram on one of the occasions—"one long step nearer our 'ideal'—a 'Universal System.' "[103] The development of radio research was yet another reinforcing strand in Bell's now well-established research program, illustrating once more the operation of our principles of the development of industrial research; it demonstrated to Bell and its subscribers, again, the tremendous value of "properly directed industrial research, conducted on scientific principles."[104]

During and after World War I, the climate of opinion in the United States became increasingly supportive of industrial research. The denial of European products during the war sensitized Americans to their dependence on German industrial goods and to the important role that scientists had played in Germany's industrial development.[105] Following industry's considerable contribution to the war

[102] Fagen, p. 363; Jewett to Blackman, "Radio Research and Development in the Bell System," March 9, 1932, Exhibit 1951-B, app. E-1; Espenshied to Fuller; and Reich, pp. 26–27; Espenshied to Fuller, p. 4.

[103] On September 29, 1915, Vail's telephone message to Carty was sent from New York to Arlington, Virginia, by long-distance wire and then to Mare Island, near San Francisco, by wireless (*Telephone Review Supplement,* October 1915). Also see L. Espenshied, "1915 Transoceanic Radio-Telephone Experiments—Account of Lloyd Espenshied's Participation," December 27, 1937, BLC.

[104] J. J. Carty, "The Relation of Pure Science to Industrial Research," *Science* 44 (October 1916): 511–18, see esp. 512.

[105] Americans arguing for industrial research would frequently cite the model of Germany; American scientists (including Noyes, Reese, and Whitney) who had spent time studying in Germany, or who (like Pupin) had immigrated to America from Europe, frequently noted the close ties between science and German industry. See

effort,[106] the United States government began a policy of supporting industrial research (including dye and chemical manufacture, scientific instruments, and communications devices for national defense). It was therefore reasonable for Bell to continue to promote its public image as "research oriented"; a pamphlet prepared in 1924 for internal use stresses the great "benefits for the Bell System" which could be obtained "by deliberately continuing to promote the public recognition of these laboratories as the greatest industrial scientific organization in the country."[107] Research, which the Bell System initially viewed only as a means of improving and creating devices, had come to be recognized, as Jewett put it in 1938, "as a currency which is more potent than gold in insuring complete freedom of action and in some cases even the right to live . . . which can be used in a bartering operation of cross-license and as an aid to charting out progress in the future, enabling better and more economical planning."[108]

On New Year's Day, 1925, a new corporation, the "Bell Telephone Laboratories," took over the engineering department of the Western Electric Company, which by now had over 3,000 persons working on its staff.[109] The Bell System thereby formally institutionalized its

---

Arthur D. Little, "Industrial Research in America," *Science* 38 (November 1913): 643–45, 648–53, 655–56; John J. Beer, "Coal-Tar Dye Manufacture and the Origin of the Modern Industrial Research Laboratory," *Isis* 49 (June 1958): 123–31; Willis Whitney, *Scientific American* (December 1921), p. 89; Robert Hilton, "The Maintenance and Preservation of Our Dyestuff Industries," *Transactions of the American Institute of Chemical Engineering* 11 (1918): 281–84; Michael Pupin, "Science and the Industries," *Columbia Alumni News* 15, no. 21 (March 1924): 314–16.

[106] Western Electric and AT&T both contributed strongly to the war effort by developing a nationwide communications network between ships and airplanes and points on the continent. Vail chaired the Committee on Telegraphs and Telephones, which acted in coordination with the Council on National Defense to coordinate the electrical communications web during the war. Jewett took a leading role in the formation of the National Research Council and in efforts of the Signal Corps to develop communications devices for the Army and Navy (see Rhodes, pp. 120–23; Lovette, p. 280; file labeled "War File Summary of Development Work in Connection with War Orders, 1917–1918," ATTL, esp. "The Bell Companies in the War," 1919," ATTL, box 2016; and W. Gifford, "The Activities of the Bell System during the World War," *Signal Corps Bulletin*, no. 102 [October–December 1936], pp. 1–9).

[107] "Notes on Research and Development Policy," ATTL, box 53.

[108] Jewett, "The Laboratory."

[109] Bell Telephone Laboratories was to be responsible to AT&T for fundamental research and development and to Western Electric for development, design, and engineering; AT&T and Western Electric were each to own 50 percent of the common stock. Carty was named chairman of the Board of Directors, Jewett director and president, and Colpitts became one of three vice-presidents (memorandum entitled "Organization Change Planned to Be Effective on or before January 1, 1923," ATTL, box 53).

"continuous program of research and development necessary to the progress of the Bell Telephone System," which, as we have seen, had been developing almost from the time the telephone was invented.

Summary and Conclusions

The establishment of research at Bell was neither sudden nor inevitable; it developed gradually, over the decades from the 1870s through the 1910s, out of a series of business decisions motivated by the increasing number of complex technological problems that the company faced as the telephone system grew larger. Important basic studies were carried out in the 1890s, well before the start of Bell's organized program of in-house research. Although the company's support for science and original invention remained cautious in 1890–1903, as Hayes's official reports in this period suggest, the context for research was already changing through Hayes's recruitment of individual scientists (Stone, Campbell, Colpitts, and Jewett), who would form the core of Bell's subsequent research program. While these men were first employed to solve practical problems, their presence on the staff and the technological applications of their successful individual scientific studies (such as Campbell's loading coil and Arnold's high vacuum tube) strengthened the basis for the company's policy of supporting basic research; and their suggestions often led to the employment of additional scientists (such as Jewett on Campbell's recommendation, and Arnold on Jewett's).

On being reinstated at AT&T in 1907 as president, Vail formally adopted and publicized as his motto the corporate goal of establishing a "universal telephone system," thus opening up a multitude of new technological problems, which underlined the company's need to develop a "repeater" suitable for use on transcontinental telephone lines. The usefulness of a focused researched effort by well-trained scientists on the staff of the company's newly consolidated engineering division in New York City was now transparent to Vail, as well as Carty and Jewett, and in 1911 the company reached the landmark of establishing in its organization a branch devoted specifically to research. Thus, the immediate motivation for the establishment of research came from internal technological needs whose urgency and relevance was fostered by the external economic, social, and political context (including competitive outside developments in wireless, the growing availability of trained scientists in America, and the wartime activities of industry) which helped to accelerate and broaden, but did not alone initiate, Bell's research program.

This case lends support, through our analysis of a series of examples—the events in 1897–1900 leading to the loading coil; those in 1908–15 leading to the high vacuum tube that made the transcontinental line feasible; and radio research, which by 1915 resulted in transoceanic and transcontinental wireless telephone communication—to the three principles put forth in the introduction, which provide a preliminary model for explaining the emergence of scientific research in an industry. In this paper, we have most fully documented the interactions between science and technology; the relationships between corporate goals, decisions, and technology require more detailed research and analysis in future studies of the Bell System. The mechanisms by which the research changed in character during the subsequent decades remains to be explained, as does the influence which this research exerted both on Bell and on the larger society. Why, for example, did a portion of Bell's research program during the late 1930s branch out to explore very general solid-state problems relating only peripherally to telephone use? How did the existence of successful industrial laboratories, such as Bell, influence the development of similar laboratories elsewhere?

Our model for the emergence of Bell's research opens up many questions concerning the development of research in industry generally, including: Can the development of research for industries other than telephone communications be described within the same framework? To what extent do the principles apply in earlier periods and in other countries, and how may we account for differences? How must the principles be extended to account for business factors, such as patents, or larger social, economical, and political factors, such as the increasing professionalization of science, government support of industries, and wartime developments? How are we to incorporate into the principles the influence of special interest groups? (For example, from industrial management's point of view, research derives from the possibility of increasing technological output, while from the scientists' point of view it often derives from the satisfaction of personal curiosity or from financial awards which management supplies.) How can we expand the principles to fit later stages in the growth of an industry's research program?

Let us close by reformulating our principles in a way that admits generalization to contexts other than technology-based industry. We picture, in analogy to Hugh Aitken's "field of influence" model,[110] the relationship as a flow of influence among the three spheres of activity: the nonscientific, the technological, and the scientific. For the case of the emergence of research at Bell, we observed flow originating in the

[110] Hugh Aitken discusses this model on pp. 298–336 of *Syntony and Spark: The Origins of Radio* (New York, 1976).

nonscientific sphere, streaming to the technological, and finally to the scientific, with feedback from the scientific to the nonscientific via the technological.[111] The more general historical problem of studying the relationships between science, technology, and society becomes that of determining the characteristic influence flow patterns for each particular context. Out of such an array of flow patterns, a small corner of which this paper has filled in, historians may begin to identify the mechanisms underlying the more general relationships between science, technology, and society.

[111] It would seem that the principles operated in the traditionally stressed reverse direction for the case of radio, where Hertz, Lodge, and Marconi's application of Maxwell's model of the electromagnetic field led to radio technology, which in turn led to the radio industry with its impact on politics, literature, art, and other aspects of modern life. But, as Aitken notes, the traditional model overlooks highly important and influential exogenous factors, including Marconi's sense of the commercial market and his entrepreneurship, which were both crucial in creating the radio industry. Aitken feels, rather, that "the appearance of radio was without question a response to the rising demand for long-distance communications in the closing decades of the nineteenth century."

Flexibility and Mass Production at War: Aircraft Manufacture in Britain, the United States, and Germany, 1939–1945

JONATHAN ZEITLIN

The relationship between military enterprise and mass production is a central theme in the history of technology.[1] Military demand, military finance, and military ideology, it is widely agreed, provided a crucial stimulus for the technological breakthroughs necessary for mass production. The origins of the "American system of manufactures" during the 1830s and 1840s, as Merritt Roe Smith has shown, lay in the U.S. Army's determined quest for a rifle with interchangeable parts, itself inspired by the earlier visions and experiments of military engineers in 18th-century France.[2] During the late 19th and early 20th centuries, as David Hounshell has demonstrated, American industrialists successfully adapted the techniques of "armory practice"—notably, sequential manufacture of interchangeable parts using special-purpose machinery, jigs, fixtures, and gauges—to high-volume production of light, standard-

DR. ZEITLIN, professor of history, sociology, and industrial relations at the University of Wisconsin—Madison, is completing a book on flexibility and mass production in the British metalworking industries, 1830–1990, with support from the Guggenheim Foundation and the German Marshall Fund of the United States. **He is grateful to the following friends and colleagues for generous provision of advice, comments on previous drafts, research materials, and unpublished manuscripts: Herrick Chapman, David Edgerton, John Guilmartin, Howell Harris, Gary Herrigel, Paul Hirst, I. B. Holley, Jr., David Hounshell, Nelson Lichtenstein, Ned Lorenz, Eric Schatzberg, Jim Tomlinson, and Jacob Vander Meulen, The usual disclaimers apply.**

[1]For a general survey, see Merritt Roe Smith, "Introduction," in Merritt Roe Smith, ed., *Military Enterprise and Technological Change: Perspectives on the American Experience* (Cambridge, Mass., 1985), pp. 1–38.

[2]Merritt Roe Smith, *Harpers Ferry Armory and the New Technology* (Ithaca, N.Y., 1977), and "Army Ordnance and the 'American System' of Manufacturing, 1815–1861," in Smith, ed., pp. 39–86; Ken Alder, "Terror and Technocracy: The French Origins of Interchangeable Parts Manufacturing" (paper presented at the Society for the History of Technology meeting, Uppsala, Sweden, August 16–20, 1992). For a dissenting view that argues for the independent origins of the American system in the private sector, see Donald R. Hoke, *Ingenious Yankees: The Rise of the American System of Manufactures in the Private Sector* (New York, 1990).

ized equipment such as sewing machines, typewriters, agricultural implements, bicycles, and ultimately automobiles.[3] Both military and economic historians also concur that the resulting capacity of civilian firms to manufacture large numbers of standardized weapons became increasingly central to the conduct of industrialized warfare, while the great munitions drives of World War I in particular played a key part in diffusing mass-production methods from the United States to other combatant powers such as Britain, France, and Germany.[4] World War II, by all accounts, marks the apogee of this symbiosis between mass production and military prowess. For the Allied victory surely depended, as contemporaries themselves believed, on the ability of American (and Soviet) mass-production industry to turn out military aircraft, tanks, and other weapons systems in unprecedented quantities.[5]

Historians of technology are well aware that international military rivalries have often centered on complex weapons systems such as naval battleships produced to customer specifications in small batches or single units using skill-intensive methods. And they have likewise documented the importance of military objectives and material support for the development of flexible, general-purpose technologies such as computers and numerically controlled machine tools.[6]

Yet the prevailing view has been that the military's overriding concern with what Roe Smith calls "uniformity and order" has shaped and reinforced two central premises of modern industrial management: control and standardization. Insofar as military procurement objectives are recognized to diverge from the mass-production model—for example, by emphasizing performance over costs—this has often been seen as an institutional distortion of normal economic processes that inhibits defense contractors from competing successfully in civilian product markets. For contemporary critics of the "baroque" technolo-

[3]David A. Hounshell, *From the American System to Mass Production, 1800–1932: The Development of Manufacturing Technology in the United States* (Baltimore, 1984).

[4]On the importance of interchangeable-parts manufacture for national military power after 1850, see William H. McNeill, *The Pursuit of Power: Technology, Armed Force, and Society since A.D. 1000* (Chicago, 1982), chaps. 7–9, esp. pp. 232–36, 330–31, 355, 358–59.

[5]For overviews, see Alan S. Milward, *War, Economy and Society, 1939–1945* (London, 1977), chap. 6; R. J. Overy, *The Air War, 1939–1945* (London, 1980), chap. 7.

[6]On the naval arms race before World War I, see McNeill, chap. 8; Jon Tetsuro Sumida, *In Defence of Naval Supremacy: Finance, Technology, and British Naval Policy, 1889–1914* (London, 1990); Clive Trebilcock, *The Vickers Brothers: Armaments and Enterprise, 1854–1914* (London, 1977). On the military role in the development of computers and numerically controlled machine tools, see Kenneth Flamm, *Creating the Computer: Government, Industry, and High Technology* (Washington, D.C., 1988), chap. 3; David F. Noble, "Social Choice in Machine Design: The Case of Automatically Controlled Machine Tools," *Politics and Society* 8 (1978): 313–47, and *Forces of Production: A Social History of Industrial Automation* (New York, 1984).

gies nurtured by postwar patterns of military demand, World War II likewise represents the high-water mark of convergence with civilian production methods from which subsequent procurement policies have systematically deviated.[7]

This view of the relationship between military and civilian technology clearly rests on normative assumptions about the superiority of mass production as a model or paradigm of modern industrial efficiency. But as a growing body of literature has argued, mass-production firms have often been successfully out-competed by practitioners of an alternative strategy—sometimes termed "flexible specialization"—based on the manufacture of a changing array of customized goods using flexible, general-purpose equipment and skilled, adaptable workers. Where demand patterns are uncertain and technologies rapidly changing—as in much present-day industry—such flexible production methods give manufacturers a vital edge in adjusting swiftly to shifting competitive conditions and bringing new products to market. As in the past, too, many firms now seek to avoid the potential disadvantages associated with exclusive reliance on either mass or flexible production by combining elements from each model in varying proportions.[8]

This article reexamines aircraft production in Britain, the United States, and Germany during World War II from a perspective informed by current upheavals in markets, technologies, and industrial organization but attentive to contemporary debates among the historical actors themselves. In each of these countries, I argue, aircraft production, like that of other weapons systems such as tanks or naval warships, required not only the capacity to manufacture complex products in vast numbers, but also considerable flexibility. As fighters, bombers, and other types of aircraft were tested against one another under real battlefield conditions, military authorities demanded continual modifications to overcome design flaws and incorporate new tactical and technical ideas.

[7]For the influence of military concerns with uniformity and order on modern management, and the primacy of performance over price in weapons performance, see Smith, "Introduction" (n. 1 above), pp. 7, 17, 20–22. For contemporary criticisms of military production as "baroque technology" by contrast to civilian mass production, see Mary Kaldor, *The Baroque Arsenal* (London, 1982); Ann Markusen and Joel Yudken, *Dismantling the Cold War Economy* (New York, 1992).

[8]On flexible specialization and hybrid forms of productive organization, see Michael J. Piore and Charles F. Sabel, *The Second Industrial Divide* (New York, 1984); Charles F. Sabel and Jonathan Zeitlin, "Historical Alternatives to Mass Production: Politics, Markets, and Technology in Nineteenth-Century Industrialization," *Past and Present,* no. 108 (1985), pp. 133–76; Paul Hirst and Jonathan Zeitlin, "Flexible Specialization versus Post-Fordism: Theory, Evidence and Policy Implications," *Economy and Society* 18 (1991): 1–55; Charles F. Sabel and Jonathan Zeitlin, eds., *Worlds of Possibility: Flexibility and Mass Production in Western Industrialization* (Cambridge, 1996, in press).

Pervasive but unpredictable shortages of raw materials, components, and labor likewise involved constant improvisation to avoid costly bottlenecks. Established mass-production methods, such as those pioneered by the automobile industry, typically proved too rigid for the high level of uncertainty and rapid pace of innovation imposed by the war economy. Successful aircraft manufacturers therefore needed to find new ways of reconciling the high throughput of mass production with the adaptability of the craft workshop. The result was a proliferation of hybrid forms of productive organization, which anticipated in many respects more recent innovations in flexible manufacturing such as those developed by the Japanese. At the same time, however, the methods used to balance the quality and quantity of aircraft output varied widely across the three countries, as did their relative effectiveness, reflecting deeper national differences in geopolitical position, resource endowments, productive capabilities, and civil-military relations, among other factors.

Britain: Quantity Production and Continuous Improvement

In Britain, as in each of the major combatant powers of the 1930s and 1940s, rearmament and total war demanded vast quantities of munitions, from shells, guns, and small arms to ships, tanks, and transport vehicles, along with the machine tools to produce them. Such a massive increase in output inevitably required both the expansion of established arms manufacturers and the conversion of civilian engineering firms to military production. Nowhere was the transformation more dramatic and far-reaching than in the production of aircraft, the weapons system accorded the highest priority in British strategic planning: on the eve of rearmament in 1933, the industry employed some 20,000 people; at its peak in 1944, the workforce involved had mushroomed to 1.8 million, or more than one-third of the entire manufacturing sector.[9]

The interwar aircraft industry, as these figures suggest, was very much a small-scale affair. From its wartime high of 268,000 employees in 1918, aircraft manufacturing in Britain contracted sharply during the 1920s. With commercial demand limited by the relatively slow development of civil aviation within the British Empire, the industry was largely sustained by the Air Ministry, which sought to maintain its military potential during a period of relative disarmament by sharing out small orders among a "ring" of approved contractors. As a result, the Royal Air Force (RAF) in 1931 was using forty-four different types of airframe and

[9]For these figures, see Peter Fearon, "Aircraft Manufacturing," in *British Industry between the Wars,* ed. Neil K. Buxton and Derek H. Aldcroft (London, 1979), p. 216; M. M. Postan, *British War Production* (London, 1952), p. 310; David Edgerton, *England and the Aeroplane* (London, 1991), p. 72.

thirty-five different types of engine, though a smaller number of more successful designs predominated. Airframe manufacture was thus fragmented into fifteen to twenty small and medium-sized firms, operating primarily as defense contractors, while aeronautical research was conducted by the Royal Aeronautical Establishment and the National Physical Laboratory, which functioned as collective resources for the industry and the RAF.

Before 1935, the airframe manufacturers remained principally a collection of design teams and jobbing workshops, with limited experience of working with jigs and fixtures or other volume-production methods, although orders for individual models might run as high as fifty to 100 in any given year. Aero engine production was concentrated in the hands of a smaller number of larger firms, several of which also made luxury cars. But even here, where interchangeability was more important for technical performance, production runs were short and opportunities for standardization limited. On both airframes and engines, therefore, "bench methods" based on general-purpose tools and extensive hand fitting remained the rule, and skill requirements ranked among the highest in British engineering.[10]

All this began to change significantly with Prime Minister Stanley Baldwin's commitment to achieve air parity with Germany in 1934 and the decision to reequip the RAF with a new generation of modern aircraft in 1936. A key step in the expansion of aircraft production was to be the construction of a series of purpose-built "shadow factories," financed by the state but managed by outside firms, notably, automobile and electrical manufacturers, working to designs supplied by established airframe and engine companies. This scheme, devised by Lord Weir, the air minister's industrial advisor and a longstanding advocate of mass production, was based explicitly on the assumption that the efficiency of aircraft manufacture could be substantially increased by the infusion of management skills and high-volume techniques from the shadow firms. The shadow scheme was forced through by the Air Ministry over the vehement opposition of the Society of British Aircraft Constructors (SBAC), which argued that manufacturers of standardized goods such

[10]On the development of the British aircraft industry to 1935, see Edgerton, *England and the Aeroplane,* chaps. 1–3, who substantially revises the negative view of its performance during this period advanced by Fearon, "Aircraft Manufacturing," pp. 216–40, "The Formative Years of the British Aircraft Industry, 1913–24," *Business History Review* 43 (1969): 476–95, "The British Aircraft Industry and the State, 1918–35," *Economic History Review* 27 (1974): 236–51, and exchange with A. J. Robertson in *Economic History Review* 28 (1975): 648–62, and "The Vissicitudes of a British Aircraft Company: Handley Page Ltd. between the Wars," *Business History* 20 (1978): 63–85. I am grateful to David Edgerton for new information about the size of RAF orders for particular models during this period.

as automobiles lacked the necessary flexibility to accommodate the frequent modifications in design inherent in aircraft production. Despite their claims to superior adaptability, however, the SBAC's own members soon showed themselves only too anxious to slow down the introduction of new types of planes in hopes of obtaining lucrative continuation orders on well-established models.[11]

Whereas some shadow firms like Standard Motors and (later) English Electric adapted easily to the close tolerances, continual design changes, and complex coordination tasks involved in aircraft production, other prominent volume manufacturers proved less successful. Thus Austin was demoted from its preeminent position as the sole assembler of complete aero engines in the second shadow scheme of 1938, while the management of the giant Spitfire shadow factory at Castle Bromwich was transferred from the Nuffield (Morris) Organisation to Vickers-Armstrong in 1940. As a result of such experiences, the Air Ministry and its advisors reluctantly acknowledged that despite the automobile industry's vital importance for the expansion program, its methods and equipment were "far less appropriate" for aero engine manufacture than had been originally anticipated. Too much reliance had been placed on the organizational skills of the motor vehicle firms, argued Sir Charles Craven of Vickers—not without a strong measure of self-interest—since aircraft manufacture was "more of a quantity production type than a mass production problem." Automobile industry practice, it appeared, could not be directly transferred to aircraft manufacture, though shadow firms like Austin, Rover, and Standard did eventually become more adept at adapting techniques such as "straight flow" machining, assembly conveyors, and time study to the production of aircraft components and subassemblies.[12]

Subsequent procurement policy accordingly favored extensions to established aircraft firms over management by outside contractors, with few major exceptions such as Ford's Merlin engine plant. By the end of the war, government expenditure on such extensions exceeded that on shadow factories by a margin of £225 million to £146 million, respectively. Even more striking, only 22 percent of wartime aircraft by weight

[11]On the planning of air rearmament, the development of the shadow scheme, and conflicts between the Air Ministry and SBAC, see Postan, chaps. 2–3; William Hornby, *Factories and Plant* (London, 1958), chaps. 6–8; Edgerton, *England and the Aeroplane,* pp. 68–77, and "State Intervention in British Manufacturing Industry, 1931–51: A Comparative Study of Policy for the Military Aircraft and Cotton Textile Industries" (Ph.D. diss., University of London, 1986), pp. 106–42, 193–202; David Thoms, *War, Industry, and Society: The Midlands, 1939–45* (London, 1989), chap. 1; Robert P. Shay, *British Rearmament in the 1930s: Politics and Profits* (Princeton, N.J., 1977).

[12]Thoms, pp. 9, 17–20, 29–30, 52–53, 56–58, quotations from p. 57.

TABLE 1

British, American, and German Aircraft Production, 1939–45

	1939	1940	1941	1942	1943	1944	1945*
A. Number of aircraft:							
United Kingdom	7,940	15,049	20,094	23,672	26,263	26,461	12,070
United States	5,856	12,864	26,277	47,836	85,898	96,318	49,761
Germany	8,295	10,247	11,776	15,049	24,807	39,807	7,540
B. Structure weight (millions of pounds):							
United Kingdom	29	59	87	134	185	208	95
United States	...	...	82	275	651	952	430
Germany	...	...	88	114	163	199	...

Source:—R. J. Overy, *The Air War, 1939–1945* (London, 1980), p. 150.
*United Kingdom = January–September; United States = January–June.

was produced outside the air firms themselves, though the proportion was substantially higher for certain important types such as heavy bombers. Two large groups, Vickers-Armstrong (Supermarine and Vickers Aviation) and Hawker-Siddeley (Armstrong-Siddeley, Armstrong-Whitworth, Gloster, Hawker, A. V. Roe), alone accounted for nearly half the total output of British aircraft during World War II.[13]

While the initial phases of rearmament proceeded slowly, not least because of the reluctance of Baldwin's successor Neville Chamberlain and his advisors to place the economy on a war footing and to negotiate over dilution (the substitution of unskilled men and women for skilled craftsmen) with the engineering unions, the expansion programs launched in 1936–38 did ultimately generate large increases of output. Thus by 1940, as table 1 shows, Britain was producing 15,000 military aircraft, fifteen times as many as five years earlier (thirty times more in terms of weight) and 50 percent more than Germany.[14]

After the fall of France, aircraft procurement programs were drastically stepped up, and a much wider range of engineering firms were drawn into the industry as subcontractors and component suppliers. This vast manufacturing archipelago was overseen by a separate Ministry of Aircraft Production (MAP), which developed a complex and sophisticated statistical planning apparatus for coordinating the flow of raw materials, components, and labor among some 12,000 firms while

[13]Edgerton, *England and the Aeroplane*, pp. 72–73, 75; Overy, *Air War* (n. 5 above), p. 165; Hornby, *Factories and Plant*, pp. 214, 222–23; Postan, p. 388.

[14]On the slow pace of prewar rearmament and the problem of dilution, see also Peggy Inman, *Labour in the Munitions Industries* (London, 1957), chap. 2; R. A. C. Parker, "British Rearmament 1936–9: Treasury, Trade Unions and Skilled Labour," *English Historical Review* 96 (1981): 306–43. For the quantitative expansion of production and the comparison with Germany, see Edgerton, *England and the Aeroplane*, p. 71; Overy, *Air War*, pp. 21, 150.

constantly adjusting production schedules for different types of aircraft to take account of supply bottlenecks and changing operational requirements. Sir Stafford Cripps, who became minister of aircraft production in 1942, was an enthusiastic exponent of mass production and scientific management who had run a government munitions factory during World War I. Under his direction, MAP deployed a variety of instruments to encourage the adoption of high-volume methods in aircraft manufacture, notably, the Production Efficiency Board and the Technical Costs Branch, which advised firms on factory layout, tooling, rate fixing, quality control, and labor utilization, placing particular emphasis on the diffusion of motion-study techniques (overseen by a representative of the Metropolitan-Vickers Electrical Co., its leading domestic practitioner). Together with the Labour Supply Inspectorate of the Ministry of Labour, MAP pressed its suppliers to increase dilution, and the proportion of female labor in the industry as a whole reached 36.5 percent by mid-1943. Where aircraft manufacturers did not measure up to these standards of efficiency, finally, MAP used its powers to force through changes directly, as in cases such as Napier's, whose management was transferred to English Electric; Fairey Aviation, where a government controller was installed as managing director and deputy chairman; or, most radically, Short Bros., which was nationalized after repeated failures to meet agreed production schedules.[15]

The central dilemma of wartime aircraft manufacture was the need to balance the qualitative gains obtainable through design modifications against the quantitative losses resulting from interruptions to continuous production runs. Since the development of new aircraft types took some four to seven years, any design might prove strategically obsolete by the time it reached the stage of large-scale manufacture, and periodic revisions were essential to take account of battlefield experience. Insofar as the development of new types could be accelerated through abridged procedures such as ordering "off the drawing board" and what would now be called "simultaneous engineering" of design and tooling, modifications during the course of production became even more crucial in removing bugs from untested models. But the more radical the design change, the longer the time required for its introduction, and the greater the loss in current output arising from alterations to

[15]On MAP and the organization of wartime aircraft production, see Postan (n. 9 above), pp. 303–22; Inman, *Labour in the Munitions Industries,* pp. 42–81, 428–33; Alec Cairncross, *Planning in Wartime: Aircraft Production in Britain, Germany and the USA* (London, 1991), pt. 1; Edgerton, "State Intervention" (n. 11 above), pp. 202–11, and "Technical Innovation, Industrial Capacity and Efficiency: Public Ownership and the British Military Aircraft Industry, 1935–48," *Business History* 26 (1984): 247–79; Nick Tiratsoo and Jim Tomlinson, *Industrial Efficiency and State Intervention: Labour, 1939–1951* (London, 1993), chap. 2.

tooling and falling labor productivity on unfamiliar operations. The solution arrived at by the British was a policy of incremental improvements and continuous modifications that could be "spliced in" more quickly and with less disruption to production schedules through careful coordination among airframe, engine, and accessory manufacturers than could a completely new design. Only at moments of extreme military urgency, as at the onset of rearmament in 1934, after the Munich crisis in 1938, and during the Battle of Britain in 1940, was the RAF's so-called doctrine of quality relaxed in order to secure larger quantities of existing types, while current designs were temporarily frozen.[16]

This policy of incremental improvement and continuous modification allowed the British to maintain a very high average quality of aircraft and leapfrog over German designs by constantly incorporating new tactical and technical ideas. Between 1938 and 1944, for example, the Supermarine Spitfire, one of the most successful and versatile fighters of the war, went through more than twenty revisions significant enough to be denoted by a separate mark number, none of which required more than a fraction—typically a small one—of the person-hours spent in developing and tooling up the original design. These revisions, in turn, were responsible for dramatic advances in the Spitfire's performance, which increased its maximum speed from 356 to 460 miles per hour, its altitude ceiling from 34,000 to 42,000 feet, and its rate of climb from 2,500 to 5,000 feet per minute, while also enhancing its firepower, maneuverability, flying range, strength, and suitability for specialized tasks such as photographic reconnaissance, naval work, and even bombing. Similar gains in performance and adaptability were achieved through piecemeal modification of other major fighter and bomber designs such as the Hawker Hurricane, the de Havilland Mosquito, the Avro Lancaster, and the Vickers Wellington. At the same time, moreover, output continued to rise steadily if not at the rate envisaged by the overambitious programs of the early war years: between 1940 and 1944, as table 1 shows, the weight of aircraft produced more than tripled, from 59 to 208 million pounds per year, notwithstanding the radical shift in its composition from fighters to bombers during 1942–43.[17]

Despite the beneficial impact of design improvements on the quality of British aircraft, such changes were greatly disliked by aircraft manufacturers and those within MAP responsible for meeting output targets,

[16]Postan, pp. 322–45; M. M. Postan, D. Hay, and J. D. Scott, *Design and Development of Weapons* (London, 1964), pp. 1–9, 139–53, 159–74. For a parallel analysis of "the dilemma of mass production: more airplanes or better?" see Irving Brinton Holley, Jr., *Buying Aircraft: Matériel Procurement for the Army Air Forces* (Washington, D.C., 1964), pp. 312–18.

[17]Postan, pp. 326–27, 339–41; Postan, Hay, and Scott, pp. 159–68, 535–36; Edgerton, *England and the Aeroplane* (n. 9 above), pp. 71–72; Overy, *Air War,* p. 150.

since they undercut the effectiveness of mass-production techniques. Thus a 1942 study conducted by Sir Ernest Lemon, MAP's director-general of production and a pioneer of high-volume rolling-stock manufacture on the prewar London, Midland and Scottish Railway, showed that while full-scale jigging and tooling-up could be justified on an uninterrupted series of 1,500 components, the continuous spate of modifications meant that aircraft like the Spitfire were largely produced in batches of 500 or less, on which bench methods would have been more economical. Similarly, prewar manufacturers of standardized consumer durables like the American-owned Hoover Ltd. (vacuum cleaners) complained bitterly that fluctuating order quantities and frequent changes in the design and specification of aircraft components—which MAP itself regarded as unavoidable—largely undermined their ability to realize the benefits of flow production. Even where mass-production firms were able to concentrate uninterruptedly on long runs of standardized equipment for the aircraft program, the inflexibility of their approach could nonetheless prove a serious handicap. Thus Ford's shadow factory turned out single-stage Merlin engines faster and more cheaply than Rolls-Royce's home plant at Derby but unlike the latter could not change over to the two-stage version (which was responsible for major improvements in speed and altitude on the Spitfire and other aircraft) without complete reequipment of both machining and assembly shops.[18]

The United States: Mass Production and Retrospective Modification

These trade-offs between quality and quantity of wartime aircraft production were resolved rather differently across the Atlantic. Despite the growth of civil aviation in the United States during the 1920s and 1930s (stimulated by government-subsidized airmail contracts), military demand nonetheless remained predominant as in Britain, and the prewar American industry, like its transatlantic counterpart, "was still making airplanes largely by hand," in the words of I. B. Holley, the historian of U.S. air force procurement.[19] As in Britain, too, automobile

[18]Postan, pp. 341–42, 410–11; Postan, Hay, and Scott, pp. 168–69; Edgerton, "State Intervention" (n. 11 above), pp. 128–30, 133–35; Cairncross, pp. 30–31, 64, 68; Sir Ernest Lemon, "Reduction in the Time of Change-Over to a New Type of Aircraft," August 1942, London, Public Record Office (PRO), BT 28/423/IC/114/2; C. B. Colston (chairman and managing director of Hoover Ltd.), "Simplified or Rationalised Production in the Light Engineering Industry," report to MAP, February 1942, and comments of MAP officials in PRO, BT 28/410/IC/105; Ian S. Lloyd, *Rolls Royce: The Merlin at War* (London, 1978), pp. 120–23.

[19]On the prewar development of the U.S. aircraft industry and the nature of its production methods, see Jacob Vander Meulen, *The Politics of Aircraft: Building an American Military Industry* (Lawrence, Kans., 1991); Holley, *Buying Aircraft,* chap. 2, and "A Detroit

manufacturers played a key part in the expansion of U.S. aircraft production after President Roosevelt's 1940 call for 50,000 planes within two years. But the application of automotive methods to aircraft manufacture was carried much further in the United States, with its greater experience of mass production, its larger reserves of labor and industrial capacity, and its geographical insulation from enemy bombing attacks. Although the Army Air Corps had decided in 1938 to employ only established aircraft firms as prime contractors for final assembly of airframes, influenced by Britain's disappointing experience with the intial phases of the shadow factory scheme, automobile manufacturers and other outside subcontractors ultimately accounted for some 40 percent of wartime aircraft production by weight, including 48 percent of all aero engines and some two-thirds of all combat engines. Automobile company executives, reluctant to forgo current car sales and preparations for new models in the booming market of 1940–41, claimed initially that only 10–15 percent of their machine tools could be converted to aircraft production, but by June 1942 the effective proportion redeployed had reached 66 percent.[20]

The most extreme experiment with the application of Detroit-style mass production to military aircraft was Ford's vast greenfield plant at Willow Run, Michigan, constructed for the manufacture of the Consolidated B-24 bomber. In June 1940, Henry Ford and his associates had proclaimed their ability to produce one thousand fighter planes a day within eight months, provided that the design was frozen at the outset and changes during manufacture were "tabooed." Despite the urgent need for vast quantities of combat aircraft—more than a year later the U.S. Army's aerial striking force was still judged to be at "zero strength"—Ford's remarkable offer was rejected by the military authorities because none of the designs earmarked for production had yet passed beyond the experimental stage. Undeterred by this setback, the company accepted a contract to supply sets of knocked-down parts for the heavy, four-engined, B-24 bomber; risked its own financial and engineering resources to tool up for complete assembly; and retrospectively obtained official sanction—including reimbursement—for the resulting fait accompli. From the

Dream of Mass-Produced Fighter Aircraft: The XP-75 Fiasco," *Technology and Culture* 28 (1987): 578–79, from which the quotation is drawn; Tom Lilley et al., *Problems of Accelerating Aircraft Production during World War II* (Boston, 1947), pp. 5–7, 39–40, 53–56.

[20]On the mobilization of the U.S. automobile industry and its contribution to wartime aircraft production, see Holley, *Buying Aircraft,* pp. 290–92, 304–16, 560–63; W. F. Craven and J. L. Cate, eds., *The Army Air Forces in World War II,* vol. 6, *Men and Planes* (Chicago, 1955; reprint, Washington, D.C.: Office of Air Force History, 1983), pp. 319–30; Lilley et al., pp. 32–36.

outset, Ford engineers planned for mechanization and subdivision of labor on the vastest possible scale. The B-24 was broken down into 20,000 separate operations, while seventy major component sections were to be prefabricated in special areas and moved on conveyors to join the final assembly line. Some 29,000 dies and 21,000 jigs and fixtures (including special fixtures for all assembly operations) were developed at a total cost of $75–$100 million in order to reduce costs, increase accuracy, and ensure interchangeability.[21]

Despite its bold conception and innovative execution—both of which greatly impressed contemporaries—Willow Run was at best a Pyrrhic triumph, as Holley has elegantly demonstrated. As the British had discovered earlier, the difficulties of transferring automotive industry practice to aircraft manufacture proved much greater than Ford had envisaged. In 1941, the B-24 was still a "shop-engineered" aircraft, and all 30,000 of Consolidated's original drawings had to be redrawn to compensate for the more limited technical skills of Ford workers and supervisors. Aluminum airframes turned out to require very different pressing techniques than steel auto bodies, forcing the company to scrap some 14,000 dies and extensively rework 2,400 others before achieving satisfactory results. Nor could the original design long remain stable once exposed to combat conditions: according to Holley, "the last B-24 turned off the Willow Run production line was an entirely different aircraft from the original item" as a result of 130 major and thousands of minor changes to engines, fuel tanks, propellers, armament, and paint which increased its gross weight from 41,000 to 60,000 pounds. The unavoidability of such modifications, coupled with frequent material shortages, blocked the full exploitation of planned economies of scale, and little more than half of all jigs and fixtures constructed by the company at vast expense were ultimately used in production.[22]

By the end of 1944, person-hours per aircraft and manufacturing costs at Willow Run were well below the industry average, while its dedicated assembly fixtures and multioperation machines yielded significant advances in accuracy as well as dramatic reductions in direct labor requirements. But Willow Run was also the slowest of all plants making the Consolidated B-24 to achieve full quantity output—so slow, in fact, that it became widely known as "Will It Run?" Not least among the sources of delay was the problem of attracting a vast new labor force to rural Michigan in the absence of adequate housing, and Ford never succeeded in recruiting more than 40,000 of the 72,000 employees originally planned for the site. By the time Willow Run reached peak

[21]Holley, *Buying Aircraft,* pp. 518–21, 523–24, 526.

[22]Ibid., pp. 521–27; Lilley et al., pp. 49–50; Cairncross (n. 15 above), p. 177.

production, the B-24 itself was partially obsolescent and the factory's potential cost advantage could not be fully realized because of the air force's unwillingness to absorb more than two-thirds of its potential capacity. When military officers came to evaluate the project immediately after the war, they understandably concluded that the Ford approach was "not flexible enough" to use on a product that had not been "completely engineered and ready for mass production," though the B-24 was no different in this respect than other similar aircraft such as the B-25.[23]

To be sure, Ford's experience was by no means representative of American aircraft production during the Second World War. More than 90 percent of all airframes, nearly half of all engines, and a substantial proportion of components and accessories were manufactured by the established aircraft constructors in government-financed extensions and branch facilities, as well as their original home plants.[24] Despite the relatively short runs of individual planes predominant before the war, American aircraft firms, Jacob Vander Meulen argues, were "poised for mass production" because of congressional procurement policies based on price competition and the denial of intellectual property rights in new designs, together with the ideological influence of the automobile industry as a paradigm of modern manufacturing practice.[25] Once massive orders for military aircraft became available on financially attractive terms, therefore, these firms moved rapidly to apply line production methods such as progressive assembly, special-purpose machinery, systematic production control and scheduling, and careful balancing of individual operations, especially at their new purpose-built branch plants. At the same time, however, the aircraft manufacturers remained much more sensitive than the automobile firms and other outside subcontractors to the inevitability of frequent design changes, and, according to a postwar study conducted by the Harvard Business School, they were "ultimately able to introduce a high degree of flexibility" into line-production techniques. Among the most important such innovations was the development of "multi-line assembly" at the Boeing plant in Wichita, Kansas, whereby heavy bombers were broken down into an ensemble of major sections, each of which could be

[23]Holley, *Buying Aircraft*, pp. 326–27, 527–29; Lilley et al., p. 82–83; Overy, *Air War* (n. 5 above), pp. 164, 178. Ford's archrival General Motors failed completely in its attempt to mass produce a new fighter plane from preexisting standardized components: see Holley, "A Detroit Dream of Mass-Produced Fighter Aircraft."

[24]Holley, *Buying Aircraft*, pp. 560–62; Craven and Cate, eds. (n. 20 above), pp. 328–30, 354–56; Lilley et al., pp. 33–36.

[25]Vander Meulen, *Politics of Aircraft* (n. 19 above), esp. chaps. 7–8.

"pre-completed" (and therefore more easily modified) in separate areas of the factory rather than taking shape along a single moving "chassis."[26]

Although American aircraft manufacturers, like their British counterparts, accepted the necessity of periodic design modifications during the course of production, U.S. procedures for accommodating such changes differed significantly from those current on the other side of the Atlantic until relatively late in the war. In most American aircraft factories, designs were temporarily frozen to allow the production of uninterrupted batches of up to 1,500 planes at a time, with indispensable changes introduced retrospectively in twenty special "modification centers" scattered around the country. An entire center at Birmingham, Alabama, for example, was occupied nearly full-time with modifications to B-24s from Willow Run. Such retrospective modification was normally "quick and dirty," extremely costly, and highly labor-intensive: the number of person-hours involved could approach those spent in constructing the original plane, and Holley estimates that "anywhere from 25 to 50 percent of the total labor spent in turning out military aircraft was actually performed at the centers." By 1944, therefore, the U.S. military authorities had become increasingly concerned to rationalize the procedures governing design changes and to reduce the volume of retrospective modifications by making aircraft manufacturers responsible for the costs and administration of the modification centers on the one hand and by helping them to develop more effective techniques of production control on the other. Such reforms, which brought American aircraft manufacturers closer to contemporary British practice, were apparently so successful that during the Korean War a firm like Republic Aviation proved able to incorporate an average of 315 modifications per week in manufacturing the P-84.[27]

[26]Holley, *Buying Aircraft* (n. 16 above), pp. 290–304, 320–24; Craven and Cate, eds., pp. 306–18, 332–34; Lilley et al. (n. 19 above), pp. 18–20, 39–48, 53–56, quotation from p. 41. For "multi-line assembly" at Boeing Wichita and its diffusion to other B-29 plants, see Jacob Vander Meulen, "Flexible Mass Production at War: Heavy Bombers and Large Aircraft Engines in the United States, 1940–45" (paper presented at the Society for the History of Technology meeting, Washington, D.C., October 14–17, 1993), pp. 10, 14.

[27]Holley, *Buying Aircraft,* pp. 528–38; Craven and Cate, eds., pp. 334–37; Air Historical Office, *The Modification of Army Aircraft in the United States, 1939–1945,* Army Air Forces Historical Studies no. 62 (Washington, D.C., August 1947). Better control of production modifications was also necessary to avoid "flooding the tactical units with a heterogenous collection of equipment," which could unduly complicate supply, maintenance, and training procedures. During the autumn of 1943, for example, local commanders frantically demanded a host of design modifications to the P-40 fighter to improve its tactical effectiveness but later complained that "the logistical difficulties imposed by these changes nullified the gains anticipated." See Holley, *Buying Aircraft,* pp. 517–18.

In sheer quantitative terms, the accomplishments of the American mass-production approach to wartime aircraft were nothing short of extraordinary. In numerical terms, as table 1 shows, U.S. output of military aircraft overtook the British in 1941; by 1944 it was nearly four times that of Britain and more than twice that of any other belligerent. Measured in terms of structure weight, the British lead lasted until 1942, but within two years U.S. output was four-and-a-half times greater than that of Britain and nearly five times greater than that of Germany. Willow Run alone turned out one bomber every hour, an output equivalent in weight to 50 percent of the entire German aircraft industry at its peak.[28]

In qualitative terms, however, the results were less impressive. On the eve of Pearl Harbor in December 1941, as an internal air forces account put it, "not one of the new plants authorized after June 1940 . . . had yet produced a single plane; and none of them was destined to get into full production until 1943," by which time the military tide had already turned in Europe. Heavy, four-engined bombers took even longer to reach quantity output, and deliveries of the long-range B-29 needed for the war with Japan had barely begun by the spring of 1944. Willow Run, as we saw, was so slow in reaching full stride on the B-24 that the military value of its vast output was significantly reduced, while the fastest transitions to quantity output of heavy bombers were achieved by the home plants of Boeing and Consolidated, where, as in the case of aero engine manufacturers like Pratt and Whitney or Curtiss-Wright, skill levels remained higher and production methods closer to prewar job shop practice than in the branch and licensee factories.[29]

The American system of temporarily frozen designs and retrospective modifications rather than continuous improvement during the course of production likewise exerted a negative impact on both the quality and quantity of combat-ready aircraft, as Sir Michael Postan argued in his official history of British war production. Where tactical experience was accumulating rapidly, as in the case of bombers in the European theater in 1942–43, Postan observed, the "United States Air Force demanded urgent improvements all the time, and the 'modification centres' were soon choked up with aircraft." The flow of planes to squadrons was thus reduced well below the remarkable numbers

[28]Overy, *Air War* (n. 5 above), pp. 150, 164; Holley, *Buying Aircraft,* pp. 552–56; Craven and Cate, eds., pp. 329–31, 350–51.

[29]Air Historical Office, *Expansion of Industrial Facilities under AAF Auspices,* Army Air Forces Historical Study no. 40 (Washington, D.C., n.d.), p. 107, quoted in Craven and Cate, eds., p. 314; Holley, *Buying Aircraft,* pp. 324–25, 552–53; Lilley et al., pp. 1, 15–20, 25–30, 54, 81–83.

leaving the factory gates, as too was the average quality of aircraft in operation.[30]

British military planners and industrialists were well aware of the U.S. achievement in increasing aircraft output, and a series of technical missions toured American plants during the war to assess the applicability of their methods for securing productivity improvements under domestic conditions. Three central contrasts stood out among the findings of these missions: the larger average size and greater vertical integration of American aircraft factories; their bigger design and planning staffs predominately composed of college-educated engineers; and their more extensive use of moving conveyors, time study, and other line-production methods. Although the British missions were greatly impressed with the efficiency of American practice, recommending both the expansion of domestic technical education and the introduction of moving assembly lines "where output warrants it," they saw little opportunity for the direct transfer of U.S. methods to domestic aircraft production during the war itself. A 1942 mission concluded:

> The fundamental difference in the average size of the plants—large units in U.S.A. and dispersal into small units forced upon us by conditions in U.K.—vitally affects the methods of production, and the extent to which American ideas can be copied in U.K. Some methods of construction and layout of shops in America which are superior to those used in the U.K., are only possible in the relatively unlimited space available and on the scale on which they can then be employed; they could not be used in U.K. except perhaps in a few of the larger plants.[31]

As subsequent studies showed, larger plants and longer production runs meant that labor productivity in American airframe manufacture—measured in terms of structure weight per person-day—was 75 percent higher than that of Britain in 1943–44. But in those relatively few cases where British production was maintained for long enough to take

[30]Postan (n. 9 above), pp. 342–43. The proportion of planes subject to retrospective modification varied significantly by type: thus nearly 100 percent of heavy bombers and transports passed through the modification centers before delivery to the armed forces but only 30–50 percent of fighters. See Holley, *Buying Aircraft*, p. 537.

[31]"Report of British Mission to United States of America to Study Production Methods," September–October 1942, PRO, AVIA 10/104, esp. pp. 3, 9–10, from which quotations are drawn; "Report of the Fedden Mission to the United States," June 1943, PRO, AVIA 10/99, and précis of its findings, April 22, 1943, PRO, AVIA 10/106. For discussions of the findings of these missions, see Postan, pp. 335–36; Cairncross (n. 15 above), chap. 8; and, more tendentiously, Corelli Barnett, *The Audit of War: The Illusion and Reality of Britain as a Great Nation* (London, 1986), pp. 148–54.

advantage of the "learning curve"—fifty to sixty fighters or twenty to thirty bombers per week—output per worker was roughly equivalent and overall costs lower than in the United States.[32] Corelli Barnett claims in his influential polemic *The Audit of War* that the British produced 1.19 pounds of aircraft structure weight per man-day in 1944, compared to 2.76 pounds in the United States, implying a substantially larger productivity gap between the two countries. But Barnett's British calculations are based on employment figures for November 1943, rather than the actual number of man-days worked in 1944, although the labor force in the aircraft industry declined by more than 14 percent during this period.[33] Even more misleading, the figure given by Barnett for American labor productivity is actually a synthetic index of relative efficiency devised by the U.S. Strategic Bombing Survey (USSBS) in which raw data on pounds of weight per employee per day were corrected for variations between countries in product mix (types of aircraft) and scale of production (units per day). Were it possible to calculate a similar index of relative efficiency using British data, such corrections would undoubtedly reduce the initial disparity in person-days per pound of structure with the U.S., as the USSBS found in the cases of both Germany and Japan.[34]

In other respects, too, such aggregate statistical comparisons arguably exaggerate the American productivity advantage. Rapidly rising output per worker in U.S. aircraft factories—direct person-hours per pound of airframe accepted in July 1944 had fallen to 43 percent of those required in January 1943—were partially offset by subsequent reworking in modification centers, which as we have seen could add a further 25–50 percent to the total labor inputs involved.[35] In Britain, unlike the

[32]Eric Mensforth, "Airframe Production," *Proceedings of the Institution of Mechanical Engineers* 156 (1947): esp. 35–37. Subsequent analyses of the "learning curve" have tended to reduce the contribution of longer production runs per se to productivity gains as opposed to related improvements in product design, equipment, tooling, materials, and management: see Louis E. Yelle, "The Learning Curve: Historical Review and Comprehensive Survey," *Decision Sciences* 10 (1979): 302–28; R. M. Bell and D. Scott-Kemmis, "The Mythology of Learning-by-Doing in World War II Airframe and Ship Production," *Industrial and Corporate Change* 4 (1994), in press. I am grateful to David Hounshell for these references.

[33]See Barnett, pp. 145–46; Edgerton, *England and the Aeroplane* (n. 9 above), pp. 79–80; Overy, *Air War* (n. 5 above), p. 171.

[34]For the method used in constructing the original efficiency index, see USSBS, Aircraft Division, *The Japanese Aircraft Industry* (Washington, D.C., May 1947), reprinted in *The United States Strategic Bombing Survey*, ed. David MacIsaac (New York, 1976), 7:27–29, and *Aircraft Division Industry Report: Strategic Bombing of the German Aircraft Industry*, 2d ed. (Washington, D.C., November 1947) (hereafter cited as *Strategic Bombing of the German Aircraft Industry*), reprinted in MacIsaac, ed., vol. 2, figs. 6-11 and 6-12; Jerome B. Cohen, *Japan's Economy in War and Reconstruction* (Minneapolis, 1949), pp. 218–19. The origins of Barnett's error can be traced to Overy, *Air War*, table 15, p. 168, quoted in Barnett, p. 146.

[35]Holley, *Buying Aircraft*, pp. 564, 532.

United States, repair of damaged aircraft was performed on the same facilities as new construction, reducing the apparent productivity of the latter but increasing the overall effectiveness of war production: 48 percent of all planes delivered to the Metropolitan Air Force (that portion of the RAF based in Britain itself) were repaired during the war, using one-third to one-fifth of the raw materials and labor needed for equivalent new construction.[36] Heavy bombers and large transport planes constituted a much larger proportion of American than British output during the final years of the war—40 percent of all planes delivered to the army air forces in 1944 compared to 24 percent of those received by the RAF, and an even larger disparity in terms of weight—but required fewer person-hours per pound of structure than lighter aircraft such as fighters.[37] Structure weight per person-hour, as American aircraft manufacturers and British military planners both insisted, was in any case a highly imperfect indicator of relative efficiency since it failed to capture the impact of variations between firms—and national economies—in factors like labor quality, degree of vertical integration, frequency of modifications, and manufacturability of individual designs, not to mention the tactical value of the planes themselves.[38]

Germany: Quality and Quantity Out of Balance

Germany, the Allies' main aerial antagonist, appears by all accounts to have managed this trade-off between quality and quantity of aircraft less successfully than either Britain or the United States. All the major combatant powers oscillated between defensive phases, in which design changes were severely restricted in order to overcome a numerical disadvantage by maximizing output of current types, and offensive phases, in which greater emphasis was placed on the incorporation of modifications aimed at achieving tactical superiority over enemy air-

[36]"Report of the British Mission" (n. 31 above), p. 16; Postan (n. 9 above), pp. 316–22. Forty-five percent of American combat aircraft were nonetheless undergoing repair on any given day: see Holley, *Buying Aircraft,* p. 246.

[37]Holley, *Buying Aircraft,* pp. 553–55; Postan, app. 4, p. 485. According to 1940 calculations by the British Air Ministry quoted in Barnett (n. 31 above), p. 146, "a ton of heavy aircraft represented less added value and a smaller industrial effort than a ton of lighter aircraft." The USSBS reckoned that the "unit cost or production hours per airplane varies inversely as the weight to the one-third power." See USSBS, *Japanese Aircraft Industry,* p. 27.

[38]For the views of American aircraft manufacturers, see East Coast and Central Aircraft War Production Councils, letter to T. P. Wright re Indices of Airplane Production Efficiency, May 8, 1944, National Archives, Washington, D.C., Truman Committee Papers, Box 685. I am grateful to Jacob Vander Meulen for supplying a copy of this document. For the views of a leading British aircraft planner, see Ely Devons, "Statistics and Planning," in his *Planning in Practice: Essays in Aircraft Planning in War-Time* (Cambridge, 1950), esp. p. 150: "nobody was really interested in the weight of aircraft produced, but in its fighting power," quoted in Edgerton, *England and the Aeroplane* (n. 9 above), p. 80.

craft. In Germany, however, these swings were particularly violent, largely for political and administrative rather than industrial reasons. After freezing existing models for serial production at the onset of aerial rearmament in 1933–34 and again in the wake of the international crises of 1938, the German military authorities were given free rein to demand frequent and radical design changes with little regard for their impact on aircraft output during the crucial period between 1939 and 1942 when the Allies were building up their vast expansion programs. From mid-1941 on, Field Marshall Erhard Milch as technical director at the Air Ministry sought to reorient the industry toward mass production of existing types at the expense of new models, but his efforts were largely frustrated by Hermann Goering's insistence on the development of heavy bombers to support a renewed offensive against Britain, Russia, and ultimately the United States. In 1944, as Allied forces closed in on the Nazis' European empire, Milch and Albert Speer as minister of munitions finally achieved a quantum leap in output of fighter planes for the defense of Germany, but only by concentrating production on a small number of standard types and imposing a ruthless freeze on design modifications.[39]

These violent oscillations and conflicts within German production policy adversely affected both the quantity and the quality of aircraft output. Despite the remarkable accomplishments of the German aircraft industry and the Luftwaffe during the late 1930s, by the time of the Battle of Britain in 1940, U.K. output was 50 percent higher, as we have seen, while the quality of fighters on the two sides was roughly equal.[40] Between 1939 and 1941, the human and material resources pumped into the German aircraft industry increased by 100 percent, but, as table 1 shows, output rose by only 30 percent, compared to planned growth of 100–400 percent. As Richard Overy has argued, the inefficiency of German aircraft production during the early years of the war can be gauged by the fact that with no more labor and fewer raw materials than in 1941, the industry turned out 31 percent more planes in 1942 and 111 percent more in 1943 (30 percent and 85 percent more in terms of

[39]This sketch of the evolution of German aircraft policy and production during the 1930s and 1940s draws on the following major secondary accounts: Edward L. Homze, *Arming the Luftwaffe: The Reich Air Ministry and the German Aircraft Industry, 1919–39* (Lincoln, Neb., 1976); Overy, *Air War* (n. 5 above), esp. chaps. 1 and 7, *Goering: The "Iron Man"* (London, 1984), esp. chaps. 6–7, and "German Aircraft Production, 1939–1942" (Ph.D. diss., University of Cambridge, 1977); Edward R. Zilbert, *Albert Speer and the Nazi Ministry of Arms: Economic Institutions and Industrial Production in the German War Economy* (London, 1981); Cairncross (n. 15 above), chap. 7; Hans-Joachim Braun, "Fertigungsprozesse im deutschen Flugzeugbau, 1926–1945," *Technikgeschichte* 57 (1990): 111–35, and "Aero-Engine Production in the Third Reich," *History of Technology* 14 (1992): 1–15.

[40]For the qualitative parity of British and German fighters in 1940, see Edgerton, *England and the Aeroplane*, p. 37.

weight); in 1944, with virtually the same resources as in 1940, the Germans produced nearly four times the number of planes (238 percent more than 1941 in terms of numbers and 126 percent more in terms of weight). During the last full year of the war, the Germans overtook the British in numbers of planes manufactured and had nearly caught up with them in terms of structure weight as well. By that time, however, matching the British was no longer a relevant target, as the German "production miracle" of 1944 could not prevent the Allies from securing overwhelming air superiority given their vast combined output.[41] In terms of productivity, too, the German aircraft industry fell further behind the United States even as it improved dramatically on its own past performance. Even though output per person-hour in the German aircraft industry increased by some 200–300 percent between 1941 and 1944, the USSBS's index of relative efficiency (which made allowances for variations in the composition and scale of output) put the American advantage at 110 percent in 1944, compared to 20–40 percent in 1941–43.[42]

Equally important was the relative decline in the quality of German aircraft during the war. Of the new generation of aircraft designed to come on stream in 1942—the Heinkel 177 and Junkers 288 heavy bombers and the Messerschmitt 210 heavy fighter—none proved successful enough for large-scale manufacture, and the Air Ministry was therefore obliged to maintain older types in production much longer than originally planned. Even where the Germans managed to develop a radical new design ahead of the Allies, as in the case of the Messerschmitt 262 jet-engined fighter, they failed to bring it into production fast enough to exploit the plane's military potential, in part because of personal interference from Hitler himself, who wanted it converted into a fighter-bomber. Conversely, where the Germans sought to accelerate development of new types through shortcuts such as making several prototypes simultaneously or using a high proportion of preexisting parts, problems constantly cropped up, as in the case of the Heinkel 162 single-engined jet fighter of which some 100 were built before the end of the war: the firm's chief test pilot was killed on its second flight and the design was plagued by persistent technical weaknesses. By 1943–44, this inability to introduce new models, together with growing restrictions on modifications to existing types, meant that the Luftwaffe had lost any claim to qualitative superiority over the Allies even on conventional aircraft.[43]

[41]Overy, *Goering,* pp. 102, 148–50, 176–77, 185, 190–93, 201–4.

[42]Ibid., pp. 150, 190–91; USSBS, *Strategic Bombing of the German Aircraft Industry* (n. 34 above), p. 85.

[43]Overy, *Goering,* pp. 166–68, 191–93, 197; Cairncross, pp. 130–31, 137–45, 149–53.

The relative failure of the German aircraft industry during World War II has often been attributed to the limitations of craft as opposed to mass production. Thus historians like Richard Overy and Hans-Joachim Braun argue that German industry's failure to match—or even approach—the volume of aircraft turned out by the Allies was partially due to the persistence of "artisanal" (*handwerkerliche*) methods and production ideals. Particularly during the early years of the war, these historians contend, skilled workers, craft-trained foremen or *Meister*, and company managers—like the military authorities themselves—resisted attempts to bypass apprenticeship and dilute the workforce through the introduction of semiskilled labor and large-scale manufacturing techniques. A skilled, adaptable workforce and flexible production methods were also necessary to cope with the frequent design changes demanded by the Luftwaffe. In many firms, therefore, aircraft continued to be built by craftsmen using general-purpose equipment, rather than by less skilled workers using specialized machines, which were themselves in short supply because of the predominance of similar "conservative" attitudes in the German machine tool industry itself. Under these conditions, Overy maintains, "Batch production gave way only slowly to line production, while time-and-motion studies on which American practice rested were either not introduced or where they were, proved unworkable because of the traditional methods of work payment and use of skills. Conveyor belt production and rational factory organization developed only slowly and many factories still had little evidence of modern methods even by 1945."[44]

This portrait of German aircraft production and its shortcomings draws heavily on contemporary criticisms by domestic rationalization advocates and by Allied intelligence officers who surveyed the industry immediately after the war. Thus, for example, a British Intelligence Objectives Sub-Committee (BIOS) team charged with investigating methods of production control in German factories concluded that "management was interested much more with [*sic*] the technical development of their products rather than with techniques of control. . . . With the higher proportion of skilled workers in German industry and the greater effort which they were prepared to give there was not so much demand for good internal production organization."

[44]Overy, *Air War* (n. 5 above), pp. 170 (from which the quotation is drawn), 172–74, *Goering*, pp. 159–62, 184–87, 190–91, and "German Aircraft Production" (n. 39 above), pp. 151–70, 205–19; Braun, "Fertigungsprozesse im deutschen Flugzeugbau," esp. pp. 116–19, 127, which speaks of the continuing predominance of "handwerklichen Produktionsideal orientierte Fertigungsmethoden" even at the end of the war, "Aero-Engine Production in the Third Reich" (n. 39 above), pp. 5–6, 11–13, and *The German Economy in the 20th Century* (London, 1990), pp. 134–38.

Of all the aircraft firms visited, only Heinkel's had created a central office for production control and progress chasing as was increasingly the practice in large British concerns. Even there, control of parts movement within a shop remained the sole responsibility of the superintendant or foreman, while irregular supplies of materials and "constant modifications in program and design" meant that "controlled production was hardly to be thought of, let alone line production," detailed planning, or the use of special-purpose machinery: "Indeed factories had to be in production so quickly to meet Air Ministry demands, that output could only be achieved through flexible organisation and the provision of general-purpose machine tools." Another BIOS report on the German aircraft instrument industry similarly found much less dilution and use of simplified methods (including time-and-motion study) than were common in the United Kingdom; German instrument firms, in contrast to their British and particularly American counterparts, "avoided elaborate mechanisation by relying on the great skill of the labour employed." Although the USSBS placed less emphasis than the British BIOS teams on the distinctiveness of German production methods, it too ascribed much of the productivity gap between the two countries to "inefficient labor utilization" in Germany, notably, lesser use of women workers and "high production, special purpose tools."[45]

Some elements of this contrast between productive organization in German and Allied aircraft factories are clearly well-founded. There is general agreement, for example, among historians and contemporary observers alike that until the later years of the war German aircraft firms, like most of the metalworking sector, relied more heavily than their Allied counterparts on the skills of their workforce. Such policies reflected the fact that, as the USSBS observed, "the supply of competent mechanics in the German labor market was much greater than that available to American aircraft manufacturers," as a result of the broad coverage achieved by a revitalized system of craft apprenticeship combining in-plant manual training with technical instruction in public vocational schools.[46] It is also true that both before and during the war

[45]BIOS, *Investigation of Production Control and Organisation in German Factories,* Final Report 537 (London, 1946), pp. 1, 7–8, and "Appendix: Production Control in the Heinkel Aircraft Organisation," pp. 2, 12–13, 19–20, 23–25; BIOS, *Conditions in the German Aircraft Instrument Industry,* Final Report 881 (London, 1947), pp. 8, 10–11, 13; USSBS, *Strategic Bombing of the German Aircraft Industry,* pp. 85–86. For criticisms by German rationalization advocates, see Overy, *Goering,* pp. 162, 184, and "German Aircraft Production," pp. 154–55, 255–56.

[46]USSBS, *Strategic Bombing of the German Aircraft Industry,* p. 86; cf. also Combined Intelligence Objectives Sub-Committee (CIOS), *A Survey of Production Techniques Used in the*

many German industrialists and military planners were critical of American-style mass-production methods for a combination of economic, technological, and ideological reasons, favoring instead production strategies based on customization, flexibility, and "quality work."[47]

In other respects, however, this opposition between German craft traditionalism and Allied mass-production modernism is seriously overdrawn and misleading. Design modifications and program changes, as Overy himself recognizes, were not an optional luxury irrationally insisted on by technologically fastidious air force officers, but rather an indispensable means of adapting equipment with long development times to rapidly shifting tactical conditions and strategic priorities.[48] Even in the United States, therefore, program and engineering changes in wartime aircraft production were sufficiently common that the USSBS claimed—probably mistakenly—that there was no substantial difference between the two countries in this regard.[49] In Britain, the incorporation of design modifications seems, if anything, to have been more continuous than in Germany, at least for certain sections of the industry. Thus British fighters retained a power edge throughout the war, as Sir Alec Cairncross of MAP found when he interviewed German designers in

---

German Aircraft Industry, report XXV-42 (Washington, D.C., July 1945), pp. 4–6. For the development of the German apprenticeship system between the wars, see Albin Gladen, "Die berufliche Aus- und Weiterbildung in der deutschen Wirtschaft 1918–45," in *Berufliche Aus- und Weiterbildung in der deutschen Wirtschaft seit dem 19. Jahrhundert,* ed. Hans Pohl (Wiesbaden, 1979), pp. 53–73. From 1941–42 on, however, foreign workers, prisoners of war, and concentration camp inmates came to form a growing proportion of the workforce, particularly in the engine plants: see Overy, "German Aircraft Production," pp. 207, 212–13, 270–72; Braun, "Aero-Engine Production in the Third Reich, p. 9; CIOS, *A Survey of Production Techniques in the German Aircraft Industry,* pp. 6–7, 11–12.

[47]For German criticisms of American mass-production methods before World War II, see Heidrun Homburg, "Anfänge des Taylorsystems in Deutschland vor dem Ersten Weltkrieg," *Geschichte und Gesellschaft* 4 (1978): 170–94, esp. 174; Hans-Liudger Dienel, "German Opposition against the American System of Production" (paper presented to the Society for the History of Technology meeting, Uppsala, Sweden, August 16–20, 1992); Gary B. Herrigel, "Industrial Organization and the Politics of Industry: Centralized and Decentralized Production in Germany" (Ph.D. diss., Massachusetts Institute of Technology, 1990), esp. pp. 277–79, 439–40, and "Industry as a Form of Order: A Comparison of the Historical Development of the Machine Tool Industries in the United States and Germany," in *Governing Capitalist Economies: Performance and Control of Economic Sectors,* ed. J. Rogers Hollingsworth, Philippe C. Schmitter, and Wolfgang Streeck (Oxford, 1994), pp. 97–128. For the German military's commitment to the doctrine of "qualitative superiority" in armaments production, see also Alan S. Milward, *The German War Economy* (London, 1965), chap. 5.

[48]Overy, *Air War,* pp. 177–78, and *Goering* (n. 39 above), pp. 180–81; but cf. also Milward, *German War Economy,* pp. 159–60.

[49]USSBS, *Strategic Bombing of the German Aircraft Industry* (n. 34 above), pp. 87–88.

1945, mainly because Rolls-Royce "had found ways of modifying the Merlin [engine] on the production line, with or without a change of Mark number, to yield more and more engine thrust," while firms like BMW "had relied on discontinuous changes in specification as one mark of engine replaced another," so that "any lead they established . . . did not last long."[50]

In both Anglo-Saxon powers, similarly, frequent design changes demanded a substantial measure of adaptability in aircraft manufacturing methods. In the United States, as we have seen, aircraft firms modified standard line-production techniques to obtain greater flexibility, while in Britain a large proportion of aircraft components were turned out in batches small enough to make bench methods more economical than full mass-production tooling.[51] Nor was opposition to rationalization and deskilling a German peculiarity. Many British craftsmen opposed dilution throughout the war, while even in the United States, as a Harvard Business School team observed, old-line aircraft manufacturers experienced considerable difficulty in "persuading foremen and other supervisors, who had grown up under job shop conditions, to accept the engineering discipline that large-scale production . . . required."[52] More fundamentally still, as I have sought to show elsewhere, many British engineers and industrialists, like their German counterparts, remained skeptical before the war about the suitability of American mass-production methods in catering to the diverse and fluctuating markets characteristic of domestic metalworking.[53]

The received contrast between craft traditionalism and mass-production modernism in wartime aircraft manufacture is misleading from the German as well as the Allied perspective. During the 1930s, as the USSBS noted, "German manufacturers had made an intensive study of aircraft tooling . . . and had evolved techniques of 'series' or line production which were advanced for that time." According to Edward

[50]Cairncross (n. 15 above), pp. 16, 148–49; on Rolls-Royce's commitment to flexibility and continuous improvement, see also Lloyd, *Rolls Royce* (n. 18 above), pp. 23, 72–74, 118–25.

[51]The CIOS survey rated German manufacturing methods superior to those of the U.S. insofar as "new products could be put into production or designs changed more quickly and easily than they could in America," while also noting that "it is possible that British methods were even more flexible than the German": see *A Survey of Production Techniques Used in the German Aircraft Industry* (n. 46 above), p. 5.

[52]For British opposition to dilution, see Parker, "British Rearmament" (n. 14 above); Inman (n. 14 above), chaps. 2–3. For difficulties in reorienting job-shop supervisors in American aircraft plants, see Lilley et al. (n. 19 above), p. 54.

[53]See Jonathan Zeitlin, *Between Flexibility and Mass Production: Strategic Debate and Industrial Reorganization in British Engineering, 1830–1990* (Oxford, 1996, in press). An article covering the period before 1914 will appear in Sabel and Zeitlin, eds., *Worlds of Possibility* (n. 8 above).

Homze, a leading historian of German aerial rearmament, one such technique, the *Baukastenflugzeuge,* a standardized basic airframe which could be combined with different engines and equipment to serve a variety of missions, was carried so far as to inhibit the search for new design solutions within the industry. Another innovation of the 1930s, the so-called hole system (*Lochbauweise*), which avoided the need for elaborate assembly jigs to secure interchangeability by mass prepunching of airframe sections, anticipating the methods introduced by Ford at Willow Run, proved so inflexible during the war that large quantities of special tooling had to be scrapped every time there was a significant change in production programs.[54]

At the end of the war, too, the USSBS claimed that German aircraft manufacturing techniques "corresponded rather closely with American practice," including the use of "moving belts, with the work along them carefully planned at definite stations"; the main difference was that "operations were broken up among buildings in the same location or dispersed over a wide area" as a defensive measure against bombing attacks. From 1941 on, as Braun has documented, leading industrialists like William Werner (who had worked for Chrysler in Detroit before becoming technical director of Auto Union) and the noted designer Willy Messerschmitt formulated ambitious plans for expanding output of aero engines and airframes, respectively, by specializing suppliers on individual components and by extending the use of modular construction on the *Baukasten* or "Meccano-set" principle. Each of these proposals was eventually rejected, however, principally because they would have increased the vulnerability of aircraft production to disruption by Allied bombing.[55]

Finally, like their British and American counterparts, German aircraft manufacturers devised innovative techniques for realizing some of the benefits of large-scale production without excessive loss of flexibility. Prominent among these were the widespread use of "universal" assem-

[54]For the general point about serial production in the 1930s, see USSBS, *Strategic Bombing of the German Aircraft Industry* (n. 34 above), p. 21. On the *Baukastenflugzeuge,* see Homze (n. 39 above), p. 119; on the "hole system" and its disadvantages, see Braun, "Fertigungsprozesse im deutschen Flugzeugbau" (n. 39 above), pp. 115–16, and *German Economy in the 20th Century* (n. 44 above), pp. 137–38; Zilbert, *Albert Speer* (n. 39 above), pp. 226–27.

[55]USSBS, *Strategic Bombing of the German Aircraft Industry,* pp. 19–20, 97; Braun, "Fertigungsprozesse im deutschen Flugzeugbau," pp. 121–31, and "Aero-Engine Production in the Third Reich" (n. 39 above), pp. 7–8. British investigators, on the other hand, claimed that "strict line production was certainly not possible in Germany" during the war, with the partial exception of Opel, General Motors' prewar German subsidiary: see BIOS, *Investigation of Production Control and Organisation in German Factories* (n. 45 above), p. 7, and "Appendix: Production Control in the Heinkel Aircraft Organisation" (n. 45 above), pp. 13–14.

bly jigs, which "could be adjusted within certain limits to permit fairly wide changes in design," and Heinkel's combination of a modified form of line production (known as "repetition" or "rhythm" work) in the earlier and later stages of airframe manufacture with batch production in other shops.[56]

The central weakness of German aircraft production, on this analysis, lay not in the fact of program changes and design modifications, or even their frequency, but rather in the way they were planned and administered. Unlike in Britain or the United States, alterations in programs and designs were not screened for their potential impact on aircraft output by officials with detailed knowledge of current industrial conditions. Nor did the Germans devote comparable effort to minimizing the impact of such changes on aircraft output by careful phasing of their introduction (whether continuously as in Britain or discontinuously as in the United States) and by systematic coordination among manufacturers of airframes, engines, components, and accessories. Product selection decisions, too, were taken without realistic appraisal of the lead times necessary for the development of new weapons systems, while output targets were often based on mistaken estimates or even deliberate misrepresentations of German and Allied productive capacity. As a special commission established in 1943 to review the problems of the German aircraft industry concluded, the most important cause of lost production during the course of the war was the lack of "clear planning" and "precise allocation of tasks" within the Air Ministry.[57]

This failure of German aircraft planning, as Overy above all has demonstrated, can be traced directly to the decision-making structure of the Nazi regime. Responsibility for aircraft development and production was dispersed among a shifting array of overlapping and competing centers of power—from individual firms, military officers, and government ministers on up to Hitler himself. The punitive response to bad news or pessimistic forecasts within the Nazi political system likewise discouraged upward transmission of accurate but unwelcome informa-

[56]USSBS, *Strategic Bombing of the German Aircraft Industry,* pp. 20, 88; BIOS, "Appendix: Production Control in the Heinkel Aircraft Organisation," pp. 12–13. For experiments at Daimler-Benz during the 1930s with hybrid forms of modified flow production based on mixed-model assembly and conveyor-belt "islands," see Michael Stahlman, "Management, Modernisierungs- und Arbeitpolitik bei der Daimler-Benz AG und ihren Vorläuferunternehmen von der Jahrhundertwende bis zum Zweiten Weltkrieg," *Zeitschrift für Unternehmensgeschichte* 37 (1992): 161–63.

[57]See Overy, *Air War* (n. 5 above), pp. 154–59, 177–80; *Goering* (n. 39 above), pp. 105–6, 159–62, 169–72, 179, 187–88 (from which the quotation is taken), 192–93, 197; Cairncross (n. 15 above), pp. 117–22, 129–36, 139, 147–49; USSBS, *Strategic Bombing of the German Aircraft Industry,* pp. 105, 108; CIOS, *A Survey of Production Techniques Used in the German Aircraft Industry* (n. 46 above), p. 11.

tion by subordinates. In contrast to Britain and the United States, no independent civilian ministry or department was established for the administration of German aircraft production to temper the military's demands for new and improved designs with a parallel concern for increased output; even inside the Air Ministry itself air strategy and aircraft development were largely planned in isolation from one another until midway through the war. As in the other Axis powers, the military retained a dominant influence over the procurement and specification of aircraft, while little effort was made to involve industry in the planning and modification of production programs. All this began to change from 1941–42 with the reforms introduced by Milch and Speer, which included the representation of firms on a series of councils, committees, and "rings" aimed at improving the coordination of aircraft production. But internal conflicts and contradictory objectives within the planning apparatus continued to inhibit the Germans from achieving a satisfactory balance between the quality and the quantity of aircraft output until the scales tipped decisively toward the latter in the desperate conditions of 1944–45. If the air war of 1939–45 was among other things a contest between competing models of productive organization, this contest was overshadowed and overdetermined, as contemporaries were all too acutely aware, by a larger contest between competing political regimes with sharply different patterns of civil-military relations.[58]

Conclusions

Aircraft manufacture did not exhaust the full range of military production during World War II. Analysis of other major weapons systems would show different balances between quality and quantity across the three countries. In the case of tanks, for example, the Germans appear to have been rather more successful in maintaining qualitative superiority over the Western Allies through frequent innovations in technology and design, while also sustaining higher levels of output than the British (but not the Americans) throughout the conflict.[59] The central element of German qualitative superiority in

[58]This is Overy's central argument in *Goering,* chaps. 6–7, *Air War,* chap. 7, and "German Aircraft Production" (n. 39 above), passim. For similar judgments about the primacy of political and administrative factors in the failure of the German aircraft industry during the war, see also Braun, "Fertigungsprozesse im deutschen Flugzeugbau," pp. 119, 127, and "Aero-Engine Production in the Third Reich," pp. 6–7; USSBS, *Strategic Bombing of the German Aircraft Industry,* p. 97; "Final Report of the Farren Mission to Germany: A Survey of a Cross Section of German Aircraft, Aircraft Engine and Armament Industries," 1945, PRO, AVIA 9/85.

[59]For brief comparative overviews of tank production during World War II, see Milward, *War, Economy and Society* (n. 5 above), pp. 181–82; R. A. C. Parker, *Struggle for Survival: The History of the Second World War* (Oxford, 1989), pp. 133–34; György Ránki, *The Economics of the Second World War* (Vienna, 1993), chap. 3.

tanks was the rapid introduction of new heavy types such as the Panther and the Tiger after encountering the Soviet T-34s in 1941–42. Rationalization and coordination of production were also more effective in tank manufacture than in that of aircraft, in part because of its higher priority in German military planning during the late 1930s and early 1940s.[60] Although the British had invented the tank during World War I and pioneered in its development through the early 1930s, their production during the subsequent decade was impeded by a combination of unresolved ambiguities about the army's military role, the low priority accorded to its rearmament, initial selection of weak firms as shadow manufacturers, and a misconceived attempt to combine drawing-board orders with severe restrictions on design modifications—a near-total contrast with aircraft procurement policies.[61] The Americans were more successful than the British in converting locomotive and motor vehicle firms to tank production after 1940, in part because they could profit from the latter's experience, but they too encountered major problems in both design and production, leaving U.S. armor clearly inferior to that of Germany at the time of the D-day invasion of 1944.[62]

But aircraft production was the largest single sector of the war economy in all three countries, involving vast numbers of civilian manufacturers as subcontractors and licensees, and Allied air superiority was widely regarded as a decisive factor in defeating the Axis powers. Hence wartime experience of aircraft manufacture—or rather a particular reading of its lessons—exerted a powerful influence on postwar reconstruction planning beyond the industry itself, especially in Britain.

British industrialists and government officials, as we saw earlier, toured both American and German aircraft factories during and immediately after the war, studying the organization of production and design with an eye to the reform of domestic practice. Despite concerns about the inflexibility of American aircraft manufacture and the adequacy of its system of retrospective design modifications, British wartime missions to the United States were more impressed by the enormous productivity gains obtainable through mass production—if only the necessary eco-

[60]Harmut H. Knittel, *Panzerfertigung im Zweiten Weltkrieg: Industrieproduktion für die deutsche Wehrmacht* (Herford, 1988); Milward, *The German War Economy* (n. 47 above), pp. 100–102, 148.

[61]Postan (n. 9 above), pp. 27–34, 160, 183–95; Postan, Hay, and Scott (n. 16 above), chaps. 13–14; Brian Bond, *British Military Policy between the Wars* (Oxford, 1980), chaps. 5–6; David Fletcher, *The Great Tank Scandal: British Armour in the Second World War* (London, 1989), pt. 1.

[62]See Constance McLaughlin Green, Harry C. Thompson, and Peter Roots, *The Ordnance Department: Planning Munitions for War* (Washington, D.C., 1953), pp. 189–204, 236–39, 250–59, 267–72, 275–304; Harry C. Thompson and Lida Mayo, *The Ordnance Department: Procurement and Supply* (Washington, D.C., 1960), chaps. 10–11.

nomic scale of operation could be achieved at home. Postwar missions to Germany, conversely, were impressed by the skills and training of the labor force at all levels, from engineers and technicians to foremen and manual workers, but not by production methods and organization, which they regarded as inferior not only to that of American firms but also of well-run British companies. "Thus as regards the internal organisation of German industrial concerns we have very little to learn," concluded the BIOS investigation of production control; and this finding was echoed in the reports of other BIOS teams dealing with individual metalworking sectors such as aircraft instruments, machine tools, and power presses.[63]

Hence British government planners took U.S. mass production and the "three charmed 'S's" of simplification, standardization, and specialization as their model for the reconstruction of the domestic metalworking industries, a message reinforced by the postwar Anglo-American productivity missions organized in the context of the Marshall Plan. Some domestic manufacturers eagerly embraced this agenda, as in the case of Standard Motors, which reconverted its wartime shadow factory for volume production of standardized cars and tractors with interchangeable engines. But others, including the leadership of the Federation of British Industries, were less enthusiastic about the Labour government's productivity drive for a combination of economic and political reasons.[64]

Yet as the mergers and reorganizations which swept the metalworking industry during the 1950s and 1960s would demonstrate, British businessmen and public officials alike increasingly accepted mass production and American management methods as a transnational standard of manufacturing efficiency. Far from reviving its competitive fortunes, however, this putative Americanization of British industry was associated instead with a rapid loss of market share both at home and abroad, resulting in a steep decline of domestic production and employment. By

[63]BIOS, *Investigation of Production Control and Organisation in German Factories* (n. 45 above), p. 8; cf. also BIOS, *Conditions in the German Aircraft Instrument Industry* (n. 45 above), p. 9, *The German Machine Tool Industry,* Final Report 641 (London, 1947), pp. 18–20; *German Gauge and Tool Industry,* Final Report 632 (London, 1947), p. 2, and *German Power Press Industry,* Final Report 1826 (London, 1949), pp. 3–6.

[64]On reconstruction planning, the postwar productivity drive, and their limitations, see Jim Tomlinson, "Productivity Policy," in Helen Mercer, Neil Rollings, and Jim Tomlinson, eds., *Labour Governments and Private Industry: The Experience of 1945–1951* (Edinburgh, 1992), pp. 37–54, and "Mr Attlee's Supply-Side Socialism," *Economic History Review,* 2d ser., 46 (1993): 1–22; Tiratsoo and Tomlinson, *Industrial Efficiency and State Intervention* (n. 15 above), chaps. 2–7; Zeitlin, *Between Flexibility and Mass Production* (n. 53 above), chap. 10. For the specific case of Standard Motors, see Nick Tiratsoo, "The Motor Car Industry," in Mercer, Rollings, and Tomlinson, eds., pp. 162–85.

the 1980s, ironically, the competitive difficulties of British metalworking firms, like those of the Americans themselves, were frequently attributed to their inability to match the standards of product innovation and productive flexibility set by the Germans and the Japanese in meeting the demands of increasingly diverse and volatile international markets.[65]

In the United States, by contrast, wartime aircraft production appears to have had little impact on domestic manufacturing practice outside the industry itself. Before the war, as a 1947 report by a Harvard Business School team observed, "standardization of design and steadiness of scheduled rate had been considered fundamental prerequisites to line production. The success of the mass production industries was built in large measure on their ability to call a halt to changes prior to large-scale output so that carefully developed plans could be put into operation without any serious interruption." Hence they concluded: "The fact that the aircraft industry was ultimately able to introduce a high degree of flexibility into production procedures, and thereby to make effective use of line production techniques in spite of change, constituted an outstanding contribution to production management."[66]

Yet there is little evidence that this contribution had much subsequent influence on production management in postwar civilian industry. In automobiles, for example, the "Big Three" manufacturers pursued productivity gains at the expense of product innovation through standardization of major components across models, specialization of production facilities, and the introduction of dedicated automation equipment such as transfer machines. Product competition focused increasingly on annual changes in body styling and superficial variations between models in outfitting and trim, while the resulting proliferation of options was accommodated through greater labor intensity and high buffer inventories within the assembly process. By the 1970s and 1980s, these strategies had left American automobile firms extremely vulnerable to competition from foreign manufacturers—above all the Japanese—able to combine lower costs and higher quality with more flexible response to market trends and more rapid introduction of new models.[67]

[65]For preliminary accounts of the transformation of British industrial structure and managerial strategy since the 1950s and contemporary contrasts with German and Japanese practice, see Steven Tolliday and Jonathan Zeitlin, eds., *The Power to Manage? Employers and Industrial Relations in Comparative-Historical Perspective* (London, 1991), esp. pp. 282–86; Paul Hirst and Jonathan Zeitlin, eds., *Reversing Industrial Decline? Industrial Structure and Policy in Britain and Her Competitors* (Oxford, 1989), and "Flexible Specialization and the Competitive Failure of UK Manufacturing," *Political Quarterly* 60 (1989): 164–78.

[66]Lilley et al. (n. 19 above), p. 41.

[67]For synthetic overviews, see William J. Abernathy, *The Productivity Dilemma: Roadblock to Innovation in the Automobile Industry* (Baltimore, 1978); William J. Abernathy, Kim B. Clark, and Alan B. Kantrow, *Industrial Renaissance: Producing a Competitive Future for America* (New

The case of Germany is the most ambiguous of the three countries. Some metalworking firms, inspired by wartime rationalization programs as well as by the American model, moved aggressively into high-volume manufacture during the 1950s and 1960s, restricting and standardizing their product ranges, reorganizing production on flow lines, introducing special-purpose machinery and dedicated automation equipment, and reducing their dependence on skilled labor. The clearest example is that of Volkswagen, whose low-cost Beetle had been planned before the war but was not manufactured on a large scale until after 1945. Less dramatic but equally significant shifts in this direction could be observed among more established motor vehicle manufacturers, as well as among large concerns in other metalworking sectors such as electrical goods, office equipment, agricultural implements, and construction machinery. At the same time, however, sectors like machine tools and textile machinery continued to be dominated by networks of small and medium-sized firms using skilled craftsmen and universal equipment to turn out specialized niche products; and these rapidly expanded their position in world markets during the 1950s at the expense of foreign competitors such as the British. Vehicle manufacturers such as Daimler-Benz who moved some distance toward mass production also maintained some elements of their prewar flexibility, manufacturing a wide range of trucks and buses as well as luxury cars with a combination of standardized and custom parts. At other volume manufacturers, too, the proportion of skilled workers remained higher than in comparable factories abroad, partly because of the abundant supply produced by the German system of craft apprenticeship. Hence German metalworking firms in many sectors adjusted more easily than their British or American counterparts to the turbulent markets of the 1970s and 1980s, while also responding more successfully to the competitive challenge posed by the Japanese.[68]

York, 1983); Steven Tolliday and Jonathan Zeitlin, eds., *Between Fordism and Flexibility: The Automobile Industry and Its Workers,* 2d ed. (Oxford, 1992), esp. introduction. For British accounts of production control methods and restriction of design changes in American metalworking factories during the early 1950s, see British Productivity Council, *Production Control: Report of a Visit to the USA in 1951 of a Specialist Team on Production Planning and Control* (London, 1953); B. E. Stokes, "The Organisation of Production Administration for Higher Productivity," *Journal of the Institution of Production Engineers* 31 (1952): 194–218.

[68]This sketch of postwar German developments draws heavily on Herrigel, "Industrial Organization and the Politics of Industry" (n. 47 above), chaps. 4–7, "Industry as a Form of Order" (n. 47 above), and "The Case of the West German Machine-Tool Industry," in *Industry and Politics in West Germany: Toward the Third Republic,* ed. Peter Katzenstein (Ithaca, N.Y., 1989), pp. 185–220. For the success of German machinery and automobile firms in world markets at British expense during the 1950s, see also Alan Kramer, *The West Germany Economy, 1945–1955* (Oxford, 1991), pp. 182–95; Alan S. Milward, *The European Rescue of the Nation State* (London, 1992), pp. 134–67, 396–424. On the performance of the German

Despite the relative success of wartime aircraft manufacturers in Britain—as well as to a lesser extent the United States and Germany—in modifying mass-production techniques to facilitate frequent changes in models and design, these innovations rarely appear to have been taken up by civilian manufacturers in other industries. Insofar as the wartime experience of aircraft manufacture did influence postwar reconstruction planning, its effect was instead to dazzle government officials and progressive industrialists with the vista of extraordinary productivity gains which could be achieved if continuous output of standard models were not interrupted by "abnormal" disruptions to large-scale production programs.

Ironically, however, the flexible production methods developed by wartime aircraft manufacturers, particularly in Britain, anticipated in many respects those deployed so effectively by Japanese firms during the 1970s and 1980s in civilian sectors such as automobiles and consumer electronics. Like wartime British aircraft manufacturers, for example, Japanese firms in these industries have gained competitive advantages in product development through continuous modification and incremental improvement, using fewer off-the-shelf parts on new models while redesigning them less radically than their American or European counterparts. Like wartime British aircraft manufacturers, too, Japanese firms have enhanced their flexibility and improved the quality of their products by turning out components in small batches "just in time" to meet the demands of rapidly changing production programs. And like wartime aircraft manufacturers in all three countries, finally, Japanese firms have also constructed complex mechanisms for orchestrating collaboration with suppliers and subcontractors in product development, process improvement, and logistical coordination.[69]

Such flexible manufacturing methods reminiscent of wartime aircraft production have proved successful not only in consumer-oriented sectors such as automobiles and electronic equipment but also in the

---

automobile industry through the 1980s, see also Wolfgang Streeck, "Successful Adjustment to Turbulent Markets: The Automobile Industry," in Katzenstein, ed., pp. 113–56.

[69]For this view of Japanese manufacturing and product development methods, see Richard J. Schonberger, *Japanese Manufacturing Techniques* (New York, 1982); Michael A. Cusumano, *The Japanese Automobile Industry* (Cambridge, Mass., 1985); Kim B. Clark and Takahiro Fujimoto, *Product Development Performance: Strategy, Organization, and Management in the World Auto Industry* (Cambridge, Mass., 1991); Tolliday and Zeitlin, eds., *Between Fordism and Flexibility*, esp. introduction and chaps. 8–9. For the role of group organization, coordination committees, complexes, and rings in orchestrating cooperation between firms engaged on common projects in the wartime British, American, and German aircraft industries, respectively, see Postan (n. 9 above), pp. 417–22; Holley, *Buying Aircraft* (n. 16 above), pp. 544–46; Homze (n. 39 above), pp. 77–78; Overy, *Goering* (n. 39 above), pp. 149–50, 190, and "German Aircraft Production" (n. 39 above), pp. 254–64; Cairncross (n. 15 above), pp. 124, 131–32, 135–36.

aerospace industry itself. As David Friedman and Richard Samuels have demonstrated, although Japan's aircraft industry, like that of Germany, was dismantled by the Allies after the war, over the past few decades Japanese manufacturers have built up a rapidly growing business as suppliers of aerospace components. Starting out as licensees and subcontractors for Western firms, Japanese aerospace manufacturers have steadily increased their autonomous design capabilities and expanded their role in each successive international collaborative project. Unlike their American or British counterparts, however, aerospace manufacture is not isolated in specialist firms heavily oriented toward defense contracting, but rather integrated alongside civilian production in large, diversified metalworking groups such as Mitsubishi, Kawasaki, and Fuji Heavy Industries. Both civil and military aircraft components are manufactured alongside other commercial products up to the phase of final assembly using easily redeployed general-purpose equipment such as computer numerically controlled machine tools, while design and engineering knowledge are systematically diffused across the group through mechanisms such as project teams, corporation-wide study groups, and technology focus centers for functional area specialists. The result, Friedman and Samuels convincingly argue, has been a high rate not only of technology "spin-off" from aerospace across the Japanese economy but also of "spin-on" of innovations in manufacturing techniques from other sectors to aerospace itself.[70]

The experience of aircraft manufacture during World War II, like that of Japanese industry today, suggests that the imperatives of military and civilian production are more similar than most historians and policy analysts have assumed, but in different ways than others have argued. The evidence presented here supports neither a sharp distinction between performance-oriented flexible production of military goods on the one hand and cost-oriented mass production of civilian goods on the other, nor the subsumption of both sectors under a universal model of productive efficiency based on a common ideological commitment to order and control. Rather it substantiates the view that under conditions of pervasive environmental uncertainty and rapid technological inno-

[70]David B. Friedman and Richard J. Samuels, "How to Succeed without Really Flying: The Japanese Aircraft Industry and Japan's Technology Ideology," in *Regionalism and Rivalry: Japan and the United States in Pacific Asia,* ed. Jeffrey A. Frankel and Miles Kahler (Chicago, 1993), pp. 251–320; Richard J. Samuels, *"Rich Nation, Strong Army": National Security and the Technological Transformation of Japan* (Ithaca, N.Y., 1994). For the isolation of defense contracting from civilian production in the United States and Britain, see also Jacques S. Gansler, *Affording Defense* (Cambridge, Mass., 1989), pp. 244–45; Markusen and Yudken (n. 7 above), chap. 4; AnnaLee Saxenian, "The Cheshire Cat's Grin: Innovation, Regional Development and the Cambridge Case," *Economy and Society* 18 (1989): 448–77.

vation, such as those prevalent today as well as during World War II, military and civilian manufacturers alike must find ways of balancing countervailing objectives—quality and quantity, design change and production continuity, standardization and customization—in adapting their products to shifting competitive conditions at acceptable costs. The more similar the challenges facing military and civilian producers, finally, the more likely it becomes that the most successful firms—or national economies—will be those in which the institutional boundaries between the two sectors remain most fluid.

Space-Age Europe: Gaullism, Euro-Gaullism, and the American Dilemma

WALTER A. MC DOUGALL

"It is a far cry from Cape Kennedy," wrote a correspondent for the *New York Times*. "There are no neon signs, no drive-ins—and no night clubs. There are only some scattered huts and towers, lost in a desolate flatland as big as New Jersey, its pebbly floor covered with a pale green haze after a spell of rain. In the huts, which are filled with electronic equipment, one can hear, almost any morning, a calm young voice on a loudspeaker saying 'dix, neuf, huit, sept . . . ' In the distance a needle with a tail of fire slowly rises above the desert and roars into the sky."[1]

The site was Hammaguir, an adobe village where sheep, goats, and a few dromedaries nosed about in the brittle weeds. Colomb-Béchar, the nearest town, lay 80 miles to the north, itself 700 miles into the Sahara from Algiers. Nearby, the parallel lines of an abandoned railroad vanished into the dunes, perhaps to meet at infinity, an artifact of France's first stab at a colonial dream, the trans-Saharan railway. In 1965, the imperative of international competition had brought France's finest engineers back into a desolation that proved congenial to the most advanced technology even as it swallowed the remains of an earlier industrial revolution. For the Algerian civil war prepared the return of Charles de Gaulle, who forestalled a military threat to overthrow the Fourth Republic by overthrowing it himself and pledged to

DR. MCDOUGALL is professor of history and Alloy-Ansin Professor of International Relations at the University of Pennsylvania. Formerly of the University of California, Berkeley, he wrote this article while researching . . . *the Heavens and the Earth: A Political History of the Space Age* (New York, 1985), which won the Pulitzer Prize for History in 1986. His research was aided greatly by the then-NASA historian Dr. Monte Wright and his assistant Dr. Alex Roland, now of Duke University. For information on the European space program, McDougall benefitted from the help of Michel Bourely, Alain Dupas, and the staff of the European Space Agency, Paris; J. L. Blonstein of EUROSPACE, Paris; the International Institute of Strategic Studies, London; Arnold Frutkin, former director of International Programs, NASA; Wreatham Gathright, formerly of the Department of State; and Dr. Hans Mark, deputy administrator of NASA and later Secretary of the Air Force and chancellor of the University of Texas.

[1]John L. Hess, "The Last Countdown," *New York Times,* February 11, 1967.

restore French greatness through technology, not empire. The Treaty of Évian ended French rule in Algeria in 1962 but reserved to the metropole, for a time, its proving grounds at Hammaguir, whence Gaullist France would become the world's third space power.

French technicians—some in burnooses like a cosmic foreign legion—were mainly men of the *Société pour l'Étude et la Réalisation d'Engins Ballistiques* (SEREB) and the *Centre National d'Études Spatiales* (CNES). Back home the SEREB shared in the design of intermediate-range ballistic missiles to cradle the bombs of the world's fourth nuclear deterrent, and the CNES designed satellites for cooperative programs with the United States and European space agencies. But the Algerian task was final checkout of Diamant, a French-made space booster, and the goal was to place a French satellite in orbit before *FR-1*, another Gallic spacecraft, went aloft aboard an American Scout rocket from Vandenberg Air Force Base. From the start, France's national program was competing with her own cooperative programs, with the Americans, and with other Europeans in the race to become the third nation in space.

The Diamant booster was a three-stage configuration composed of engines developed in previous rocket programs, one propelled by the exotic mix of nitric acid and turpentine, the others solid-fueled. Together they developed 107,000 pounds of thrust, roughly equivalent to that of the Jupiter-C that had launched the first American satellite seven years before. By mid November of 1965, the NASA launch of *FR-1* and a French presidential election were both three weeks away. Forty-three months before, the design for Diamant had been frozen and airframe construction commenced at the Nord Aviation plant for SEREB. Now the identity of Gaullist France, wedded to the prestige and power of technological dynamism more consciously even than Kennedy's America, rode on the outcome. "Trois, deux, un . . . ," the countdown ended on November 26. Preset charges exploded the bolts holding down the sleek cylinder, and its own large exhaust nozzle fired up to full thrust. Soon the tracking stations reported in: *Asterix-1*, a modest 42-kg satellite named for the red-whiskered barbarian of French comics, was transmitting from orbit. Its chemical batteries quit after just two days, but Diamant had glistened, and *Le Monde* proudly proclaimed "La France Troisième 'Puissance Spatiale'!"

* * *

Today, over a quarter-century after *Sputnik*, the political patterns of the space age have undergone a radical shift. Where two superpowers vied alone for prestige and military advantage, now seven nations have launched satellites on homemade boosters, and dozens have participated in cooperative satellite programs for commercial, scientific, and technological motives. Where once international cooperation and "space for peace" were universally touted, at least in rhetoric, now vigorous competition obtains, not only between the United States and the USSR but between the United States and its industrial allies as well. Where once government arsenals monopolized spaceflight, now a spectrum of institutions—public, semipublic, and private, military and civilian, national, bilateral, and multinational—adapt to the demands of space development, operations, and marketing. In the 1980s, the surprising conclusion is that the space age, defined not only by revolutionary technologies but also by mobilization of national resources for the force-feeding of technological change, will be shaped in years to come as much by developments in the "second tier" of European and Asian states as in the Big Two. For the political history of space technology has validated neither the early hopes of a "humanity united in space" nor the fears of a yawning "technology gap" stemming from economies of scale in the United States and the USSR to the detriment of all others. This article examines, from the point of view of the "others," how both these expected outcomes of the space age were forestalled and why the Gaullist model, rather than the American or Soviet, has come to shape the international politics of technology in the space age.[2]

The wartime hero de Gaulle rose to power just eight months after *Sputnik 1*. His mission, brooded over for twelve years, was to save France. This meant military independence, without which no state was truly sovereign; economic independence, without which no state was master of its own house; and technological revolution, without which no modern society could maintain the first two conditions. Sensing that the colonial mission drained French resources and earned opprobrium rather than prestige, de Gaulle liquidated imperial France. Bristling under the Anglo-American "special relationship," he withdrew from NATO command, blocked Britain's entry into the Common Market, and thus proscribed an Atlanticist France. Contemptuous of integration and fearful of German power, he capped progress toward a European France. Needless to say, de Gaulle also abhorred the Left's vision of a Socialist France. Instead, de Gaulle launched a revolution

[2]Some of the ideas herein were raised in two short papers I was asked to give during the early stages of my research on the political history of the space age: "Space-Age Europe 1957–1980," NASA-Yale Conference on the History of Space Activity (February 1981), and "The Struggle for Space," *Wilson Quarterly* 4 (Autumn 1980): 66, 71–82.

from above to reify his "certaine idée" of Technocratic France, the R&D state.[3]

The unabashed theme of Gaullism was *grandeur, la gloire*—for "la France ne peut être la France sans la grandeur." But "glory" is not a policy any more than "peace" is, and in the case of Gaullist France, *grandeur* was an axiom or self-definition implying that any "France" not willing or able to play the role of a Great Power was not France at all. And if de Gaulle had cherished such beliefs ever since 1940, two events just prior to his return sufficed to persuade his countrymen. The first was the 1957 British White Paper in which Defense Minister Duncan Sandys argued that economic decline, social demands, and superpower dynamism forced Britain thenceforth to rely on cheap nuclear deterrence. Though directed against the Soviets, a beefed-up deterrent would, as Harold MacMillan admitted the following year, also increase British leverage vis-à-vis the United States.

The second event was *Sputnik*. Now that the Soviets were capable of threatening the U.S. homeland with hydrogen bombs and intercontinental ballistic missiles, was the American nuclear umbrella still credible? Would America risk New York to save Paris? Such imponderables reinforced French determination to press on with their own nuclear *force de frappe*. But *Sputnik* gave another ironic twist to Franco-American relations, for the Eisenhower administration, in the post-*Sputnik* panic, expanded strategic cooperation with Britain, while Congress amended the McMahon Act to enable more nuclear secrets to be passed to friendly nuclear powers. In the interest of nonproliferation, friendly *non*nuclear powers received no such aid. When de Gaulle sought to purchase KC-135 tankers for inflight refueling of his Mirage IV jets, the U.S. government hindered the sale. When France concluded contracts with Boeing for missile components, the State Department withheld approval. The French concluded that U.S. policy was designed to keep France a *nation secondaire* for all time, and when de

[3]This notion of modern technocracy as the R&D state, "the institutionalization of technological change for state purposes," is the theme of my article, "Technocracy and Statecraft in the Space Age: Toward the History of a Saltation," *American Historical Review* 87 (Oct. 1982): 1010–40. For Gaullist ideas on politics and technology, see esp. his *Memoirs of Hope: Renewal and Endeavor* (New York, 1970) and his collections of speeches in Ambassade de France, *Major Addresses, Statements, and Press Conferences of General Charles de Gaulle* (New York, 1964), as well as *De Gaulle parle*, 2 vols., ed. André Passeron (Paris, 1962–66) and the following works on Gaullist foreign policy: Paul-Marie de la Gorce, *De Gaulle entre deux mondes* (Paris, 1964) and *La France contre les empires* (Paris, 1969); W. W. Kulski, *De Gaulle and the World* (Syracuse, N.Y., 1966); John Newhouse, *De Gaulle and the Anglo-Saxons* (New York, 1970); Paul Reynaud, *The Foreign Policy of Charles de Gaulle*, trans. Mervyn Savill (New York, 1964).

Gaulle pronounced on NATO and military matters, his rhetoric aimed, in every case, not at Moscow but at Washington.[4]

But technical independence, the mark of a Great Power abroad, dictated a revolution at home, and, after seven years of the Fifth Republic, France was scarcely familiar to those who had known her in the 1950s. For 150 years French business had distinguished itself by jealousy, traditionalism, and acrimony with labor; and state policy by vacillation between nationalization and laisser-faire. But the constitution of the Fifth Republic enhanced the power of the executive, which in turn reformed the universities, folded small industrial concerns into mighty semipublic corporations, and linked them to state agencies in a coordinated national team for the force-feeding of technological change, with the state itself as managerial czar. In space technology, de Gaulle's technocrats combined the air force's office for aeronautical research and the private firms of Nord Aviation, Sud Aviation, Engins MATRA, and Dassault into the new SEREB. Gradually French public and private aerospace concerns became a single team, with contracts drawn from the Defense Ministry and CNES, distributed among firms, overseen by the tough, ubiquitous *inspecteurs des finances*, and the results exploited by bureaucratic managers. Thanks to this national complex for R&D, solid-fueled IRBMs and submarine-launched missiles entered flight testing as early as 1967, and the first nine nuclear-tipped missiles were put into silos in Haute Provence in 1971. Nuclear-armed submarines entered service the following year, and, together with the Mirage jet bombers, completed France's little triad of nuclear forces.[5]

[4]For instance, de Gaulle declared early in 1958 that "I would quit NATO if I were running France. . . . NATO is no longer an alliance, it is a subordination" (C. L. Sulzberger, *The Last of the Giants* [New York, 1970], pp. 61–62). Although the official justification of the *force de frappe* was to provide France with a modern deterrent, Gaullist ministers invariably spoke of it as the only way for France to rejoin the ranks of the Great Powers, make herself heard in world councils, receive equal treatment in the Western alliance, and qualify for American nuclear aid. See Wilfrid Kohl, *French Nuclear Diplomacy* (Princeton, N.J., 1971), pp. 98–100. Even Raymond Aron, a friend of NATO, saw the *force de frappe* as a "political trump" in dealings with the United States.

[5]On the domestic revolution promoted by de Gaulle in the name of technological dynamism, see especially Robert Gilpin, *France in the Age of the Scientific State* (Princeton, N.J., 1968). On the evolution of the *force de frappe*, see: Kohl, *French Nuclear Diplomacy*; Wolf Mendl, *Deterrence and Persuasion: French Nuclear Armament in the Context of National Policy 1945–1969* (London, 1970); Bertrand Goldschmidt, *L'Aventure atomique* (Paris, 1962); Lawrence Scheinmann, *Atomic Energy Policy in France under the Fourth Republic* (Princeton, N.J., 1965); Charles Ailleret, *L'Aventure atomique française* (Paris, 1968). French nuclear research was well advanced in 1940 when the German conquest put a halt to the work of the Curies. The Fourth Republic founded the French atomic energy commission, which worked steadily toward the fabrication of weapons-grade plutonium,

De Gaulle also announced plans for an orbital space program in 1959. The Hammaguir proving ground, home for France's share of captured V-2s since 1947, became the most active rocket range outside the United States and the USSR. In 1961, the state combined its various research groups, and CNES emerged as a full-fledged space agency and joined forces with SEREB to build a space-launch capacity. Unlike the American NASA, CNES made no artificial distinction between military and civilian rocketry. Its launchers were developed under military aegis, and its director-general, Robert Aubinière, was an air force general and an advocate of the military uses of space. In the early 1960s the SEREB crept up on orbital capacity with a series of ever more precious stones: the Agate, Topaze, and Rubis solid-fueled stages, the Émeraude liquid-fueled first stage, the Saphir two-stage configuration, and finally the Diamant-A.[6]

Why a French space program? First, if prestige were a primary aim of Gaullist policy, then space beckoned irresistibly. Second, orbital flight could be pursued relatively cheaply as an offshoot of the planned military missile program. Third, a mature nuclear strike force would itself someday require satellite support systems for geodesy, targeting, surveillance, communications, and meteorology. Indeed, French military theorists such as Aubinière, Pierre Gaullois, and Colonel Petkovsek argued the inevitability of space militarization in the missile age more candidly than American officials (who treated the military space program as a public relations albatross).[7] But the fourth and fundamental reason for a French space program was the apparent centrality of space-related technologies in the Gaullist drive for permanent technological revolution.

The traditional "stalemate society" that was France had never adequately adjusted even to the industrial age.[8] But the advent of elec-

---

often without official blessing. In 1954 the cabinet of Pierre Mendès-France approved continuation of work leading to a bomb test. By that time French strategists already justified "going nuclear" as an economy move, "more force for the franc," in imitation of Eisenhower's New Look (see, e.g., Charles Ailleret, "L'Arme atomique, arme à bon marché," *Revue de défense national* 10 [1954]: 315–25). Hence, when de Gaulle took over in 1958 he had only to make public France's intention of building its own nuclear force and vastly increase the funding.

[6]On French rocket development, see U.S. Congress, Committee on Science and Technology, *World Wide Space Programs*, 95th Cong., 2d sess. (1977), pp. 142–57.

[7]See, e.g., Petkovsek, "L'Utilisation militaire des engins spatiaux," *Revue militaire générale*, July 1961.

[8]A "stalemate society" (the evocative phrase is Stanley Hoffman's) is one in which conflicting socioeconomic interest groups are strong enough to block implementation of the programs of others but not strong enough to realize their own through a weak, fragmented parliamentary system. Such a society is incapable of major reforms. One

tronics, atomic power, computers, and space technology in the 1950s ushered in a postindustrial age with still stiffer requirements. "It is no longer enough," wrote de Gaulle, "for industry, agriculture, and trade to manufacture, harvest, and exchange more and more. It is not enough to do what one does well; one must do it better than anyone else. . . . Expansion, productivity, competition, concentration—such, clearly, were the rules which the French economy, traditionally cautious, conservative, protected, and scattered, must henceforth adopt." How could such a revitalization come about? First, through state leadership under the French Economic Plan. Second, through priority for international competition, "the lever which could activate our business world, compel it to increase productivity, encourage it to merge, persuade it to do battle abroad. . . . " Thus, the Common Market, in de Gaulle's view, was not a means to submerge France into Europe but a means to expand the market in which French industry might achieve dominance. Third, through statist stimulation of advanced R&D in the fields of nuclear power, aviation, computers, and space "because their labs and their inventions provide a spur to progress throughout the whole of industry."[9]

De Gaulle envisioned a hybrid economy uniquely adapted to an age of continuous technological revolution. He rejected laissez-faire capitalism, for that model carried within it "the seeds of a gigantic and perennial dissatisfaction. It is true that the excesses of a system based on laisser-faire are now mitigated by certain palliatives, but they do not cure its moral sickness." Communism, on the other hand, theoretically "prevents the exploitation of men by men, [but] involves the imposition of an odious tyranny and plunges life into the lugubrious atmosphere of totalitarianism without achieving anything like the results, in terms of living standards, working conditions, distribution of goods, and technological progress which are obtainable in freedom."[10]

---

solution to such stalemate is "corporatist" decision making by which labor, business, political parties, and bureaucracies, e.g., compromise to bring about centralized social progress. The French Third Republic proved singularly incapable of effecting such compromise even under the threat of foreign competition or internal disruption. In perverse fashion, it was left to the collaborationist Vichy regime to foster a number of reforms—in industrial organization, labor relations, and scientific research—amounting to a certain "modernization" of the French state. The Fourth Republic then established new research institutes and the Economic Plan after the war, presaging in many ways the Gaullist era. See, e.g., Stanley Hoffman, *In Search of France* (Cambridge, Mass., and London, 1963); Gilpin, *France in the Age of the Scientific State* (n. 5 above); Charles S. Maier, *Recasting Bourgeois Europe: Stabilization in France, Italy, and Germany in the Decade after World War I* (Princeton, 1974); and Robert O. Paxton, *Vichy France: Old Guard and New Order, 1940–1944* (New York, 1972).

[9]De Gaulle, *Memoirs of Hope*, pp. 133–35.

[10]Ibid., p. 136.

Both "world systems" were unsuitable to the space age; de Gaulle sought a *juste milieu.* Competition was indeed the engine of progress but also the solvent of community. Hence the competitive stimulus must be international, while at home French institutions combined in a dynamic unity. The initial results were stupendous: real growth of 7 percent per annum in the early 1960s, zero unemployment despite the influx of demobilized soldiers and *pieds noirs* from North Africa, a fivefold increase in state R&D funding from 1959 to 1964—until the government subsidized three-quarters of all R&D performed in France. To an even greater degree than in post-*Sputnik* America, R&D in France was nationalized. But unlike space-age America, Gaullist France subjected its national effort to a centralized plan. As Michel Debré explained the Five-Year Plan for R&D in 1961, the additional funds were to constitute a "masse de manoeuvre" which the state could target on carefully selected sectors whose "spin-off" effects would advance national technology across-the-board. Master planning fell to various standing committees reporting directly to the prime minister, like the *Comité Consultatif de la Recherche* (known as "The Wise Men") or the *Délégation Générale à la Recherche.* Together with the Ministry of Science, they plotted strategy for the conscious invention of the future.[11]

Despite fundamental restructuring of French political, academic, and industrial life and the huge strides made in the first decade of the Fifth Republic, the evident explosion of technology in America symbolized by Project Apollo seemed only to widen the "technology gap" across the Atlantic. By 1964, de Gaulle was warning of "bitter mediocrity" and the "colonization" of France if she did not push her technology forward even more relentlessly. One economist believed Kennedy's America had found "the keys to power" in command R&D. By means of its favorable "technological balance of payments," superiority in "point sectors," and direct investment abroad through multinational corporations, the United States had adjusted first to the new technological age and threatened to dominate the world. The Fifth Republic, therefore, embraced the assumptions of such American enthusiasts as NASA administrator James Webb that (1) basic research is the cutting edge of national competitiveness; (2) there is a direct relation in R&D between scale and results; (3) multinational entities threaten national

[11]Debré in "Le Programme pour la recherche scientifique," *Figaro,* May 4, 1961. Generally, see Gilpin, *France in the Age of the Scientific State,* and C. Freeman and A. Young, *The Research and Development Effort in Western Europe, North America, and the Soviet Union* (Paris: OECD, 1965).

independence; and (4) priority of invention is self-perpetuating, that is, leading nations tend to increase their lead.[12]

How could France hope to compete if scale and priority were critical in advanced technology? Many Europeans concluded that the appropriate response to the technology gap was more vigorous integration. Only by pooling their resources and talent might Europeans hope to forestall U.S. hegemony. But this was not the Gaullist conception. France did not flee dependence on America only to become dependent on a European mélange. Rather, France's cooperative programs in nuclear technology, space, and aviation (e.g., the SST) were fashioned so as to draw on the resources of others in the interest of her national programs rather than to donate French expertise in the interest of multilateral progress. In space, as in Euratom, French contributions to Europe were a fraction of the efforts made at home. Cooperative programs were of interest insofar as they channeled foreign funds, ideas, and markets into a technology flow irrigating France's own garden.

The French Five-Year Plan for space, approved in 1961, made room for cooperation with NASA and France's European partners, but the announced goals of CNES were (1) to create a French technological base capable of original experimentation in space and (2) to put French industry in a favorable position vis-à-vis the competition certain to develop in Europe. The first goal meant that France must not merely duplicate, later and on a smaller scale, what the superpowers had done but select technological targets of opportunity in which France might someday compete. The second goal assumed eventual European independence from the United States but that competition within Europe would also obtain. Such goals demanded vigor, not only the world's third largest space program but a precise strategy to guide it.

"La méthode assez française," according to Aubinière, was to fix objectives from the outset, then create the instrument needed to fulfill them. Rather than hasten to "do something about space," letting existing institutions stumble forward, France shaped her institutions to her goals. Aubinière and Pierre Auger, scientific chief of CNES, together forged an "infrastructure technique très importante," a government-industry-university team of the sort James Webb and NASA would soon promote in America. In the early years the French

[12]Michel Drancourt, *Les Clés de pouvoir* (Paris, 1964). These assumptions buttressed NASA budgetary appeals throughout the 1960s but were challenged by American critics as early as 1962. For the general French adherence to them, see Gilpin, *France in the Age of the Scientific State*, pp. 32–71, esp. pp. 56–57.

still relied heavily on imported American technology. Forty percent of the *FR-1* satellite, for instance, consisted of U.S.-made components. But once in possession of such subsystems, French technicians replicated them at home and gained an advantage over other Europeans. One result of this "competition through cooperation" was the almost total European dominance enjoyed by France in solar-cell systems in the late 1960s.[13]

Space technology was to be a force for global unity, said the academics. But the new age, in de Gaulle's intuition, would be one of heightened self-sufficiency and competition, even neomercantilism, for any state considering itself a Great Power. The cost and complexity of space-age technologies rendered the free market obsolete—capitalism at home was no longer "competitive" in the global arena. French policy on space, in sum, amounted to statist cooperation in science and statist competition in engineering. This was the France that sat down with the other European states in 1962 to build a joint European space program.

* * *

Suddenly, after *Sputnik*, the airy zealots of the British Interplanetary Society no longer appeared to be candidates for Bedlam. For a half-century they had predicted the coming of spaceflight, and now their pleas for a British space policy resounded in Parliament itself. David Price, a Tory backbencher, intoned, "We are now in the space age, whether we like it or not. All public policy must be shaped to accommodate this sudden change in the human environment. . . . Viewed historically, Europe dare not stand apart from the space race." But European states could not compete by themselves, said Price, or be content to lean on the superpowers, or expect a United Nations space program. The only solution was a pooling of resources—and they need not even start from scratch, for the British Blue Streak intermediate-range

[13]On French competitive strategy for the European market, see Aubinière, "Réalisations et projêts de la recherche spatiale française," *Revue de défense nationale*, November 1967, pp. 1736–49. On the strategy and execution of the French space program in the 1960s, see: U.S. Congress, *World Wide Space Programs*, pp. 139–70; Georges L. Thomson, *La Politique spatiale de l'Europe*, 2 vols. (Dijon, 1976), vol. 1, *Les Actions nationales*, chap. 1; Michiel Schwartz, "European Policies on Space Science and Technology 1960–1978," *Research Policy* 8 (1979): 204–43; "Programme spatiale français jusqu'en 1965," *Figaro*, July 28, 1961; "Le Débat sur le centre d'études spatiales," *Le Monde*, October 19, 1961; "More French Satellites after 'Diamond,'" *Daily Telegraph*, June 1, 1962; Kenneth Owen, "France's Space Programme: The Reasons Why," *Flight International*, July 12, 1962; L. Germain, "Le Recherche spatiale en France," *Revue militaire d'information*, January 1963; Charles Cristofini (president of SEREB), "Planned Cooperation Is France's Aim," *Financial Times*, June 10, 1963.

missile, headed for cancellation before flight testing, could survive as the first stage of an all-European satellite launcher. Price also foresaw coordinated space research in firms and laboratories across Europe; manufacture of components and whole spacecraft in European plants; joint launch facilities; communications and other commercial satellites; nuclear and solar power for spacecraft; hypersonic, reusable winged vehicles; space medicine; and even nuclear, ion, or plasma "space drives." None of this was fantastic, he insisted. The Common Market states plus Britain, Norway, and Switzerland had combined gross national products greater than the Soviet and over half the American. Without the burden of military or manned programs, Europe could surely compete in selected technologies of scientific and economic potential.[14]

Price's assumptions met a willing audience in a Europe searching for its place in the postwar, postimperial world. It seemed the old continent, the cradle of the Industrial Revolution, must decay by the 21st century into a global backwater unless she joined the new technological revolution. By 1959 both British political parties were sponsoring bills for a ministry of science or technology, Gaullist France was embarked on an R&D boom, and the West Germans were eager to master ancillary space technologies (despite a shyness about missilery stemming from the V-2 heritage). So the Council of Europe and a committee of experts at Strasbourg in 1960 endorsed the principle of a European space program. When Minister of Aviation Peter Thorneycroft offered the Blue Streak to Europe the following year, the European Launch Development Organization (ELDO) was born. Britain would perfect the Blue Streak, France would provide a second stage called Coralie, Germany the Astris third stage, Italy the test satellite, the Netherlands the telemetry, Belgium the guidance station, and Australia its test site at Woomera in the outback.

Here was an enterprise in multinational technological cooperation on an unprecedented scale, and a mission for Europe—space-age Europe. In the initial enthusiasm, potential difficulties were brushed aside or unappreciated. While ELDO-financed research was to be shared openly, the French insisted that members not be required to share data acquired in national research. The French also insisted that no restrictions be placed on national military application of ELDO-derived technology, thus killing chances of American aid. The big states insisted that voting power in ELDO reflect contributions, the smaller states feared being pawns. So the convention required that the

[14]"European Cooperation in Space," *Spaceflight*, January 1961; David Price, "Political and Economic Factors Relating to European Space Cooperation," *Spaceflight*, January 1962; Kenneth Owen, "Europe's Future in Space," *Flight*, July 6, 1961.

annual budgets be approved by both a two-thirds majority and by countries whose total contributions constituted 85 percent of the budget. But how would contributions be distributed? Tortuous costing of the planned rocket stages produced a £70 million project, 38.8 percent of which was Britain's responsibility, 23.9 for France, 22.0 for Germany, 9.8 for Italy, and the remainder for the others. But the costs of each stage were as uncertain as cost overruns were predictable. These and other sources of discord hung over the ELDO convention signed in 1962.

Meanwhile, European scientists led by Eduardi Amalfi, Pierre Auger, and Sir Harrie Massie took the initiative in space science. The "brain drain" of European talent to the United States was the academic equivalent of the "technology gap" that the scientists hoped to stem through a European space science program. Working separately from those discussing launch-vehicle development, delegations from ten countries (Belgium, Denmark, France, Germany, Italy, the Netherlands, Spain, Sweden, Switzerland, United Kingdom), founded the European Space Research Organization (ESRO), also in 1962. ESRO dedicated itself to peaceful purposes only, free exchange of information, and joint construction of satellites and experiments for launch by NASA and eventually by ELDO. Again, at French insistence, members were released from the obligation of sharing data "obtained outside the organization." The ESRO convention provided for a European Space Technology Center (eventually based at Noordwijk, the Netherlands), a European Space Data Center (later the Space Operations Center) for telemetry and tracking (Darmstadt, West Germany), a sounding rocket range (Kiruna, Sweden), and a headquarters (Paris). ESRO had less difficulty than ELDO with procedure: each state received one vote, with most issues decided by simple majority. The budget would be voted by a two-thirds majority every three years, with national contributions fixed in proportion to the net national income of the member state. Projected spending for the first eight years was a mere $306 million.[15]

Politically, these numbers were acceptable: some half a billion dollars for ESRO and ELDO divided among several countries over six to eight years. This was hardly an excessive entry fee into the postindustrial world. But was it enough? By the middle of the decade the United States would be spending $5 billion per year on civilian space technol-

[15]On the origins of ESRO and ELDO, see: U.S. Senate, Committee on Aeronautical and Space Sciences, *International Cooperation and Organization for Outer Space*, 89th Cong., 1st sess. (1965), pp. 103–17; U.S. Congress, *World Wide Space Programs*, pp. 237–77; Thomson, *La Politique spatiale de l'Europe*, vol. 2, *La Coopération européenne*, chap. 3; Alain Dupas, *La Lutte pour l'espace* (Paris, 1977), chap. 10.

ogy alone. European aerospace firms, the most enthusiastic but also the most discerning of observers, understood better than the politicians the cost and frustrations of large-scale R&D. Hawker-Siddeley and SEREB accordingly gathered about them an industrial lobby of ninety-nine companies called EUROSPACE to educate and influence the bureaucrats. Almost half the member firms were French, as were the president, Jean Delorme, and secretary-general, Yves Demerliac. In the words of the former, "Unless the European countries wish to join the ranks of the backward and underdeveloped countries within the next fifty years, they must take immediate steps to enter these new fields." A low-orbit launcher, scientific satellites, and half a billion dollars did not suffice for what Delorme called "a matter of survival."[16]

How so? Space technology scarcely promised big profits in the near future—the motives for the superpowers were defense and prestige. Even communications satellites, which held immediate promise, were hardly "a matter of survival." But EUROSPACE took a larger view that might be termed Euro-Gaullism. It advised against importing U.S. systems, even if permitted to do so, in order that Europeans might gain experience in R&D. The payoff was in the means, not just the ends. "European industry," recalled Demerliac, "never considered space as a money-making activity. [Its] main initial motive was to improve its technology so as to remain competitive in world markets. Space was a means of forming or retaining qualified teams capable of delivering advanced items of equipment and also—perhaps above all—to manage the joint development of complex systems or sub-systems. . . . The target for European industry is clearly to acquire prime contractor ability for all space applications systems."[17]

"Prime contractor ability for all space applications"! Even as the French hoped to target specific markets and achieve technological primacy within Europe, so the European aerospace industry as a whole sought competitiveness in targeted world space markets. The impact of

[16]SEREB and Hawker-Siddeley Aviation, *L'Industrie et l'espace* (Paris and London, 1961); Jean Delorme in EUROSPACE, *Proposals for a European Space Program* (Fontenay-s-bois, 1963), pp. 11–13, 96–97.

[17]Yves Demerliac, "European Industrial Views on NASA's Plans for the '70s," AAS Goddard Memorial Symposium (Washington, D.C., March 1971). See also the testimony of Eilene Galloway, congressional staff expert on space policy, after interviews with European officials in April 1967: Library of Congress, Clinton Anderson Papers, Box 919. EUROSPACE also proposed an ambitious program of space R&D including a reusable "space transporter" or "shuttle." See EUROSPACE, *Proposals for a European Space Program*, pp. 17–67, and *Aerospace Transporter* (Fontenay-s-bois, 1964). The preface to the latter was written by an aging Eugen Sänger, German rocket engineer who designed a reusable "boost-glide" space vehicle during World War II and can be considered the progenitor of the Space Shuttle.

such technological strategies on trans-Atlantic cooperation was profound. NASA was eager to cooperate in space science; the United States might, in its generous moods, even welcome Europeans as subcontractors in expensive missions. But it was not likely to transfer technology sufficient to create full-scale competition for American aerospace firms. NASA was forthcoming with proposals to train foreign scientists and launch scientific payloads on a reimbursable basis. But the State Department tried to discourage the Europeans from forming ELDO and turned a cold shoulder when the Europeans sought help for their Europa-1 booster. De Gaulle insisted, and Europeans listened, that the day might come when American willingness even to launch foreign satellites might cease. Europe must have her own space booster.[18]

Nevertheless, the EUROSPACE plea for an additional £218 million for space failed to persuade European parliaments committed to expanding social welfare in the 1960s. So ELDO and ESRO, underfunded and poorly conceived, came to exemplify all the risks of multilateral R&D. First, the governments delayed ratification of the conventions until 1964, by which time the Americans and Soviets had pulled much further ahead. Then the brick and mortar work of building the centers, especially for ESRO, absorbed several more years, while most of the first triennial budget went for overhead. Not until 1967 did the first experimental satellite *ESRO-1* reach orbit, courtesy of NASA.

The ESRO members also quarreled over disproportionate distribution of contracts, the issue of *juste retour.* National responsibilities in ESRO projects were not fixed in advance, so contracts flowed to the most competitive firms. France, true to her intent, garnered a percentage of contracts up to twice the level of her contribution. Efficiency demanded that business go to the most qualified firms, but politics demanded "affirmative action" for countries playing "technological catch-up." Either the poor subsidized the rich, or the rich subsidized mediocrity in the short run and new competition in the long run. Even as the French deplored American dominance vis-à-vis Europe, they themselves exploited a dominant position within Europe.

Yet American progress obliged Europeans to press on, despite their growing organizational troubles. In 1962 the U.S. Congress passed the Communications Satellite Act. The following year NASA launched *Syncom 1*, the world's first geosynchronous communications satellite,

[18]A. V. Cleaver, "European Space Activities since the War: A Personal View," British Interplanetary Society Paper (March 1974). The United States cooperated heartily with ESRO, including providing a tracking station in Alaska. When ESRO turned increasingly to commercial applications of space technology at the end of the 1960s, however, American enthusiasm cooled.

while President Kennedy embarked on a hurried campaign for a global communications network. This first commercial application in space was precisely the sort of enterprise in which Europeans hoped to specialize, yet the United States moved so quickly that negotiations ensued for an international telecommunications satellite consortium long before the Europeans had any technical leverage whatever. INTELSAT, founded by nineteen nations in 1964, fell under exclusive American leadership. The United States controlled 61 percent of the voting authority, and virtually 100 percent of the necessary technology, and the U.S. COMSAT Corporation was the only entity in the world capable of deploying and managing the global system. The European Conference on Satellite Communications, formed to provide the Europeans with a united front in negotiations with the American giant, succeeded in making the INTELSAT accord temporary, but for the time being the Americans had a monopoly. All contracts necessarily went to U.S. firms, only NASA had the means to launch the satellites, Americans managed the system (sometimes, it seemed, in the interest of the American common carriers like AT&T and ITT), and NASA was told to refuse launch service for potentially competitive foreign satellites.[19]

After 1966 European parliaments began to grasp what EUROSPACE had understood from the beginning. The entry cost to the space market would be far higher than conceived in the sanguine moments of 1962, and European space spending, to be effective and politically acceptable, must be targeted narrowly on "practical" applications rather than basic science. ELDO's Europa booster, therefore, underwent several upgrades, turning the low-orbit launcher into one capable of launching heavier payloads into geosynchronous orbit 22,000 miles above the earth. Confusion attending these redesigns, made before *Europa-1* had achieved a single success, meant more delay and waste. By 1969 ELDO had still not made a launch despite a budget

[19]On background and negotiation of the INTELSAT Convention, see: Murray L. Schwartz and Joseph M. Goldsen, *Foreign Participation in Communications Satellite Systems: Implications of the Communications Satellite Act of 1962*, RAND RM 3484-RC (1963); U.S. Congress, Committee on Government Operations, *Satellite Communications*, 88th Cong., 2d sess. (1964), pt. 2, pp. 661–65; Jonathon F. Galloway, *The Politics and Technology of Satellite Communications* (Lexington, Mass., 1972); Delbert D. Smith, *Communications via Satellite: A Vision in Retrospect* (Boston, 1976); Michael Kinsley, *Outer Space and Inner Sanctums: Government, Business, and Satellite Communication* (New York, 1976); U.S. Senate, *International Cooperation and Organization for Outer Space*, pp. 117–20. On the European Conference on Satellite Communications, see Smith, *Communications via Satellite*, pp. 135–41; for European acquiescence in temporary U.S. domination in hopes of gaining future influence over INTELSAT, see U.S. Congress, *Satellite Communications*, pt. 1, p. 28.

three and one-half times the initial estimate. The European Space Conference, at French insistence, debated plans to shift priorities in ESRO from scientific to commercial satellites, while Britain and Italy, pleading straitened finances, threatened to pull out of the European space effort altogether.

Insufficient capital, political disputation, the problem of *juste retour*—any of these handicaps might alone have crippled such an awkward venture in multinational command R&D. But there was more. Systems integration for an international space booster was a boondoggle. Every technical hurdle had to be surmounted by an international committee whose babble of tongues only exacerbated the habitual lack of communication among scientists, engineers, and bureaucrats. One of the more tolerant veterans of those days recalled that whenever an improvisation was called for, the French stubbornly refused to violate any hard-won procedural principle; the Germans endorsed the principle, then listed all conceivable exceptions; the Italians excitedly urged renegotiation of the principle to accommodate the offending contingency; while the British cheerfully accepted any improvisation so long as under no circumstances would it serve as a precedent! Nor did the French send their best men to ESRO and ELDO, reserving them for the national effort, while others were accused of loading the space agencies with deadwood personnel.[20]

By the end of the decade the European space program was a shambles—and this at the peak of concern over the technology gap, brain drain, and "industrial helotry," all presumably products of explosive American technocracy. EUROSPACE tried to capitalize on this mood, best expressed by Jean-Jacques Servan-Schreiber's *The American Challenge*, by warning its cultured countrymen against their tendency to sniff at the technical achievements of boorish Americans: Carthage's flourishing culture did not save it from the Romans, nor did Rome's superior culture fend off the barbarians. Echoing NASA, EUROSPACE identified the real value of Apollo not in lunar exploration itself but in the perfection of techniques for large-scale R&D, national mobilization, and technological spin-offs. If Europe did not steel itself to make the necessary effort toward technological independence, it would soon be too late. The Germans expressed this as *Torschlusspanik*: Europe must jump through the door to the space age before the door

[20]On the problems of the European space program in the 1960s, see especially: *ELDO, 1960–1965: First Annual Report* (Brussels, 1965) and *Annual Reports (1966–)*; *ESRO, First General Report, 1964–65* (Paris, 1966) and *Annual Reports (1967–)*. Books on the frustrations of the 1960s include Jacques Tassin, *Vers l'Europe spatiale* (Paris, 1970) and Orio Giarini, *L'Europe et l'espace* (Lausanne, 1968). Anecdote on national temperaments from Tassin, pp. 98–99.

slammed shut. The Italian government called for a "technological Marshall plan" and British Prime Minister Harold Wilson for a "European technological community" to supplement the Common Market.[21]

Americans at the time, entering the home stretch in the race for the moon, naturally believed their own advertising about the superiority of the American system for generating high technology. The Atlantic Institute, the Organization for Economic Cooperation and Development, and other Euro-American institutions earnestly inquired into how to bridge the technology gap. Robert McNamara and James Webb argued that the real gap was not in hardware but in management techniques and systems analysis as practiced in the United States. Zbigniew Brzezinski emphasized the importance of scale: "All inventions for a long time will be made in the U.S. because we are moving so fast in technology and large-scale efforts produce inventions."[22]

Dismay and discouragement made 1968–72 years of confusion and cautious rebirth for space-age Europe. The Europa-2 booster failed four times to launch a satellite, and ELDO finally collapsed.[23] The Nixon administration, absorbed in planning for the post-Apollo period, invited the Europeans to collaborate in the proposed Space

[21]EUROSPACE, *Towards a European Space Program* (Fontenay-s-bois, 1966). On the technology gap generally, see: Jean-Jacques Servan-Schreiber, *Le Défi américain* (Paris, 1967); Norman Vig, *Science and Technology in British Politics* (Oxford, 1968); Pierre Vellas, *L'Europe face à la révolution technologique américaine* (Paris, 1969); Klaus-Heinrich Standke, *Europäische Forschungspolitik im Wettbewerb* (Baden-Baden, 1970).

[22]Atlantic Institute, *The Technology Gap: United States and Europe* (London, 1970); Richard R. Nelson, *The Technology Gap: Analysis and Appraisal*, RAND P-3694-1 (1967); Roger Williams, *European Technology: The Politics of Collaboration* (London, 1973). McNamara cited by Williams, p. 25; Brzezinski cited by James Webb, memo to Arnold Frutkin, June 22, 1967, NASA History Office.

[23]Events in Britain determined the final fate of ELDO. Despite the role of Price, Massie, Thorneycroft, and other Britons in the founding of the organization, British cabinets exhibited a lasting confusion about space and technology policy. Historically, the United Kingdom had the third-highest R&D budget in the world, while its industrial decline periodically raised alarms about the need for new technology. Nevertheless, budgetary pressures and bungling seemed always to prevent a coherent policy on the French model. Officials responsible for space suffered from a bureaucratic minuet that shifted them among nine different ministries over the space of a decade. Having canceled its national missile programs, the British government revived scientific rocket research in 1964 and finally launched a single, homemade satellite on the Black Knight booster in 1971. After 1957 the British depended on the United States for strategic missiles and among the Europeans were the most willing to rely on NASA for access to space, earning them in European space councils the epithet "the delegates from America." In April 1968, Anthony Wedgewood-Benn declared on behalf of the Labour government that Britain would make no further commitment to ELDO beyond her current obligations: "The effort of the Government should be directed to reinforcing the industrial potential which Europe already possesses."

Shuttle program. The Germans were especially enthusiastic, but this seemed to imply the permanent "subcontractor status" that the French in particular despised. The United States also proved accommodating in the scheduled renegotiation of the INTELSAT convention. Europeans, together with Third World members, won the right to outvote the United States in the assembly—for example, on placement of contracts—and an eventual termination of the COMSAT Corporation's management contract. But Europe could not take full advantage of such concessions without its own launch capacity and state-of-the-art comsat technology.

All was not bleak. France and Germany collaborated on a communications satellite named *Symphonie* and Britain on a geosynchronous test satellite of its own. France, of course, pursued her own military missiles, national satellite programs, and limited cooperation with the United States and, after 1966, with the USSR. After 1967, when the Hammaguir lease expired, de Gaulle also approved construction of a new equatorial spaceport in Kourou, French Guiana. Above all, the ELDO and ESRO experiences, however barren of results, gave firms and agencies the apprenticeship they needed in space technology and management. Their work on the Coralie, for instance, taught French engineers to handle hypergolic fuels and the high-energy LH_2/LOX upper stages favored for boosting heavy payloads to geosynchronous orbits. Finally, the organizational flaws that plagued ELDO and ESRO could be corrected. EUROSPACE electronics and aerospace firms formed multinational consortia with names like MESH, STAR, and COSMOS to compete for contracts and alleviate problems of *juste retour*. The European Space Conference sponsored the Bignier and Causse Reports that proposed fundamental reforms of the European space effort. They included a single European space agency, long-range program planning with guaranteed budgeting, centralization of authority for program management and systems integration, and smorgasbord participation by which member states could elect to share in some major programs and opt out of others, overall responsibility for major projects to be vested in the country paying most of the bill.[24]

[24]Michel Bourely, *La Conférence spatiale européenne* (Paris, 1970); J. Henrici, *An Overall Coherent and Long-Term European Space Program* (Munich, 1969); Théo Lefevre, *Europe and Space* (Brussels, 1972); Laurence Reed, *Ocean-Space—Europe's New Frontier* (London, 1969); C. R. Turner, "A Review of the Third EUROSPACE US-European Conference" and T. H. E. Nesbitt, "Future US-European Cooperation in Space: Possibilities and Problems," *Spaceflight*, January 1968 and May 1969; A. V. Cleaver, "The European Space Program, DISCORDE," *Aeronautics and Astronautics*, October 1968.

In December 1972, after five years of uncertainty about its own and America's future plans in space, the European Space Council adopted the above recommendations and proclaimed a new European Space Agency (ESA). It absorbed ELDO and ESRO and promised common, coordinated, long-term space and industrial policies. ESA grew out of the failures of the 1960s but also from the changed setting of the 1970s. De Gaulle was gone, Britain was in the Common Market, dynamic Germany took up the slack left by Britain and Italy, the United States–Soviet space race seemed over, and all nations were turning attention to the "practical" benefits of spaceflight. Last but not least was the U.S. space program for the 1970s. The Space Shuttle, approved by Nixon in 1972, promised to inaugurate a new era of routine, inexpensive orbital flight and yet offered the Europeans an intriguing target of commercial opportunity. For the Shuttle, a low-orbit workhorse, would not markedly improve American capability to launch payloads into high orbits. The Europeans had not only a political imperative but a technical opportunity to press on with development of a conventional heavy booster.

The ESA rested on a grand compromise. The other Europeans granted a renewed drive for the independent launch capacity demanded by France on the condition that France assume management and provide the bulk of funding. Second, ESA acceded to Germany's wish for a major cooperative program with the United States on condition that the Germans take charge and absorb most of the cost. Third, the British won approval for their pet project, a marine comsat, on the condition that they take the lead. The first program was the L3S launcher, soon to be dubbed Ariane, and 70 percent French; the second was *Spacelab*, made to fly inside the U.S. Shuttle cargo bay, and 53 percent German; the third, Marecs, was 56 percent British. Support of research centers and administration inherited from ESRO remained common responsibilities. But "big R&D" was now a mixture, not a solution, of national inputs; ESA won the loyalty of member states only through a partial nationalization of its international program. Gaullism and Euro-Gaullism coexisted, and the current era of neomercantilist competition in space was born.[25]

* * *

Europe has not escaped all the difficulties of multinational R&D. Member parliaments still have an aversion to long-range planning and

[25]On the transition to ESA, see: ESA, *Space—Part of Europe's Environment* (Paris, 1979) and *Europe's Place in Space* (Paris, 1981); U.S. Congress, *World Wide Space Programs*, pp. 285–314.

financial commitments, and budgets have remained at a level less than one-tenth that of NASA (see table 1). But ESA's first decade must be judged a success. On Christmas Eve 1979, twenty-two years after *Sputnik* and seventeen years since the birth of ELDO, the Ariane placed a European satellite in orbit from Kourou. Since then the Ariane has had a mixed record but is the first nonsuperpower booster declared operational for competitive commerce. The French (with a 59.25 percent interest) promptly incorporated a "private" company, Arianespace, and won contracts to launch comsats for INTELSAT, Arab and South American states, Australia, European Space Agency (of course), and even some U.S. firms.

In the meantime, the United States spent twelve times the cost of Ariane to develop the Space Shuttle. How is it that the American monopoly in space transportation was broken by a rocket that merely duplicated a capability (equivalent to the Thor Delta) the United States had had for two decades? According to the "méthode assez française," French engineers designed Ariane for one purpose: provide Europe with a sturdy, reliable heavy booster capable of launching satellites for the world market at a price competitive with that of American systems. What French officials foresaw as early as 1972 was that "outmoded" technology could still be commercially viable. The first two stages of Ariane, derived from the French Viking engine, are fueled by UDMH and nitrogen tetraoxide. This combination is not as powerful as LH_2 and LOX but need not be refrigerated to -250°C. The third is a high-energy stage sufficient to boost a 3,850-pound payload into trans-

TABLE 1

Global Space Budgets 1982

	Space Budget (in $Billions)	% of U.S. Budget	% of GNP
United States	12.1*	100	.40
USSR	14–28	116–227	1.0–2.0
France	.42	3.47	.08
Japan	.40	3.31	.04
West Germany	.33	3.35	.05
ESA	1.52	12.58	.06

Source.—Office of Technology Assessment, June 1984.

*The United States total includes military space spending on less classified programs (including reconnaissance satellites). The U.S. civilian space spending for 1982 amounted to some $5.7 billion, of which the majority went for Shuttle development and operations. Foreign R&D spending, therefore, though merely a fraction of American efforts, sufficed to sustain competition in a number of targeted space applications. The French agency CNES, for instance, spent 1,162.7 million francs, or 39.4 percent of its budget, on space applications, as compared to NASA's $333.8 million, or 5.8 percent of its budget, in 1982.

fer orbit, a highly eccentric path that reaches apogee at the required 22,000 miles above the earth. The final apogee motor that nudges the satellite into a circular orbit at that height is the responsibility of the customer. An upgraded Ariane-4 with a payload of over 9,000 pounds is approved and scheduled for operation by 1990.[26]

Now let us compare Ariane to the Shuttle, the only surviving project in a post-Apollo plan that originally included a space station, a lunar base, and even a manned voyage to Mars. Even to win approval of the Shuttle, NASA had to cut its initial cost estimates in half. Budgetary, technical, and military constraints, as well as the "fully reusable" feature, all dictated concentration on low orbits. The manned capacity of the Shuttle in turn meant cuts in payload and operating envelope. Finally, even the fully reusable feature was compromised, and the Shuttle evolved as an unlikely combination consisting of the orbiter, two strap-on recoverable solid rockets, and the bulky, nonrecoverable external tank.[27] The Shuttle is a spectacular tool for low-orbit operations but does not significantly increase efficiency on high-orbit launches. To meet the needs of communications customers, the Shuttle must be augmented by a perigee stage, or "inertial upper stage," carried in the cargo bay and released at an altitude of 120 miles. The perigee stage then boosts the payload to its apogee of 22,000 miles, whereupon the apogee stage connected to the satellite fires to achieve the circular geostationary orbit. It is a cumbersome process: it was this which French planners perceived in 1972.[28]

Even if the Shuttle-plus system proves reliable—and it too has had mixed results—can it match the Ariane in price? That depends entirely on government policies, for state-funded high technology is a neomercantilist controlled market, as de Gaulle sensed from the dawn of the space age. NASA claimed a cost base for geosynchronous insertion of $30 million, equal to that claimed for Ariane. Either competitor could slash prices, even below cost, in order to capture a greater market share for political purposes. But a liberal pricing policy is harder for the United States: the Shuttle cost $12 billion through 1980, the Ariane

[26]A. Dattner, "Reflections on Europe in Space—the First Two Decades and Beyond," ESA BR-10 (March 1982).

[27]On the Shuttle decision, see: U.S. Congress, Subcommittee on Space Science and Applications, *U.S. Civilian Space Programs 1958–1978*, 97th Cong., 1st sess. (1981), pp. 445–57; John M. Logsdon, "The Space Shuttle Decision: Technological and Political Choice," manuscript supplied by author. Logsdon, the director of the graduate program in Science, Technology, and Public Policy at George Washington University, is completing a book on the decision to build the Shuttle.

[28]For an excellent summary of the current launch-vehicle competition, see Alain Dupas, *Ariane et la navette spatiale* (Paris, 1981).

about $1 billion. Could not the United States continue to compete with its reliable "old-fashioned" rockets? To be sure, but the United States had such a stake in the Shuttle that it cut back production of disposable boosters in the expectation that Shuttle would revolutionize the industry. The Reagan administration, by contrast, has encouraged a renaissance of disposable launchers, but preferably through private enterprise.[29] Even as the United States pushes space technology forward into a new era, therefore, it is losing its hold on the only commercial rewards of this expensive and vital field.[30]

Yet launchers are not the only arena of competition; nor are France and ESA the only challengers in space technology. Seeking niches for themselves in these and other space markets of the future, the British, Canadians, and Japanese have forged ahead of the U.S. civilian space program in the research necessary to the next age of satellite communications, the 30/20 gigahertz or Ka band of the radio spectrum. Satellites using this broad, high-frequency band will vastly expand overall capacity, alleviate the growing shortage in geosynchronous orbital slots over the equator, and facilitate sophisticated services such as data transmission and conference calling. Meanwhile, France has shown no appreciable decline in technocratic vigor. Indeed, in matters of international technological competition, François Mitterand and the Socialists seem more Gaullist than de Gaulle, pledging to revivify national R&D and realize the old Gaullist vision of "France in the Year 2000." The French space agency CNES has promised "to consolidate our position in the principal means of applications (telecommunications, television, earth observation), to construct a solid space industry, and enlarge our penetration of the international market for launchers,

[29]Current U.S. policy options in civilian spaceflight are detailed in: U.S. Congress, *United States Civilian Space Programs 1958–1978*, pp. 5–28; Office of Technology Assessment, *Civilian Space Policy and Applications* (Washington, D.C., 1981), pp. 3–77; and the Office of Technology Assessment's new study, *Competition and Cooperation in Outer Space* (Washington, D.C., 1984).

[30]Can the market for satellite launches really be worth the effort Europe has made to break the American monopoly? In strictly commercial terms, the answer depends on how many scientific and commercial satellites will be orbited in coming decades. European analysts expect 170 missions into geosynchronous orbits alone by 1995: 110 for communications, fifty for television, and ten for meteorology. Even NASA estimates between 103 and 163 payloads by 1998 for which U.S. launchers and Ariane will compete (Battelle Columbus Laboratories, *Outside Users Payload Model* [NASw-338, June 1983]). Other analysts, however, predict a glut in communications circuits in the near future resulting from a decline in the rate of increase of demand or improvement in the capacity and durability of satellites or indeed from competition from earthbound fiber-optic circuits. But the "success" of Ariane or satellite systems is not strictly a commercial matter. Their inspiration was largely political.

satellites, and associated services and ground equipment."[31] In 1978 the French targeted remote sensing from space, developed a system called Spot designed to compete with NASA's Landsat, and founded another chartered company, Spotimage, to provide data for minerals prospecting, fishing, land-use analysis, mapping, and soil and agricultural management, especially to developing nations. Since the U.S. government has been unable to decide whether or how to market Landsat data, France threatens again to reap the rewards of a technology pioneered by the United States. And if the United States should determine to make full use of the Shuttle and exploit the prospects for space-based manufacturing in low orbit, the Europeans are ready; ESA is currently evaluating various plans for a minishuttle of its own, either the manned Hermes or the unmanned Solaris, as well as several space-station concepts.

Hence the age of Gaullism and Euro-Gaullism has spawned national and multinational, government-funded and -managed inventions of the future—not to encourage the presumed interdependence and integration stemming from global technologies but to preserve national autonomy and power. As de Gaulle perceived, however, competitiveness abroad necessitated centralization at home. Each of the European governments has absorbed or consolidated its aerospace companies into giant, semipublic behemoths: British Aerospace, France's Aerospatiale, Italy's Aerospaziale, recently the German merger of MBB (Messerschmidt) and VFW. Whatever the power of computers, nuclear weapons, jet aircraft, or space communications to make of our world a "global village," "Spaceship Earth," or a "lifeboat" in which all survive or perish as one, the space age has nonetheless sparked new political-economic fragmentation, even within the non-Communist world. The efforts of the "great" but not "super" powers to mobilize and scrap for an abiding autonomy and self-sufficiency mean that neither the Wilsonian nor the Leninist model, but rather the Gaullist model, of international order is riding the tide of technology in our time.

* * *

Where does all this leave the putative "free world leader"? It seems that the United States, the traditional, secure industrial leader and exponent of free trade, finds itself in a position reminiscent of late-19th-century Britain—not challenged in overall leadership but outmaneuvered first in one market then another by determined local

[31] Jean-Marie Luton, "La Politique spatiale française," *Les Cahiers français*, May–September 1982, p. 94.

rivals. Space technology, the very symbol of American technological hegemony just a decade ago, now reflects the bitter challenge facing the United States in the 1980s.

Why did Brzezinski's expectation of a growing U.S. lead in high technology prove false, especially since the nation still spends more on space than the rest of the free world combined? First and foremost, it is because of a nefarious division of labor that places on the United States almost sole responsibility for the strategic defense of the non-Communist world. If one counts military interest in the Shuttle, almost 60 percent of American space spending goes toward military research, while less and less money has been available for critical commercial sectors in astronautics (and aeronautics). The military requirements have an even more vexing effect—several technologies now exploited by foreigners have been developed by the Pentagon but cannot be transferred to the private sector for security reasons. These include high-resolution remote-sensing techniques and 30/20 comsat technology. United States corporations, in turn, shy away from risky markets in which foreign competition is heavily subsidized by government.

The Shuttle could still transform the space environment if vigorously exploited. The challenge of the various "Gaullists" might be transcended by a U.S. space station, space-based manufacture of drugs and crystals, or communications "platforms" with functionally limitless capacities, all built with modules boosted by the Shuttle. But to justify such expensive enterprises politically and underwrite them with tax dollars merely to maintain an image of leadership and technical virtuosity would be to adopt a Gaullist approach on our own account, while attempts to justify such a process commercially might not survive limited-range cost-benefit analysis. Nor is it natural for Americans to engage in centrally managed change, in "ten-year plans" dictated by bureaucrats, even assuming that NASA, the Pentagon, the White House, Commerce, NOAA, the aerospace industry, and Congress could agree on a long-term strategy for civilian space exploitation. Gaullism is not an expression of the American culture of technology. These considerations help to explain why Reagan aped Kennedy, in a dramatic presidential appeal, by defining a civilian space station as a national goal to be achieved "within a decade," despite the opposition of the Pentagon, the president's science adviser, and the Office of Management and Budget. Yet we cannot know whether such a station is a bold investment or a folly until decades after its completion—much less its approval.

These perplexing problems, at home and abroad, are not confined to space technology. They exist in more and more sectors as state-driven technological change pushes foreign governments further away from the free-market ideal that flourished in the merely industrial age of the 19th and early 20th centuries. Western intellectuals always assumed that the postcapitalist age would usher in socialism and integration. Instead, it ushered in Gaullism with its rejection of capitalism and communism, its enforced unity at home and Darwinist struggle abroad. The space age is an age of neomercantilism, or competing national socialisms, and for the first time Americans can neither retreat from the world nor hope to make it over. How will they adapt to life in a technocratic world they helped to make but do not like, cannot change, and cannot escape?

Nuclear Power and the Environment: The Atomic Energy Commission and Thermal Pollution, 1965–1971

J. SAMUEL WALKER

During the latter half of the 1960s, the decline of environmental quality in the United States took on growing urgency as a public policy issue. A series of controversies over the effects of substances such as DDT, mercury, and phosphates, ecological disasters such as a huge oil spill off the coast of California and the death of Lake Erie from industrial pollution, and easily visible evidence of foul air and dirty water fueled public alarm about the deterioration of the environment. At the same time that the environmental crisis commanded increasing attention, questions about the availability of electrical power triggered deepening concern. Since the early 1940s, the use of electricity in the United States had expanded by an average of 7 percent per year, which meant that it roughly doubled every decade. Utility and government planners found no indications that the pace of growth was likely to slow in the near future. A report prepared by the White House Office of Science and Technology in 1968 predicted that the nation would need about 250 "mammoth-sized" new power plants by 1990. Power blackouts and brownouts became increasingly commonplace in the late 1960s, graphically illustrating the discomfort and inconvenience that a shortage of electricity could cause.[1]

DR. WALKER is historian of the U.S. Nuclear Regulatory Commission and the author of *Containing the Atom: Nuclear Regulation in a Changing Environment, 1963–1971* (University of California Press, 1992).

[1] *Considerations Affecting Steam Power Plant Site Selection* (Washington, D.C.: Executive Office of the President, Office of Science and Technology, 1968); *Nucleonics Week*, December 12, 1968, pp. 4–5; Jeremy Main, "A Peak Load of Trouble for the Utilities," *Fortune* 80 (November 1969): 116–19 ff.; "Why Utilities Can't Meet Demand," *Business Week*, November 29, 1969, pp. 48–62; "Conservation Forces Thwart Utilities' Hunt for Power Plant Sites," *Wall Street Journal*, March 23, 1970, p. 1; "Danger of More Power 'Blackouts,' " *U.S. News and World Report*, April 20, 1970, pp. 48–50. For the best overviews of environmental issues, see Samuel P. Hays, *Beauty, Health, and Permanence: Environmental Politics in th United States, 1955–1985* (New York, 1987); Martin V. Melosi, *Coping with Abundance: Energy and Environment in Industrial America* (Philadelphia, 1985); Martin V. Melosi, "Lyndon Johnson and Environmental Policy," in Robert A. Divine,

The growing public and political concern with environmental quality and the continually increasing demand for electricity put utilities in a quandary. Electrical generating stations were major polluters; fossil fuel plants, which provided over 85 percent of the nation's electricity in the 1960s, spewed millions of tons of noxious chemicals into the atmosphere annually. Coal, by far the most commonly used fuel, placed a much greater burden on the environment than other fossil fuels. The sulfur dioxide and nitrogen oxides that coal plants released were important ingredients in air pollution, and the carbon dioxide they emitted raised the possibility of harmful climatic changes over a long period of time. The difficulties caused by burning coal, and to a lesser extent oil and natural gas, defied easy solutions. The concurrent demands for sufficient electricity and clean air created, in the words of the Conservation Foundation, "a most vexing dilemma: How do we protect the environment from further destruction and, at the same time, have the electricity we want at the flick of a switch?" An article in *Fortune* magazine depicted the problem in even starker terms: "Americans do not seem willing to let the utilities continue devouring ever increasing quantities of water, air, and land. And yet clearly they also are not willing to contemplate doing without all the electricity they want. These two wishes are incompatible. That is the dilemma faced by the utilities."[2]

After the mid-1960s, utilities increasingly viewed nuclear power as the answer to that dilemma. While conforming with their plans to achieve "economies of scale" by building larger plants, it promised the means to produce sufficient electricity without fouling the air. Envi-

---

ed., *The Johnson Years,* vol. 2: *Vietnam, the Environment, and Science* (Lawrence, Kans., 1987); Edward W. Lawless, *Technology and Social Shock* (New Brunswick, N.J., 1977); Richard H. K. Vietor, *Enviornmental Politics and the Coal Coalition* (College Station, Tex. 1980); and Roderick Nash, *Wilderness and the American Mind,* 3d ed. (New Haven, Conn., 1982).

[2]*Considerations Affecting Steam Power Plant Site Selection* (n. 1 above), chap. 4; James G. Terrill, Jr., E. D. Harward, and I. Paul Leggett, Jr., "Environmental Aspects of Nuclear and Conventional Powerplants," *Journal of Industrial Medicine and Surgery* 36 (July 1967): 412–19, reprinted in Joint Committee on Atomic Energy (JCAE), *Selected Materials on Environmental Effects of Producing Electric Power,* 91st Cong., 1st sess., 1969, pp. 121–33; *Nucleonics Week,* November 11, 1965, p. 2; *Restoring the Quality of Our Environment: Report of the Environmental Pollution Panel* (Washington, D.C.: President's Science Advisory Committee, 1965); Main, "A Peak Load of Trouble" (n. 1 above), p. 205; National Rural Electric Cooperative Association, *The Electric Power Crisis: Its Impact on Workers and Consumers,* 1971, copy in Box 35 (Reaction of Utilities, JCAE, etc.), Office Files of James T. Ramey, Atomic Energy Commission Records, Department of Energy, Germantown, Md. (hereafter cited as AEC/DOE); Conservation Foundation, *CF Letter,* March 1970, copy in Box 181 (Pollution), Office Files of Glenn T. Seaborg, AEC/DOE.

ronmental concerns were a major spur to the rapid growth of nuclear power, and industry voices emphasized the environmental benefits of nuclear generation. In a rare editorial, entitled "Let the Public Choose the Air It Breathes," the trade publication *Nucleonics Week* concluded in 1965 that, in comparison with coal, "the one issue on which nuclear power can make an invincible case is the air pollution issue." In a memorandum to senior staff members, public information officials of Atomics International, a leading nuclear vendor, itemized environmental assets of nuclear power. Describing it as "safe, clean, quiet, and odorless," they observed that atomic plants "do not release harmful amounts of pollutants to the atmosphere . . . [or] to water" and that they "assure continued supply of low cost power and conserve our natural resources." Like Atomics International, other nuclear vendors stressed the cleanliness of atomic power; it was an important selling point in their effort to expand their markets.[3]

As the buyers of generating facilities, many utilities found the case for the environmental advantages of nuclear power to be compelling. Sherman R. Knapp, chairman of the board of Connecticut Light and Power, told an American Nuclear Society meeting in February 1965: "Atomic power is bound to be increasingly attractive to communities as concern over air pollution intensifies." Other utility executives echoed the same sentiments and took actions that proved the accuracy of Knapp's prediction. Northern States Power of Minneapolis, for example, decided in 1967 to build a 550-megawatt nuclear unit because of environmental considerations, even though the estimated costs were higher than for a comparable fossil fuel plant. Richard D. Furber, vice-president of the utility, explained that Northern States had just suffered through a lengthy controversy over the construction of a coal plant and added: "Many times during the three-year controversy the opposition indicated they would lay off if we would convert this plant to a nuclear plant."[4]

This was not an isolated case in which environmentalists expressed support for nuclear power, though they clearly were less enthusiastic

[3]*Nucleonics Week*, February 25, 1965, p. 1; J. H. Wright, "Nuclear Power and the Environment," *Atomic Power Digest*, published by Westinghouse Nuclear Energy Systems, 1969, copy in Box 53 (Westinghouse Electric Co.), Seaborg Office Files, AEC/DOE; JCAE, *Environmental Effects of Producing Electric Power*, 91st Cong., 2d sess., 1970, pp. 1512–26; "Suit on Pollution to Seek A.E.C. Aid," *New York Times*, September 29, 1970, p. 25; AI-Public Information to AI Supervision, January 18, 1971, Box 194 (Anti-Nuclear Organizations), Craig Hosmer Papers, University of Southern California, Los Angeles; Richard F. Hirsh, "Conserving Kilowatts: The Electric Power Industry in Transition," *Materials and Society* 7 (1983): 295–305.

[4]"News about Industry," *Nuclear Industry* 12 (March 1965): 7–11, and "Northern States Power Puts the Accent on Environment," ibid. 14 (November 1967): 32–33.

about the technology than were industry representatives. While acknowledging the advantages of nuclear power in combating air pollution, some environmentalists cautioned that radioactive effluents could also pose a serious problem. Malcolm L. Peterson, a spokesman for the Greater St. Louis Committee for Nuclear Information, declared in 1965: "Because nuclear power plants do not pollute the air with smoke, nor produce any of the ingredients of photochemical smog, they are regarded as 'clean,' but it should not be forgotten that radioactivity, though invisible, is also a contaminant." Another prominent environmental organization, the Sierra Club, was ambivalent in its position on nuclear power. It protested plans to build a power reactor at Bodega Head on the California coast in the early 1960s, but as a policy matter it neither endorsed nor opposed the construction of nuclear plants. The attitudes of environmental groups were perhaps best summarized in the equivocal assessment of Thomas E. Dustin, president of the Izaak Walton League of America, in 1967: "I think most conservationists may welcome the oncoming of nuclear plants, though we are sure they have their own parameters of difficulty."[5]

The attitudes of the general public about the environmental effects of nuclear power were seldom evaluated. One poll published in early 1966 suggested that many members of the public lacked strong views or informed opinions about the subject. The survey, conducted with residents of Buchanan, New York (site of the Indian Point nuclear plant), Philadelphia, and Atlanta, asked, among other questions, "How 'clean' are nuclear plants in operation?" In each location, from 40 to 50 percent of the respondents had no answer. Those who did respond, however, overwhelmingly expressed a favorable outlook on the cleanliness of nuclear power.[6]

Officials of the U.S. Atomic Energy Commission (AEC) actively promoted the idea that nuclear power provided the answer to both the environmental crisis and the energy crisis. Under its statutory mandate, the AEC was responsible both for encouraging the use of atomic energy for peaceful purposes and for regulating its safety, and

[5]"Conservation Policy Guide: Abstract of Directors' Actions, 1946–1968" (Minutes), rev. ed., July 1968, Sierra Club Records, William E. Colby Memorial Library, Sierra Club, San Francisco; Thomas E. Dustin to Glenn T. Seaborg, February 17, 1967, Box 7717 (MH&S-11, Industrial Hygiene), and Thomas L. Kimball to Seaborg, November 6, 1970, Box 181 (Pollution), Seaborg Office Files, AEC/DOE; Malcolm L. Peterson, "Environmental Contamination from Nuclear Reactors," *Scientist and Citizen* 8 (November 1965): 1–11; Norman Cousins, "Breakfast with Dr. Teller," *Saturday Review,* March 19, 1966, pp. 26, 54.

[6]"Are Nuclear Plants Winning Acceptance?" *Electrical World,* January 24, 1966, pp. 115–17; "Nuclear Power and the Community: Familiarity Breeds Confidence," *Nuclear News* 9 (May 1966): 15–16.

it saw the energy/environment dilemma as an opportunity to enhance the attractiveness of nuclear power. Chairman Glenn T. Seaborg told the National Conference on Air Pollution in 1966 that, in light of expanding demand for electricity and deteriorating air quality, "we can be grateful that, historically speaking, nuclear energy arrived on the scene when it did." Although he acknowledged that nuclear power had some adverse impact on the environment, he insisted that its effects were much less harmful than those of fossil fuels. In comparison with coal, he once declared, "there can be no doubt that nuclear power comes out looking like Mr. Clean." Other AEC officials expressed the same views on numerous occasions.[7]

Other than radiation protection, the focus of its regulatory functions, the AEC did not view environmental issues as a central part of its responsibilities. Although the AEC expressed concern about environmental matters in general, it insisted that its statutory mandate for regulating its licensees did not extend beyond radiation hazards. The agency conducted numerous research projects around its own installations to seek information about the consequences of nuclear weapons tests, underground explosions, and reactor wastes for the natural environment and animal life. In nearly every case, the projects focused on the effects of radiation; the major exception was a series of studies done on heated water in the Columbia River from the plutonium reactors on the Hanford reservation.[8]

The AEC cooperated on an informal basis with other government agencies in assessing the environmental aspects of reactor licensing and operation, particularly the U.S. Public Health Service, a part of the Department of Health, Education, and Welfare, and the Fish and Wildlife Service (FWS), a part of the Department of the Interior. Relations between the AEC and other agencies were usually cordial

[7]U.S. Atomic Energy Commission (USAEC), *Nuclear Power and the Environment* (one of a series of booklets on "Understanding the Atom"), 1969, Atomic Energy Commission Records, Nuclear Regulatory Commission, Rockville, Md. (hereafter cited as AEC/NRC); Glenn T. Seaborg speeches, December 13, 1966, Box 7717 (MH&S-11, Industrial Hygiene, vol. 1), May 5, 1969, Box 25 (Pollution of the Environment), and April 20, 1970, Box 35 (Regulatory—General), Nixon Library Materials, Seaborg Office Diary, August 7, 1968, James T. Ramey speech, August 13, 1970, Box 35 (H. Peter Metzger), Ramey Office Files, AEC/DOE; "Atom and Environment," *Washington Evening Star*, June 16, 1969, p. A-10.

[8]John G. Palfrey to Donald F. Hornig, September 5, 1964, Box 168 (Office of Science and Technology), Seaborg Office Files, Glenn T. Seaborg to Edward Wenk, Jr., January 19, 1966, Box 1362 (MH&S-3-3, Contamination and Decontamination), AEC/DOE; John R. Totter speech, December 3, 1969, Box 194 (Environment—General), Hosmer Papers; USAEC, *Atoms, Nature, and Man* (one of a series on "Understanding the Atom"), 1966, and USAEC, *Nuclear Power and the Environment*, AEC/NRC.

and mutually respectful, but on occasion the arrangements led to disputes. One of those disagreements occurred when the FWS questioned the AEC's denial of regulatory authority over nonradiological environmental matters. The FWS, under an interagency understanding, reviewed power reactor applications submitted to the AEC to evaluate the effects of the proposed plant on the animal and marine environment. The FWS began to suggest in the mid-1960s that the AEC should take nonradiological environmental effects into account in licensing cases, especially the consequences of discharging large quantities of heated water for aquatic life. The AEC responded that it lacked authority to set requirements for any nonradiological effect that a nuclear plant might have on the environment.[9]

What began as a dispute between the AEC and the FWS soon flared into a major public debate over "thermal pollution." As it developed, the controversy not only embroiled the AEC in a conflict with Interior but also antagonized some prominent members of Congress, generated unfavorable publicity, and raised questions about the extent to which nuclear power was environmentally superior to fossil fuels. As a result of the thermal pollution issue, nuclear power, rather than being seen as an answer to environmental degradation from electrical production, appeared to a growing number of observers to be a part of the problem.

Thermal pollution resulted from cooling the steam that drove the turbines to produce electricity in a fossil fuel or nuclear plant. The steam was condensed by the circulation of large amounts of water, and in the process the cooling water was heated, usually by 10–20 degrees Fahrenheit, before being returned to the body of water from which it came. This problem was not unique to nuclear power plants; fossil fuel plants also discharged waste heat from their condensers. It was more acute in nuclear plants, however, for two reasons. Fossil fuel plants, unlike nuclear ones, dispelled some of their heat into the atmosphere through smokestacks. More important, fossil plants used steam heat more efficiently than nuclear ones, meaning that nuclear plants generated 40–50 percent more waste heat than did comparably sized fossil fuel plants. The cooling water that nuclear power stations

[9]U.S. Senate, Subcommittee on Air and Water Pollution, *Hearings on Clean Air,* 88th Cong., 2d sess. 1964, pp. 1069–97, and *Hearings on Thermal Pollution—1968,* 90th Cong., 2d sess., 1968, pp. 1248–62; Lester R. Rogers to Harold L. Price and others, June 13, 1963, Glenn T. Seaborg to Stewart L. Udall, March 27, 1964, John F. Newell to the Files, June 9, 1964, Troy Connor to the Separated Legal Files, July 9, 1964, Connor to Harold L. Price, May 25, 1965, Harold L. Price to Lewis A. Sigler, July 6, 1965, Price to Commissioner Ramey, February 2, 1968, L-4-1 (Memo of Understanding, AEC–Dept. of Interior and Fish and Wildlife), AEC/NRC.

released was not radioactive; it circulated in a separate loop from the water used to cool the reactor core.[10]

The problem of thermal pollution was not new in the mid-1960s, but it created more anxiety at that time because of the growing number of power plants being constructed, the greater size of those plants, and the increasing inclination of utilities to order nuclear units. Those trends combined to amplify concern about the effects of waste heat on the environment. Although the precise impact of thermal pollution was uncertain, there appeared to be ample cause to be disturbed about its implications. Some scientists suggested that waste heat deposited in lakes and rivers from steam power plants posed a grave threat both to fish and to other forms of aquatic life.[11]

The effects of thermal discharges on fish were worrisome because many species were highly sensitive to changes in temperature. A rise in water temperature could alter their reproductive cycles, respiratory rates, metabolism, and other vital functions. A drastic or a sudden shift in temperature could be lethal. Between 1962 and 1967, the Federal Water Pollution Control Administration found at least ten cases in which fish were killed by waste heat from fossil fuel power stations. The most serious incident occurred in the Sandusky River in Ohio, where over 300,000 fish died in January 1967; the others were much less severe. It was more common for fish to be killed indirectly by heat discharges. In the Hudson River around the Indian Point nuclear power station, for example, tens of thousands of bass died in 1963 after being attracted during cold weather to the warm currents coming from the plant. As nearly as experts could determine, the fish got caught in the water intake system of the plant and died from exhaustion or from contact with pumps or other equipment. Large fish kills attracted a great deal of attention, but the more subtle threats to the marine environment were at least as troubling. As one writer argued: "In the long run temperature levels that adversely affect the

[10]Federal Power Commission, *Problems in Disposal of Waste Heat from Steam-Electric Plants,* 1969, Box 7763 (MH&S-11, Bulky Package), AEC/DOE; USAEC, *Thermal Effects and U.S. Nuclear Power Stations,* WASH-1169, 1971, MH&S-3-1 (Thermal Effects, Nov. 1970–), AEC/NRC; Ralph E. Lapp, "Power and Hot Water," *New Republic,* February 6, 1971, pp. 20–23.

[11]Federal Power Commission, *Problems in Disposal of Waste Heat* (n. 10 above), pp. 1–2, 17–22; Roger Don Shull, "Thermal Discharges to Aquatic Environments," June 9, 1965, attachment to Clifford K. Beck to James T. Ramey, November 6, 1967, Legal-4-1 (Federal Water Pollution Agency), AEC/NRC; CW [Charles Weaver] memorandum, April 27, 1967, File 23-1-10 (Vermont Yankee Nuclear Power Plant Material, 1967), George D. Aiken Papers, University of Vermont, Burlington; John R. Clark, "Thermal Pollution and Aquatic Life," *Scientific American* 220 (March 1969): 19–26; Wolfgang Langewiesche, "Can Our Rivers Stand the Heat?" *Reader's Digest* 96 (April 1970): 76–80.

animals' metabolism, feeding, growth, reproduction and other vital functions may be as harmful to the fish population as outright heat death."[12]

The concern about thermal pollution extended not only to its hazards for fish but also to other potential consequences. It could disrupt the ecological balance by killing certain kinds of plant life while causing other kinds to flourish. Water warmed by thermal discharges, for example, contained relatively greater quantities of blue-green algae than of other species, and an excess of blue-green algae made water look, taste, and smell unpleasant. Rising temperatures also reduced the capacity of water to retain dissolved oxygen, which was needed to chemically convert waste matter into innocuous forms. As the amount of oxygen in the water diminished, the amount of undesirable wastes and pollutants increased.[13]

The nature and severity of the environmental damage attributable to waste heat depended on variables that differed widely from place to place, including the size and efficiency of the power plant, the type and adaptability of the fish and plant life in the affected body of water, the rate and volume of water flow, and the natural thermal characteristics of the water. While many questions about thermal pollution remained unanswered, the prospect that scores of new power plants, over half of them nuclear, would be built within two decades generated substantial alarm about its long-term effects. An article on the subject in *Scientist and Citizen,* the publication of the Committee for Environmental Information (the successor to the Committee for Nuclear Information), declared in 1968: "We cannot continue to expand our production of electric power with present generating methods without causing a major ecological crisis." Television newsman Edwin Newman informed his viewers of an even drearier prognosis. "The gloomiest forecast we know of about the future of our water resources is that by the end of the decade our

[12]Federal Power Commission, *Problems in Disposal of Waste Heat* (n. 10 above), pp. 18–21; Clark, "Thermal Pollution and Aquatic Life" (n. 10 above), pp. 19–22; JCAE, *Hearings on Participation by Small Electrical Utilities in Nuclear Power,* 90th Cong., 2d sess., 1968, pp. 89–92; "Con Ed Abolishes Fish 'Death Trap,' " *New York Times,* July 1, 1966, p. 37; Robert H. Boyle, "A Stink of Dead Stripers," *Sports Illustrated,* April 26, 1965, pp. 81–84; Allan R. Talbot, *Power along the Hudson: The Storm King Case and the Birth of Environmentalism* (New York, 1972), pp. 112–14.

[13]William M. Holden, "Hot Water: Menace and Resource," *Science News* 94 (August 17, 1968): 164–66; John Cairns, Jr., "We're in Hot Water," *Scientist and Citizen* 10 (October 1968): 187–98; Frank Graham, Jr., "Tempest in a Nuclear Teapot," *Audubon* 72 (March 1970): 13–19; Federal Power Commission, *Problems in Disposal of Waste Heat* (n. 10 above), pp. 21–22; USAEC, *Thermal Effects and U.S. Nuclear Power Stations* (n. 10 above), pp. 18–23.

rivers may have reached the boiling point," he reported in 1970. "Three decades more and they may evaporate." Newman added: "This vision of an ultimate cataclysm is based on the assumption that we will continue to discharge heat into our rivers at the rate at which we're doing it now." Most warnings about thermal pollution were far less apocalyptic than the one that Newman cited, but anxiety about the dangers of waste heat from power plants was widespread among both experts and laymen.[14]

Some observers, however, found less cause for concern. Although they acknowledged that thermal pollution was a problem, they also argued that its threat to the environment had been exaggerated. Scientists who took this point of view noted that laboratory experiments demonstrating serious effects of waste heat sometimes conflicted with actual field experience. They also showed that, contrary to the impression that newspaper and magazine articles often gave, only a small percentage of fish kills were caused by heated water from power plants. In addition, some scientists argued that heated water from generating stations could be beneficial. While certain kinds of fish were adversely affected, others throve in warmer water. The Pacific Gas and Electric Company, citing studies by the California Department of Fish and Game, asserted in 1970: "Fishermen rarely criticize utility companies for the warmer temperature of water near power plants. That's where the fishing is likely to be best." Heated water offered other potential advantages. Glenn Seaborg, for example, suggested that waste heat could be put to work irrigating fields to extend the growing season and reduce frost damage. In this regard, he and others maintained that the proper term for waste heat was not "thermal pollution" but "thermal enrichment."[15]

Yet even those who were most sanguine about the implications of waste heat recognized that its harmful effects could not be ignored. The disagreement of opinion arose over the severity of the problem, not the existence of one. In order to find out more about the

[14]Cairns, "We're in Hot Water" (n. 13 above), p. 187; Clark, "Thermal Pollution and Aquatic Life" (n. 10 above), p. 19; "Operators of Nuclear Plants Fear Heat Pollution," *Philadelphia Inquirer*, September 30, 1970, p. 1; "The Problem of Thermal Pollution," NBC television broadcast, May 24, 1970, Box 37 (Environmental Effects, including Thermal), Nixon Library Materials, AEC/DOE.

[15]Holden, "Hot Water: Menace and Resource" (n. 13 above), pp. 164–65; "Problems Associated with U.S. Thermal Effects Standards Examined," *Nuclear Industry* 17 (August 1970): 32–36; Atomic Industrial Forum, *Info: Thermal Effects*, February 20, 1970, copy in Box 18 (AIF Meeting, Thermal Considerations), Ramey Office Files, AEC/DOE; John A. Harris to the Commission, April 8, 1970, MH&S-3-1 (Thermal Effects), AEC/NRC: Glenn T. Seaborg and William R. Corliss, *Man and Atom: Building a New World through Nuclear Technology* (New York, 1971), pp. 81–83, 117–20.

consequences of thermal pollution and ways to control it, several government agencies and a number of utilities sponsored research programs. But most utilities could not wait for research to produce conclusive results about waste heat; they needed to build plants immediately to meet anticipated demand for electricity. Public concern about thermal pollution and newly established state water quality standards made it imperative for many of them to act promptly to curb thermal discharges.[16]

Technical solutions were available to deal with the problem of waste heat, but they required extra expenses in the construction and operation of steam-electric plants. Gradually, and often reluctantly, a growing number of utilities decided to pay the costs of mitigating the effects of thermal pollution. To do so, they built systems to replace their traditional, and preferred, practice of "once-through cooling," in which water was drawn into the plant, used to cool steam in the condenser, and then directly returned to its source. Utilities generally elected to use alternatives to once-through cooling because the volume and flow of the water available for cooling were insufficient, environmental groups raised vocal protests, and/or limits set by state agencies required them to reduce the temperature of waste heat discharges. The federal Water Quality Act of 1965 encouraged states to establish water quality standards for interstate streams and coastal waterways, and many states moved promptly to control water temperatures. The increasing concern about environmental quality, the imposition of state standards, the growing number and size of power stations, and the paucity of good sites for plants accelerated the trend away from once-through cooling, although utilities still employed it where they could.[17]

Utilities could choose from several options to reduce the effects of waste heat. The cheapest and easiest approach was to limit the environmental impact of heated water without building a separate system. This could be done, for example, by pumping more water through the condenser, which raised its temperature less, or by providing a long channel to discharge the heated water into different

[16]"The Effects and Control of Heated Water Discharges: A Report to the Federal Council for Science and Technology by the Committee on Water Resources Research," November 1970, copy in Box 169 (Office of Science and Technology), Seaborg Office Files, and "Thermal Effects Studies by Nuclear Power Plant Licensees and Applicants," 1969, Box 5625 (Environmental Pollution), OGM Files, AEC/DOE.

[17]R. E. Hollingsworth to Joseph D. Tydings, January 23, 1970, MH&S-3-1 (Thermal Effects), SECY-812 (December 28, 1970), AEC/NRC; Milton Shaw to Commissioner Larson, February 20, 1970, Box 7751 (ID&R-6, Hazards, vol. 2), AEC/DOE; *Nucleonics Week,* May 4, 1967, pp. 1–2.

sections of the source body of water. In many cases, however, a more elaborate system was essential. The available alternatives offered the means to resolve or greatly alleviate the problem of waste heat but also exacted significant costs. One method was to dig a cooling pond, where contact with air would cool the heated water on the surface. The primary disadvantage of a cooling pond was that it required a sizable area of land. A large plant would need a pond of several hundred acres (the rule of thumb was 2 acres for every megawatt), and, except in rural regions, the cost of that much land was prohibitive.[18]

Utilities generally found it more economical to build cooling towers. Several different designs were available, but the most commonly used were natural draft or mechanical draft towers. Either type of tower dumped waste heat into the atmosphere as warm vapor or warmed air. A natural draft tower could rise as high as a 30-story building. It worked like a chimney, drawing air warmed by contact with heated water upward and out the top of the tower. This process cooled the water, some of which evaporated; the rest either was recirculated in the condenser or returned to its source. The principal drawback to a natural draft tower was its cost, estimated in 1967 to be four thousand to ten thousand dollars per megawatt. Mechanical draft towers used fans to circulate air and cool the water from the condenser. They were less expensive to build than natural draft towers because they did not need to be nearly as high, but they were more expensive to operate. In addition to their costs, cooling towers posed other problems. They reduced the generating capacity of the plant by a small, but not negligible, amount. The water that cooling towers added to the atmosphere raised concern that they would cause localized fog and icing conditions, although there was little evidence that this was a common occurrence. Finally, natural draft towers were aesthetically objectionable to those who disliked the way they dominated the skyline for miles around.[19]

[18]Steve Elonka, "Cooling Towers: A Special Report," *Power,* March 1963, copy in Box 16 (Reactors—General), Office Files of Wilfrid E. Johnson, and P. N. Ross, "Presentation to President's Water Pollution Control Advisory Board," December 6, 1968, Box 3 (Water Pollution-Muskie Bill), Ramey Office Files, AEC/DOE; *Considerations Affecting Steam Power Plant Site Selection* (n. 1 above), chap. 4.

[19]Elonka, "Cooling Towers"; Ross, "Presentation to . . . Advisory Board"; Frank H. Rainwater, "Thermal Waste Treatment and Control," June 29, 1970, Box 18 (AIF Meeting, Thermal Considerations), Ramey Office Files; Glenn T. Seaborg to Sally Morrison, February 3, 1970, Box 7763 (MH&S-11, Environmental Studies, vol. 5), AEC/DOE; *Nucleonics Week,* May 4, 1967, pp. 1–2; "Outlook for Cooling Towers," *Nuclear Industry* 14 (October 1967): 8–14; "Wealth of New Data on Nuclear Plant Cooling Methods, Costs," ibid. 17 (July 1970): 7–15; "Utilities Burn over Cooling

The problem of cooling waste heat discharges was not peculiar to nuclear plants, but it was particularly troublesome in them. A utility that considered building a nuclear unit in the late 1960s inevitably confronted the issue of thermal pollution. In 1967, only a handful of power companies planned to use cooling towers, but, by early 1970, over half of the eighty-five plants on order or under construction were designed with cooling systems. Most of those without cooling apparatus were located on oceans, bays, or the Great Lakes, where the threat of waste heat seemed less acute. Although the trend was clear, it did not emerge without major controversies over the effects of thermal pollution and the role of the AEC in regulating them.[20]

Control over thermal pollution was, in the phrase of a writer for the trade journal *Nuclear Industry,* "a jurisdictional 'no man's land.' " The Department of the Interior, including both the Federal Water Pollution Control Administration and the Fish and Wildlife Service, took particular interest in the problem, but its statutory power extended only to advising other federal agencies and state governments on the protection of aquatic life. Enforcement of water standards remained a function of the states, but their regulations were not always adequate or uniform. Some members of Congress and officials of the FWS suggested that the AEC should assume greater responsibility over thermal discharges from nuclear plants, but it denied that it had the statutory authority to do so.[21]

The AEC's refusal to regulate thermal effects stirred private expressions of concern and, later, unusually blunt protests from the FWS. The differing views of the two agencies emerged clearly, and publicly, in a disagreement over an application for a construction permit for the Millstone Nuclear Power Station in Waterford, Connecticut. In November 1965, the AEC, as a part of its customary procedures, sent a copy of the Millstone application to the FWS for comment. The FWS, in turn, forwarded the document to one of its subdivisions, the Radiobiological Laboratory of the Bureau of Com-

Towers," *Business Week,* April 3, 1971, pp. 52–54; *Considerations Affecting Steam Power Plant Site Selection* (n. 1 above), chap. 5.

[20]Shaw to Larson, February 20, 1970, AEC/DOE; Hollingsworth to Tydings, January 23, 1970, SECY-812, H. L. Price to the Commission, July 27, 1971, Job 9, Box 19 (Legal-11 REG Litigation), AEC/NRC; "Outlook for Cooling Towers" (n. 19 above), pp. 8–12.

[21]Troy Conner to Harold L. Price, May 25, 1965, L-4-1 (Memo of Understanding, AEC–Department of Interior and Fish and Wildlife), AEC/NRC; "A Jurisdictional 'No Man's Land,' " *Nuclear Industry* 15 (March 1968): 3–5; Ellen Thro, "The Controversy over Thermal Effects," *Nuclear News* 11 (December 1968): 49–53.

mercial Fisheries. Theodore R. Rice, director of the laboratory, prepared an evaluation of the possible effects of the proposed plant on fish in the vicinity. He concluded that the reactor could be operated without radiological injury to fish. Rice appended a section cautioning that thermal discharges from the plant might have adverse consequences, but he accepted the AEC's view that its jurisdiction was "limited to matters pertaining to radiological safety."[22]

To that point, the comments of the FWS had followed well-established patterns. Clarence F. Pautzke, head of the FWS, made a major departure from routine procedures, however, when he sent Rice's report to the AEC in March 1966. Pautzke announced that, even though his agency had in the past submitted comments similar to Rice's, it had changed its position because of growing federal concern for environmental quality. "We wish to make clear that Dr. Rice's statements . . . concerning the jurisdiction and responsibility of the Atomic Energy Commission in regard to thermal pollution," he declared, "does [*sic*] not represent the policy of the Fish and Wildlife Service." Pautzke asserted that the AEC's regulatory authority covered thermal pollution and suggested that it ask the Department of Justice to review the question. If Justice supported the AEC, he thought that "legislation to provide this necessary authority should be sought by the Commission."[23]

Pautzke's letter caught the AEC by surprise. Harold L. Price, the AEC's director of regulation, complained that the FWS had not only sent it to several state agencies but also had "openly and publicly challenge[d] the position of the Commission with respect to authority over thermal effects." The Joint Committee on Atomic Energy, the AEC's congressional oversight committee, was equally startled by the implied effrontery and concerned about the possible effect of Pautzke's arguments. It had recently heard similar criticism from John D. Dingell, chairman of the House Subcommittee on Fisheries and Wildlife Conservation, who suggested that the AEC was evading the provisions and the intentions of the Fish and Wildlife Coordination Act. "The effect of this has been," he charged, "that they have

[22]"Preliminary Evaluation of Possible Effects on Fish and Shellfish of the Proposed Millstone Nuclear Reactor," December 15, 1965, Box 587 (Reactors: Millstone Point), Papers of the Joint Committee on Atomic Energy, Record Group 128 (Records of the Joint Committees of Congress), National Archives, Washington, D.C. (hereafter cited as JCAE Papers). This document is printed in U.S. Congress, House, Committee on Merchant Marine and Fisheries, Subcommittee on Fisheries and Wildlife Conservation, *Hearings on Miscellaneous Fisheries Legislation*, 89th Cong., 2d sess., 1966, pp. 191–94.

[23]Clarence F. Pautzke to Harold L. Price, March 23, 1966, Box 587 (Reactors: Millstone Point), JCAE Papers. See also Subcommittee on Fisheries and Wildlife Conservation, *Hearings on Miscellaneous Fisheries Legislation* (n. 22 above), pp. 191–92.

proceeded without due care for either the enhancement or the preservation of fish and wildlife values." In response to a request from Joint Committee Chairman Chet Holifield and for its own information, the AEC reviewed its legal stance on regulating against thermal pollution and applying the Fish and Wildlife Coordination Act to its activities.[24]

Howard K. Shapar, who, as assistant general counsel for licensing and regulation, was the AEC staff's authority on the legal aspects of regulatory issues, reaffirmed the agency's position in a lengthy analysis. He argued that the Atomic Energy Act of 1954 and its subsequent amendments restricted the AEC's regulatory power to hazards peculiar to nuclear facilities and that, therefore, its statutory mandate extended only to radiological health and safety.

Shapar further contended that the Fish and Wildlife Coordination Act, which Congress had passed in 1934 and strengthened in 1958, did not apply to AEC licensees. The act required federal agencies to consult with the FWS "with a view to the conservation of wildlife resources" if they undertook or licensed activities in which water would be "impounded, diverted, . . . controlled or modified." Shapar submitted that nuclear plants simply circulated and returned water to its source "essentially unchanged." They did not impound, divert, control, or modify it in the way that dredging, irrigation, or flood control projects did. Shapar acknowledged that a nuclear facility would raise the temperature of the water it used, but he did not view that as sufficient grounds to require AEC compliance with the Fish and Wildlife Coordination Act. Moreover, the act did not expand the regulatory authority of the AEC or any other federal agency. Consequently, even if the AEC were to agree that the law was binding, it would apply "only with respect to the radiological effects of licensed activities."[25]

Shapar's brief demonstrated that the AEC could make a strong legal case for not regulating thermal pollution. But the problem

[24]Harold L. Price to the Commission, March 28, 1966, L-4-1 (Memo of Understanding, AEC–Department of Interior and Fish and Wildlife), AEC/NRC; Chet Holifield to Glenn T. Seaborg, March 21, 1966, Box 202 (Regulatory Matters—General Files), Seaborg Office Files, AEC/DOE; John D. Dingell to Chet Holifield, March 17, 1966, John T. Conway to All Committee Members, April 2, 1966, Box 512 (Pollution: Thermal Pollution), JCAE Papers; John D. Dingell to Clinton P. Anderson, March 18, 1966, Box 845 (Joint Committee on Atomic Energy—General 1966), Clinton P. Anderson Papers, Library of Congress, Washington, D.C.

[25]Howard K. Shapar to the Files, April 18, 1966, Box 512 (Pollution: Thermal Pollution), JCAE Papers; Joseph F. Hennessey to the Commission, April 21, 1966, L-4-1 (Memo of Understanding, AEC–Department of Interior and Fish and Wildlife), AEC/NRC.

remained, and the AEC offered no alternative approaches for dealing with waste heat. When several members of Congress introduced legislation to resolve the issue by explicitly subjecting the agency to the provisions of the Fish and Wildlife Coordination Act, the AEC objected. One reason was that the bills did not grant the AEC any new regulatory authority, so that, in its view, its jurisdiction would still be limited to radiological hazards. A more important consideration was that the agency feared that the proposals, if enacted, would discriminate against nuclear power. Since fossil fuel plants were not licensed by federal agencies, they would not be required to meet the same conditions to control thermal discharges that nuclear plants would.[26]

In hearings held on May 13, 1966, Dingell grilled AEC officials about their views on thermal pollution. He opened the hearings by lamenting the "grossly inadequate protection now being afforded fish and wildlife resources," and the AEC's explanation of its position did not mitigate his anxiety. Harold Price told him that the AEC was "very much in sympathy" with programs intended to protect fish and wildlife but stressed that it opposed measures that would affect nuclear but not fossil fuel plants in doing so. When Dingell asked whether the agency assumed any responsibility for or took any interest in nonradiological environmental problems, Price replied that its authority was restricted to radiation hazards but that it was "very much interested in" preserving fish and wildlife resources. Dingell wondered whether the AEC had proposed any legislative solutions to vest "in your agency power to correct the hazard that is clearly apparent?" Price said no, that he believed the problem was "not peculiar to atomic energy plants, and it ought to be attacked more broadly." Dingell inquired about what the AEC would do if a proposed plant would obviously heat a river enough to be "enormously destructive?" Price responded that "we would be very unhappy" but that the AEC "could not, under the law, deny the license on that ground." Although Dingell was unfailingly polite to Price and other AEC representatives, he did not conceal his annoyance that their expressions of concern about thermal pollution did not convey a willingness to suggest anything they might do about it.[27]

[26]Howard K. Shapar to Mr. Trosten, May 9, 1966, John T. Conway to All Committee Members, May 10, 1966, William T. England to John T. Conway, May 17, 1966, Box 512 (Pollution: Thermal Pollution), JCAE Papers; *Nucleonics Week*, May 19, 1966, p. 5; Subcommittee on Fisheries and Wildlife Conservation, *Hearings on Miscellaneous Fisheries Legislation* (n. 22 above), pp. 112–13.

[27]Subcommittee on Fisheries and Wildlife Conservation, *Hearings on Miscellaneous Fisheries Legislation* (n. 22 above), pp. 97, 207–22; William T. England to John T. Conway, May 17, 1966, Box 512 (Pollution: Thermal Pollution), JCAE Papers.

The AEC was aware of the problem but uncertain of how to handle it. Agency officials agreed that thermal pollution required regulatory action, but they opposed any solution that would place nuclear power at a competitive disadvantage with fossil fuel plants. None of the several legislative measures proposed between 1966 and 1969 resolved that dilemma. The AEC did not want to exercise authority over thermal effects of nuclear plants unless fossil facilities had to meet the same conditions. It also objected to granting the secretary of the interior regulatory jurisdiction over thermal discharges from atomic power stations. As an alternative, it continued to consult with the FWS, which, for its part, stopped insisting that the AEC already had the necessary authority to regulate waste heat from nuclear plants. It asked that the AEC urge applicants to take action to control thermal discharges and to cooperate with interested state agencies. The AEC passed on the views of the FWS and, through it, the Federal Water Pollution Control Administration, to nuclear plant applicants as a normal part of the licensing process. But the recommendations were strictly advisory; compliance with them was not mandatory for receiving a construction permit.[28]

Meanwhile, public and congressional concern about thermal pollution continued to grow. The focal point of the enlarging controversy was the proposed Vermont Yankee Generating Station. In November 1966, the Vermont Yankee Nuclear Power Corporation, a consortium of ten utilities, applied for a construction permit for a 514–electrical megawatt plant on the Connecticut River at Vernon, Vermont. The situation in Vermont with regard to energy needs and environmental concerns reflected the national outlook in particularly sharp relief. Vermont had so little generating capacity of its own that it imported about 80 percent of its power, and its out-of-state suppliers were unable to provide for its rapidly increasing demand. At the same time, residents and state officials were committed to protecting Vermont's environmental resources from the threats posed by industrial development and population growth. The Vermont Yankee plant was intended to serve both energy and environmental requirements, but it soon aroused a sharp debate over the issue of thermal pollution.[29]

[28]James T. Ramey to Donald F. Hornig, December 14, 1966, Box 62 (Nuclear Power Reactors), Clarence F. Pautzke to Harold Price, February 8, 1968, James T. Ramey to David Black, February 9, 1968, Draft Memorandum, "Legislation on Thermal Effects," March 12, 1968, Box 64 (Thermal Effects or Pollution), Ramey Office Files, AEC/DOE; R. E. Baker to John F. Newell, March 13, 1967, Legal-4-1 (Federal Water Pollution Agency), Harold L. Price to Commissioner Ramey, October 8, 1969, Legal-4-1 (Memo of Understanding), AEC/NRC.

[29]Richard M. Klein, "Bananas in Vermont," *Natural History* 79 (February 1970): 11–18; John Walsh, "Vermont: A Power Deficit Raises Pressure for New Plants," *Science*

Officials in Vermont and adjacent states were gravely concerned about the threat of thermal pollution. James B. Oakes, attorney general of Vermont, insisted that the plant would need cooling towers to prevent ecological damage from waste heat. New Hampshire, across the Connecticut River from the proposed plant, and Massachusetts, 5 miles south of the site, expressed equally deep apprehensions about the environmental impact of Vermont Yankee. Elliot L. Richardson, attorney general of Massachusetts, complained: "Vermont will receive a million-dollar injection into its economy. Massachusetts will receive hot water." All three states protested the AEC's refusal to regulate thermal effects as a part of its licensing process.[30]

The Vermont Yankee Nuclear Power Corporation initially rebuffed suggestions that it add cooling towers to its plant by maintaining that they were unnecessary and too costly. Within a short time, however, the utility relented in the face of determined opposition from Vermont, New Hampshire, and Massachusetts. The company's concession was not enough to end the controversy. It made plans to use "open cycle" towers, in which the water from the condenser would circulate through the cooling system and then be returned to the river. This would enable the plant to meet Vermont's water standards by raising the temperature of a "mixing zone" in the river by a maximum of 4 degrees. But this was not sufficient to conform with the water standards of New Hampshire and Massachusetts, which required that even at the point of discharge the plant could not heat the river water at all. This could be done only by building a "closed cycle" system, in which the condensate water returned to the condenser after running through the cooling towers. The drawbacks of the closed cycle system were not only that it would be more expensive to build but also that it would reduce plant efficiency substantially. The issue was still unresolved in December 1967 when the AEC

---

173 (September 17, 1971): 1110–15; John Walsh, "Vermont: Forced to Figure in Big Power Picture," *Science* 174 (October 1, 1971): 44–47; Steven Ebbin and Raphael Kasper, *Citizen Groups and the Nuclear Power Controversy* (Cambridge, Mass., 1974), pp. 90–94.

[30]"They Want to 'Cool' Vernon Outflow," *Rutland Herald,* July 25, 1967; "Oakes Convinced Vernon Nuclear Plant Needs Cooling Towers," *Rutland Herald,* July 28, 1967; "Thermal Pollution Issue of A-Plant Still Up in Air," *Burlington Free Press,* October 20, 1967; "Vernon and the AEC," *Boston Globe,* November 10, 1967; *Nucleonics Week,* September 7, 1967, p. 1, September 14, 1967, p. 5; "Thermal Effects: An Acute Issue," *Nuclear Industry* 14 (September 1967): 8–14.

granted Vermont Yankee a construction permit, once again disclaiming responsibility for regulating thermal pollution.[31]

Edmund S. Muskie of Maine observed the Vermont Yankee proceedings with interest and growing impatience with the AEC's position. As chairman of the Subcommittee on Air and Water Pollution of the Senate Committee on Public Works, Muskie had already won recognition as a leading advocate of tough antipollution laws, and he took it on himself to investigate the issues involved in his neighboring states. On September 20, 1967, he wrote to Seaborg, questioning the legal basis for the AEC's refusal to consider thermal effects in its licensing actions. Muskie asserted that an executive order of July 1966, implementing sections of the Federal Water Pollution Control Act of 1965, had instructed all agency heads to combat water pollution from federal government activities. He wondered how the AEC could justify its denial of authority, and he asked for a prompt response to his query. More than a month later, Harold Price replied to Muskie. He pointed out that the executive order did not expand the AEC's regulatory jurisdiction and contended that it applied only to installations operated by federal agencies and not to licensees of the AEC.[32]

Muskie was visibly irritated by Price's letter; one of his staff members commented that the senator thought that the AEC was "thumbing its nose at the intent of Congress." He fired off another letter to Seaborg, reasserting his contention that the executive order and the Federal Water Pollution Control Act required the AEC to regulate thermal pollution. He also noted that his concern over the issue had been further piqued by the application of the Maine Yankee Atomic Power Company to build a plant in his home state. On November 4, ten days after Muskie's letter, Seaborg responded. He reiterated the AEC's standard arguments on why it believed that its authority did not extend to thermal discharges, but he promised that the agency would seek the opinion of the Department of Justice about the legal soundness of its position. In the meantime, Muskie had announced that he would hold hearings to investigate the AEC's

[31]"They Want to 'Cool' Vernon Outflow" (n. 30 above); "Nuclear Pollution Hearings Soon," *Rutland Herald,* October 31, 1967; "Power Plant Wins Federal Board Permit," *Rutland Herald,* December 9, 1967; *Nucleonics Week,* November 23, 1967, p. 6, December 14, 1967, p. 5; "Thermal Effects: An Acute Issue" (n. 30 above), pp. 8–11.

[32]Edmund S. Muskie to Glenn T. Seaborg, September 20, 1967, Harold L. Price to Edmund S. Muskie, October 23, 1967, Box 512 (Pollution: Thermal Pollution), JCAE Papers.

practice of granting licenses "without giving due consideration to the effect of waste heat."[33]

In hearings he conducted in Montpelier, Vermont, on February 14, 1968, Muskie heard representatives of Vermont, New Hampshire, and Massachusetts denounce the AEC for its refusal to exercise jurisdiction over thermal pollution. The governor of Vermont, Philip H. Hoff, after declaring that his state was "blessed with a matchless environment," went on to attack the AEC's position. "We were dismayed during the Vermont Yankee hearings when the AEC decided that thermal pollution was none of its concern," he said. "When it ignored the issue of thermal pollution . . . I think it declared itself to be a promotional agency—in effect, a publicly financed lobby." Officials of the other two states expressed similar opinions in language that was only slightly less blunt. The consensus clearly favored regulatory action by the AEC or some other federal agency. Muskie agreed, observing at one point that the AEC was "about as arbitrary in rejection of responsibility [as] I can recall in [my] experience with federal agencies."[34]

Despite the vocal objections to its denial of authority, the AEC received support for its legal stance from two important sources. The first came from Justice, which the AEC had asked, in response to Muskie's queries, to review the question of whether or not it had statutory jurisdiction over thermal discharges. In April 1968, Justice reported that it concurred with the AEC's view. After examining the provisions of the Atomic Energy Act, the Federal Water Pollution Control Act, and the executive order implementing sections of the latter act, Department of Justice attorneys concluded that the AEC did not have authority to regulate against thermal pollution.[35]

The AEC's legal claims also received support from the U.S. Court of Appeals for the First Circuit in Boston, which sustained the agency's position but viewed its policy implications with an obvious lack of enthusiasm. After the Atomic Safety and Licensing Board, a part of the AEC, granted a construction permit for the Vermont Yankee plant, the state of New Hampshire filed an appeal for a rehearing by the five AEC commissioners. The commission turned

[33]Edmund S. Muskie to Glenn T. Seaborg, October 25, 1967, John T. Conway to All Committee Members, October 31, 1967, Glenn T. Seaborg to Edmund S. Muskie, November 4, 1967, Box 512 (Pollution: Thermal Pollution), JCAE Papers; *Nucleonics Week*, November 2, 1967, p. 2; Bryce Nelson, "Thermal Pollution: Senator Muskie Tells AEC to Cool It," *Science* 158 (November 10, 1967): 755–56.

[34]U.S. Senate, Subcommittee on Air and Water Pollution, *Hearings on Thermal Pollution—1968* (n. 9 above), pp. 311–46.

[35]Joseph F. Hennessey to Frank M. Wozencraft, November 16, 1967, Box 7717, MH&S-11 (Industrial Hygiene, vol. 3), AEC/DOE; Frank M. Wozencraft to Joseph F. Hennessey, April 25, 1968, Box 512 (Pollution: Thermal Pollution), JCAE Papers.

down the request. New Hampshire then took its case to court, arguing, in terms similar to those of Muskie, that the AEC had the statutory obligation to consider thermal pollution in its decision to issue the permit to Vermont Yankee. The court of appeals denied that assertion in a ruling of January 13, 1969. It agreed with the AEC and the Department of Justice that existing legislation did not assign the AEC authority to regulate the thermal effects of licensed plants. But the court also declared: "We confront a serious gap between the dangers of modern technology and the protections afforded by law as the Commission interprets it. We have the utmost sympathy with the appellant and with the sister states of Massachusetts and Vermont." The court expressed its regret that Congress had not resolved the issue by "requiring timely and comprehensive consideration of nonradiological pollution effects." New Hampshire appealed the decision to the U.S. Supreme Court, which allowed the lower court ruling to stand by refusing to hear the case.[36]

Although the AEC won its battle in court, it was left in an uncomfortable position. It had clear judicial support for its argument that it lacked jurisdiction over thermal pollution, but it was under attack from critics who accused it of indifference to the environment. The once widely held assumption that nuclear power would provide both electricity and environmental protection was being questioned because of the emerging debate over thermal pollution. From the AEC's perspective, the best way out of this predicament was to support legislation that would clarify the roles of federal agencies in regulating waste heat discharges. But the agency favored legislation only if it did not discriminate against nuclear power or give the Interior Department final authority to decide thermal issues for nuclear plants. None of the several bills that were introduced during 1968, some granting the AEC and some the Interior Department responsibility over thermal pollution, won enough backing in Congress for passage, and the impasse continued.[37]

[36]AEC-R 141/34 (April 9, 1968), AEC-R 141/36 (April 23, 1968), AEC/NRC; *Nucleonics Week,* July 18, 1968, p. 6; "AEC Holds the Line on Jurisdictional Contention," *Nuclear News* 12 (January 1969): 8–9; "N.H. Challenge to AEC Rejected," *Bennington Banner,* June 16, 1969; State of New Hampshire v. Atomic Energy Commission, 406 F. 2d 170 (1969).

[37]Joseph F. Hennessey to the Commission, April 29, 1968, Glenn T. Seaborg to James F. C. Hyde, Jr., June 12, 1968, Seaborg to Edmund S. Muskie, October 15, 1968, Legal-4-1 (Federal Water Pollution Agency), AEC/NRC; Seaborg to Charles Schultze, December 11, 1967, AEC 783/98 (August 21, 1968), Box 64 (Thermal Effects or Pollution), Ramey Office Files, AEC/DOE; John T. Conway to John O. Pastore, August 27, 1968, Pastore to Muskie, September 5, 1968, General Files—Atomic Energy (Pastore Outgoing Mail), John O. Pastore Papers, Providence College, Providence, R.I.;

While the issue remained unresolved, criticism of the AEC became increasingly more pointed and more frequent. For a time, the attacks were sporadic and localized, largely limited to several members of Congress, a handful of environmentalists, and critics in the specific locations of a few proposed nuclear plants. But the problem of thermal pollution and the AEC's position on it captured expanding national attention after the publication of an article in the high-circulation *Sports Illustrated* in January 1969. The article was written by Robert H. Boyle, a senior editor for the magazine, devout fisherman, conservationist, and author of a book on the natural history and resources of the Hudson River.

Boyle's article, entitled "The Nukes Are in Hot Water," was a scathing attack on the AEC. "What literally may become the 'hottest' conservation fight in the history of the U.S. has begun," it opened. "The opponents are the Atomic Energy Commission and utilities versus aroused fishermen, sailors, swimmers, homeowners, and a growing number of scientists." Boyle went on to describe the threat of thermal pollution to aquatic life and to water quality. He assailed the AEC for refusing to take responsibility for the problem, attributing its inaction to a fear of the "financial investment that power companies would have to make . . . to stop nuclear plants from frying fish or cooking waterways wholesale." He gibed at Seaborg, suggesting that, even though the AEC chairman had won a Nobel Prize for finding plutonium, he had "yet to discover hot water." Boyle predicted that, since "more than 100 nuclear plants are on the drawing boards, . . . almost every major lake and river and stretches of Atlantic, Gulf and Pacific coasts are likely to become battlegrounds." The article was in many ways distorted and unfair; it misrepresented the AEC's position to the point of caricature. Yet Boyle obviously had no intention of writing a balanced scholarly treatise, and his tone of indignation and incredulity was an effective way to advance his own point of view.[38]

Although the precise impact of Boyle's article was impossible to define, it clearly broadened and called attention to the thermal pollution controversy more than any previous discussion had done. Debate over the issue was well under way before the article appeared, but after its publication, and to an appreciable degree because of its publication, thermal pollution became the subject of elevated interest

---

Nucleonics Week, July 4, 1968, p. 1, September 26, 1968, pp. 3–5, October 10, 1968, p. 6; "Thermal Effects Jurisdiction Stirs Further Controversy," *Nuclear Industry* 15 (September 1968): 36–38.

[38]Robert H. Boyle, "The Nukes Are in Hot Water," *Sports Illustrated*, January 20, 1969, pp. 24–28; Talbot, *Power along the Hudson* (n. 12 above), pp. 112–14.

and heightened concern on a national scale. One indication was the reaction and commentary that the article stirred. For three consecutive weeks *Sports Illustrated* ran letters to the editor that both commended and criticized the article. Representative Tim Lee Carter of Kentucky inserted it into the *Congressional Record*, hoping that it would "begin a rational discussion of what might be a tremendous problem in the future." Chet Holifield, worried that some of his colleagues "may have taken Mr. Boyle's utterances at face value," countered the assertions presented in "that esteemed technical journal, *Sports Illustrated*." He defended the AEC from the charge that it did not care about the effects of waste heat and maintained that nuclear power was essential for achieving the twin goals of producing sufficient electrical power and preserving the environment.[39]

Some industry spokesmen reaffirmed the same view, but the article dismayed many of their colleagues. According to one knowledgeable observer, the story "shocked many in the industry," especially since "many utilities had in fact chosen nuclear power largely *because* of environmental advantages." The AEC's director of public information concluded in March 1969 that, although "public acceptance of nuclear power remains at a high level," the "biggest problem today is the question of thermal effects." He added that "until some positive action is taken to place responsibility for thermal effects this question will continue to give us trouble." One major source of concern was that, even if the environmental effect of a single nuclear plant were relatively inoffensive, the consequences of placing several plants that discharged waste heat into the same body of water might be ruinous. This was a point that Boyle highlighted, and the projections for the rapid growth of nuclear power fed those fears.[40]

The spreading alarm about thermal pollution was evident in protests against a number of proposed nuclear plants. Although the specifics varied widely from place to place, the general patterns of the debate followed similar lines. Like the controversy over the Vermont Yankee reactor, they usually started when state officials or conservationists raised questions about the thermal effects of a plant, became matters of dispute when a utility refused to build cooling towers or take other action to mitigate waste heat discharges, and ended only after considerable acrimony and/or concessions by the power com-

[39]"19th Hole: The Readers Take Over," *Sports Illustrated*, February 3, 1969, p. 62, February 10, 1969, p. 76, February 17, 1969, p. 72; *Congressional Record*, 91st Cong., 1st sess., 1969, pp. E412, H1467–H1469.

[40]"Are the 'Nukes' In Hot Water?" *Nuclear News* 12 (March 1969): 12–15; AEC 688/62 (March 17, 1969), Box 7751 (I&P-4, Public Information), AEC/DOE; Paul Turner, "The Radiation Controversy," *Vital Speeches of the Day* 37 (September 1, 1971): 697.

pany. In several cases, what began as local issues received widespread attention as a part of growing national concern over environmental quality in general and thermal pollution in particular.

One of the first examples of this pattern was a dispute over a reactor that the New York State Electric and Gas Corporation proposed to build on Cayuga Lake, the second largest of the celebrated Finger Lakes. When a group of Cornell University faculty members raised the issue of thermal pollution and urged the utility to add a cooling system to the plant, New York State Electric and Gas refused, citing the high cost of towers. This set off a contentious controversy that ended only when the utility decided to postpone work on the plant indefinitely to study the economic and environmental effect of cooling systems. Thermal discharges also emerged as a major issue in the construction of nuclear plants on Florida's Biscayne Bay. After a series of meetings, hearings, court rulings, and negotiations involving the Florida Power and Light Company, state officials, the Federal Water Pollution Control Administration, and environmental groups over a period of two years, the utility agreed to build a cooling canal system at a cost it had initially deemed to be excessive.[41]

The thermal pollution question generated even more acrimony in the case of the Palisades nuclear plant, located on Lake Michigan in western Michigan. A group of intervenors appealed to the AEC in June 1970 to deny the application of Consumers Power Company for a low-power operating license. They charged that the plant provided insufficient protection against thermal pollution and radiation. The attorney for the intervenors, Myron M. Cherry, argued that the AEC was obligated to regulate waste heat discharges, regardless of the Vermont Yankee decision, and that its radiation regulations were outmoded and inadequate. With construction of the plant complete, Consumers was anxious to secure its operating license, but it resisted making concessions to the intervenors. Finally, in March 1971, after numerous delays, several hearings, and sharp exchanges between the

[41]Luther J. Carter, "Thermal Pollution: A Threat to Cayuga's Waters?" *Science* 162 (November 8, 1968): 649–50; Dorothy Nelkin, *Nuclear Power and Its Critics: The Cayuga Lake Controversy* (Ithaca, N.Y., 1971); Claude R. Kirk, Jr., to Glenn Seaborg, December 12, 1967, Box 180 (Pollution), Seaborg Office Files, and Seaborg to Dante Fascell, February 1, 1968, AEC 544/79 (April 5, 1968), Box 7717 (MH&S-11, Industrial Hygiene, vol. 4), AEC/DOE; Harold L. Price to the Commission, February 27, 1970, MH&S-3-1 (Thermal Effects), AEC/NRC; *Nucleonics Week*, July 10, 1969, p. 3, July 17, 1969, p. 3, February 26, 1970, p. 3, March 5, 1970, pp. 2–3, April 1, 1971, p. 5, July 8, 1971, p. 5, September 2, 1971, p. 4, December 2, 1971, p. 4; "Florida Utility Wins a Round in Thermal Effects Court Action," *Nuclear Industry* 17 (April 1970): 29–30; "Cloudy Sunshine State," *Time*, April 13, 1970, pp. 48–49.

attorneys for both sides, the utility decided that it preferred a settlement with its opponents to the prospect of further costly delays. It agreed to build cooling towers and to virtually eliminate the discharge of liquid radioactive wastes into the lake. In return, the intervenors dropped their action against the plant and opened the way for its full-power operation.[42]

Although thermal pollution assumed major proportions as an environmental concern and a regulatory issue, the controversy over it largely died out by the early 1970s. The defusing of the question occurred for a number of reasons. One was that the results of the first meticulous studies of thermal effects, though far from conclusive, were encouraging. An investigation of the Connecticut River in the vicinity of the Connecticut Yankee nuclear plant, which opened in 1967 and did not have cooling towers, demonstrated "no significant deleterious effect on the biology of the river," according to an article published in 1970. Scientists who traced the consequences of thermal discharges from the AEC's plutonium reactors on the Columbia River made a similar assessment, finding "no demonstrable effect" on the salmon or trout in the river. Neither study claimed to evaluate the long-term effects of waste heat or the implications of placing many plants on a single body of water. Their findings, therefore, played only a limited role in alleviating concern over thermal pollution.[43]

A more important influence was that, after years of fruitless efforts, Congress passed legislation that assigned federal agencies a clearly defined role in regulating water quality. Since much of the thermal pollution controversy had centered on the AEC's denial of statutory authority, congressional action removed one of the leading sources of dispute. In January 1969 Muskie introduced a bill that would require applicants for AEC construction permits or other federal licenses to present certification from appropriate state or interstate agencies that the plant could meet the water quality standards of their jurisdiction.

[42]"Cooling Tower Slated for Nuclear Plant," *New York Times*, March 17, 1971, p. 29; *Nucleonics Week*, July 2, 1970, p. 5, February 25, 1971, p. 1, March 18, 1971, p. 3; "In Government," *Nuclear Industry* 17 (July 1970): 33–39; "Cooling Tower Concession Seen as Aim of Palisades Intervenors," *Nuclear Industry* 17 (August 1970): 22–26; Frances Gendlin, "The Palisades Protest: A Pattern of Citizen Intervention," *Bulletin of the Atomic Scientists* 27 (November 1971): 53–56; James B. Graham to Edward J. Bauser, September 29, 1970, Box 512 (Pollution: Thermal Pollution), JCAE Papers; Bauser to All Committee Members, June 30, 1970, Box 181 (Reactors—Palisades), Hosmer Papers.

[43]Daniel Merriman, "The Calefaction of a River," *Scientific American* 222 (May 1970): 2–12; "Burgeoning Atomic Plants Run into Pollution Awareness," *Washington Post*, August 25, 1970, p. 4; *Nucleonics Week*, August 27, 1970, pp. 7–8; USAEC, *Thermal Effects and U.S. Nuclear Power Stations* (n. 10 above), pp. 15–36.

Members of the House proposed similar legislation. Although the measures did not extend the direct authority of the AEC, they required that it formally consider thermal effects as a part of its licensing process. They did not apply only to nuclear plants, or even only to thermal pollution, but to any activity that could lower water quality.[44]

On March 3, 1969, Commissioner James T. Ramey announced that the AEC supported the Muskie bill, a position he reaffirmed three days later when testifying on similar House proposals. He explained that, although the AEC had objected to earlier measures on the grounds that they discriminated against nuclear power, it believed that Muskie's bill addressed the problem satisfactorily, if not completely. The agency had discovered that fossil fuel plants that were at least partially constructed on navigable waters required a permit from the U.S. Army Corps of Engineers and therefore came under the provisions of the bill. In 1967, two-thirds of the large fossil fuel plants licensed would have been included in this category.

There were other reasons that the AEC found Muskie's proposed legislation attractive. Compared to previous bills, it diminished the role of the Department of the Interior, which allayed the AEC's concern that Interior would exert undue influence in its licensing actions. Most important, the AEC was anxious to see a law governing thermal effects passed, because, even though the new proposals were not drastically different than earlier ones, the political atmosphere was. As *Nuclear News* pointed out: "For the AEC, the sooner adequate and appropriate legislative control can be established over thermal effects the better. . . . The rash of adverse public opinion stirred up recently by the national news media (and by the Muskie hearings themselves) has made early and appropriate control mandatory." The AEC's endorsement of the Muskie bill was not enough in itself to ensure its enactment, but it also won the backing of key members of the Joint Committee on Atomic Energy and others who had opposed previous proposals. In March 1970, after the addition of clarifying amendments and more than five months of discussion in conference committee, Congress passed the final version of the legislation as the Water Quality Improvement Act. A broader measure, the National Environmental Policy Act, signed into law on January 1, 1970,

[44]"Muskie Thermal Effects Bill Provides State Certification," *Nuclear Industry* 16 (January 1969): 54–57; "Interior May Seek Own Pollution Plan," *Washington Post*, January 29, 1969, p. A-2.

provided further assurance that federal agencies would treat the problem of thermal pollution.[45]

The most important reason that thermal pollution ceased to be a major focus of environmental concern was that utilities increasingly took action to curb its consequences. Most nuclear plants being built on or planned for inland waterways by 1971 included cooling systems. Although power companies initially resisted the calls for cooling equipment, they soon found that the costs of responding to litigation, enduring postponements in construction or operation of new plants, or suffering loss of public esteem were less tolerable than those of adding towers or ponds. Even though a cooling system added substantially to the expense of a plant, it was still usually a small percentage of the total cost of the facility. Utilities increasingly saw it as a part of the price they had to pay to fulfill their primary objective, which was to meet the growing demand for electricity. Once they reached that conclusion and began to act on it, the issue of thermal pollution lost much of its potency and immediacy.

Even after the thermal pollution question largely faded from view, its legacy lingered on. The most important effect of the debate from the perspective of the AEC and the nuclear industry was that the image of nuclear power as an antidote for the environmental hazards of electrical production was irreversibly tarnished. The controversy played a vital role in transforming the ambivalence that environmentalists had demonstrated toward the technology into strong and vocal opposition. By the end of the 1960s environmental groups spearheaded protests against plans for many nuclear power plants, and thermal pollution was a major element in their arguments. In a similar manner, the issue wakened doubts among the general public about the environmental benefits of nuclear power. Before thermal pollution was featured in a plethora of news stories, public attitudes about the environmental impact of nuclear plants seemed to be at worst uninformed and ill defined, and at best highly favorable. As the

[45]Chet Holifield to Edmund S. Muskie, February 26, 1969, Box 46 (JCAE General Correspondence), Edward J. Bauser to All Committee Members, April 6, 1970, Box 47 (Bauser Memos), Chet Holifield Papers, University of Southern California, Los Angeles; AEC 783/106 (February 13, 1969), AEC 1318/6 (September 12, 1969), AEC/NRC; U.S. Congress, House, Committee on Public Works, *Hearings on Federal Water Pollution Control Act Amendments—1969*, 91st Cong., 1st sess., 1969, pp. 407–23; "In Government," *Nuclear Industry* 16 (March 1969): 37–42; "Muskie Bill for Thermal Effects State Control Gains Vital Support," *Nuclear News* 12 (April 1969): 22; "Conferees Approve Curb on Oil Spills," *New York Times*, March 13, 1970, p. 77; *Nucleonics Week*, April 24, 1969, pp. 4–5, March 19, 1970, p. 2–3.

public became increasingly concerned about environmental problems, however, it increasingly viewed nuclear power as one more threat to environmental integrity. Although no opinion polls on the subject were published in the late 1960s or early 1970s, a survey conducted in 1975, several years after thermal pollution had ceased to be a headline topic, indicated that 47 percent of the public thought that "the discharge of warm water into lakes and rivers" from nuclear plants was a "major problem"; another 28 percent believed it was a "minor problem."[46]

Thermal pollution was not the sole cause for the declining confidence in the environmental advantages of nuclear power that the industry and AEC emphasized. By late 1969, a new controversy was emerging over the health and environmental implications of radioactive effluents released by nuclear plants. Critics alleged that the AEC's regulations for permissible levels of radioactivity discharged to the environment by civilian power plants were inadequate. This set off a bitter debate that received a great deal of attention and eventually displaced thermal pollution as the focus of environmental concern. The thermal pollution issue laid the foundations for the radiation controversy and subsequent disputes over reactor safety. It was the first problem to raise widespread skepticism about the environmental benefits of nuclear power, and the doubts it stirred gradually expanded into other areas. It also undermined the credibility of the AEC. The agency's reluctance to take action against thermal pollution offered support to charges that it was indifferent to the environment. This fueled suspicions about its performance on other environmental issues that were apparent, for example, in the growing radiation debate.

The AEC was convinced that nuclear power offered the means to provide both ample electricity and environmental protection, and it was slow to respond to those who questioned this view. The agency came under increasing attack for failing to weigh the impact of thermal pollution in its licensing procedures, and its protestations that it lacked authority sounded like insensitivity to the environment to its growing legion of critics. In fact, the AEC was concerned with environmental quality; it funded a number of ecological studies on thermal effects and doubled its expenditures for research on the issue between 1969 and 1970. But it was even more concerned with minimizing obstacles to the use of atomic power to meet the nation's escalating energy requirements. As a matter of priorities, the AEC

[46]Louis Harris and Associates, *A Survey of Public and Leadership Attitudes toward Nuclear Power Development in the United States* (New York, 1975), pp. 39, 56.

gave greater attention to the need for power than to the problem of thermal pollution, partly because it believed that the possibility of a shortage of power was a more acute danger, and partly because it had long been inclined to emphasize the development over the regulation of nuclear electricity.[47]

The AEC's commitment to environmental quality would have been more apparent and convincing if it had acted aggressively to combat thermal pollution. It did not oppose regulating against the effects of waste heat, but it insisted that the same standards must apply to fossil plants. Its argument that imposing regulations only on nuclear plants would imperil the technology's growth by placing it at a competitive disadvantage was more of an intuitive assumption than a result of studied analysis. When quizzed about the impact of adding cooling towers on the relative economic advantages of fossil and nuclear plants, Ramey acknowledged in 1968: "I don't know that this would be a significant difference in [their] competitiveness."[48] The AEC made a strong legal case that it lacked the statutory jurisdiction to compel licensees to observe water quality standards, but a growing number of observers wondered why the AEC was so passive in its approach to the thermal pollution problem. The answer was that the agency feared that taking forceful action would discourage the growth of nuclear power. Ironically, in its view, this would lead to greater use of fossil fuels and harm the environment by causing more air pollution. In the thinking of the AEC, it was providing an important benefit to the environment by licensing new plants.

The AEC's reasoning was not clear or persuasive to those whose priorities were different. Critics portrayed it as indifferent to environmental needs and therefore loath to force its licensees to comply with water standards. The complaints were on solid ground in pointing out that the AEC's primary concern was not environmental protection, although they were often oversimplified and sometimes overwrought. Still, they sounded persuasive to many people in a time of growing concern over environmental quality and growing outrage against those who abused it. As a result of the thermal pollution controversy, the AEC and the nuclear industry frequently found themselves included among the ranks of the enemies of the environ-

[47]Walter G. Belter to Milton Shaw and G. M. Kavanagh, July 24, 1969, Box 5626 (Environmental Pollution, 1969), OGM Files, AEC/DOE. For a discussion of the AEC's efforts to balance its developmental and regulatory responsibilities, see George T. Mazuzan and J. Samuel Walker, *Controlling the Atom: The Beginnings of Nuclear Regulation 1946–1962* (Berkeley and Los Angeles, 1984).

[48]JCAE, *Hearings on Participation by Small Electrical Utilities in Nuclear Power* (n. 12 above), p. 43.

ment. In a period of a few years, the image of nuclear power was transformed from being a solution to the dilemma of producing electricity without ravaging the environment to being itself a significant threat to the environment. This perception endured long after the debate over thermal pollution ended, and it played a major role in subsequent controversies over nuclear power and the environment.

INDEX